AF718708

Shock Wave Science and Technology Reference Library

The new Springer collection, Shock Wave Science and Technology Reference Library, conceived in the style of the famous Handbuch der Physik has as its principal motivation to assemble authoritative, state-of-the-art, archival reference articles by leading scientists and engineers in the field of shock wave research and its applications. A numbered and bounded collection, this reference library will consist of specifically commissioned volumes with internationally renowned experts as editors and contributing authors. Each volume consists of a small collection of extensive, topical and independent surveys and reviews. Typical articles start at an elementary level that is accessible to non-specialists and beginners. The main part of the articles deals with the most recent advances in the field with focus on experiment, instrumentation, theory, and modeling. Finally, prospects and opportunities for new developments are examined. Last but not least, the authors offer expert advice and cautions that are valuable for both the novice and the well-seasoned specialist.

Shock Wave Science and Technology Reference Library

Collection Editors

Hans Grönig

Hans Grönig is Professor emeritus at the Shock Wave Laboratory of RWTH Aachen University, Germany. He obtained his Dr. rer. nat. degree in Mechanical Engineering and then worked as postdoctoral fellow at GALCIT, Pasadena, for one year. For more than 50 years he has been engaged in many aspects of mainly experimental shock wave research including hypersonics, gaseous and dust detonations. For about 10 years he was Editor-in-Chief of the journal Shock Waves.

Yasuyuki Horie

Professor Yasuyuki (Yuki) Horie is internationally recognized for his contributions in high-pressure shock compression of solids and energetic materials modeling. He is a co-chief editor of the Springer series on Shock Wave and High Pressure Phenomena and the Shock Wave Science and Technology Reference Library, and a Liaison editor of the journal Shock Waves. He is a Fellow of the American Physical Society, and Secretary of the International Institute of Shock Wave Research. His current interests include fundamental understanding of (a) the impact sensitivity of energetic solids and its relation to microstructure attributes such as particle size distribution and interface morphology, and (b) heterogeneous and nonequilibrium effects in shock compression of solids at the mesoscale.

Kazuyoshi Takayama

Professor Kazuyoshi Takayama obtained his doctoral degree from Tohoku University in 1970 and was then appointed lecturer at the Institute of High Speed Mechanics, Tohoku University, promoted to associate professor in 1975 and to professor in 1986. He was appointed director of the Shock Wave Research Center at the Institute of High Speed Mechanics in 1988. The Institute of High Speed Mechanics was restructured as the Institute of Fluid Science in 1989. He retired in 2004 and became emeritus professor of Tohoku University. In 1990 he launched Shock Waves, an international journal, taking on the role of managing editor and in 2002 became editor-in-chief. He was elected president of the Japan Society for Aeronautical and Space Sciences for one year in 2000 and was chairman of the Japanese Society of Shock Wave Research in 2000. He was appointed president of the International Shock Wave Institute in 2005. His research interests range from fundamental shock wave studies to the interdisciplinary application of shock wave research.

Y. Horie (Ed.)

Shock Wave Science and Technology Reference Library, Vol. 3

Solids II

With 149 Figures, 13 in Color, and 9 Tables

Springer

Yasuyuki Horie
AFRL/MNME Munitions Directorate
2306 Perimeter Road
Eglin AFB, FL 32542, USA
Email: yasuyuki.horie@eglin.af.mil

Yasuyuki Horie

Professor Yasuyuki (Yuki) Horie is internationally recognized for his contributions in high-pressure shock compression of solids and energetic materials modeling. He is a co-chief editor of the Springer series on Shock Wave and High Pressure Phenomena and the Shock Wave Science and Technology Reference Library, and a Liaison editor of the journal, Shock Waves. He is a Fellow of the American Physical Society, and Secretary of the International Institute of Shock Wave Research. His current interests include fundamental understanding of (a) the impact sensitivity of energetic solids and its relation to microstructure attributes such as particle size distribution and interface morphology, and (b) heterogeneous and nonequilibrium effects in shock compression of solids at the mesoscale.

ISBN: 978-3-540-77078-7 e-ISBN: 978-3-540-77080-0

Library of Congress Control Number: 2008921393

© 2009 Springer-Verlag Berlin Heidelberg

This work is subject to copyright. All rights are reserved, whether the whole or part of the material is concerned, specifically the rights of translation, reprinting, reuse of illustrations, recitation, broadcasting, reproduction on microfilm or in any other way, and storage in data banks. Duplication of this publication or parts thereof is permitted only under the provisions of the German Copyright Law of September 9, 1965, in its current version, and permission for use must always be obtained from Springer. Violations are liable to prosecution under the German Copyright Law.

The use of general descriptive names, registered names, trademarks, etc. in this publication does not imply, even in the absence of a specific statement, that such names are exempt from the relevant protective laws and regulations and therefore free for general use.

Cover design: eStudio Calamar

Printed on acid-free paper

9 8 7 6 5 4 3 2 1

springer.com

Preface

This book is the second volume of *Solids Volumes* in the *Shock Wave Science and Technology Reference Library.* These volumes are primarily concerned with high-pressure shock waves in solid media, including detonation and high-velocity impact and penetration events.

This volume contains four articles. The first two describe the reactive behavior of condensed-phase explosives, and the remaining two discuss the inert, mechanical response of solid materials. The articles are each self-contained, and can be read independently of each other. They offer a timely reference, for beginners as well as professional scientists and engineers, covering the foundations and the latest progress, and include burgeoning development as well as challenging unsolved problems.

The first chapter, by S. Sheffield and R. Engelke, discusses the shock initiation and detonation phenomena of solids explosives. The article is an outgrowth of two previous review articles: "Explosives" in vol. 6 of *Encyclopedia of Applied Physics* (VCH, 1993) and "Initiation and Propagation of Detonation in Condensed-Phase High Explosives" in *High-Pressure Shock Compression of Solids III* (Springer, 1998). This article is not only an updated review, but also offers a concise heuristic introduction to shock waves and condensed-phase detonation. The authors emphasize the point that detonation is not an uncontrollable, chaotic event, but that it is an orderly event that is governed by and is describable in terms of the conservation of mass, momentum, energy and certain material-specific properties of the explosive.

The article, written by two leading experimentalists in the field, is an excellent introductory or refresher reading for any class or workshop on condensed-phase detonation with focus on shock initiation and hydrodynamic phenomena. It also offers a quick reading of the most recent progress. The article compliments the chapter by M.R. Baer on "Mesoscale Modeling of Shocks in Heterogeneous Reactive Materials" in the first volume of the *Shock Wave Science and Technology Reference Library* (*High-Pressure Shock Compression of Solids II*).

The second chapter, by F. Zhang, S. Alavi, A. Hu and T.K. Wo, presents an overview of first-principles quantum-mechanical simulation of energetic materials, and select applications at high static and dynamic pressures. The latter are taken from their own studies on the dissociation of nitromethane and high-pressure, nonmolecular solid phases of polynitrogen. They are a testament not only to the progress made in inexpensive computing technology, but also to the power of the first-principles approach in gaining a basic understanding of materials at atomic scales. As is the case for the first article, it is thematically related to "What Is a Shock Wave to an Explosive Molecule" by C.M. Tarver in *High-Pressure Shock Compression of Solids VI* (Springer, 2003). In this field a few years is a long time, but the 2003 ITRI study "*Molecular Dynamics Simulations of Detonation Phenomena*", chaired by B.L. Holian, is still a valuable source of information.

The third chapter, "Combined Compression and Shear Waves," is a comprehensive topical review by two leading researchers in the field. It begins with a historical introduction, and is followed by an in-depth discussion of (1) the general theory for the combined waves in linear and nonlinear elastic solids and elastic-plastic solids and (2) a description of experimental and diagnostic methods for the combined *plane* waves. These techniques offer the fundamental value of enabling more complete characterizations of the shock-compressed state, as well as the state that is distinctly different from the principal Hugoniot. Only recently, the isentropic compression experiment (ICE) provided an alternative well-controlled method of determining off-Hugoniot states with good diagnostics. For the ICE, the reader is referred to the article by M.D. Knudson in *High-Pressure Shock Compression of Solids 1* of the *Shock Wave Science and Technology Reference Library.*

The section on applications includes investigations of rate-dependent plasticity in Al and Ti alloys, solid-state phase transformations in calcium carbonate and cadmium sulfate, high strain rate deformation of Al_2O_3, SiC and poly(methyl methacrylate), and sliding friction between WC and 4340 steel. The last example illustrates the potentials of the combined pressure and shear methods for investigating the response of dynamic materials that may not be accessible by other methods.

The fourth chapter is concerned with dynamic fragmentation of solids. The author is one of the leading authorities, if not the leading authority on the subject. He has written, among others, a review article entitled "Spall and Fragmentation in High-Temperature Metals" that appeared in *High-Pressure Shock Compression of Solids II* (Springer, 1995) and a definitive book "*Fragmentation of Rings and Shells*" (Springer, 2006). The current article, however, is a self-contained, book-length discourse on dynamic fragmentation, including many recent results from his own work, as well as insightful critiques of current thinking.

The article begins with select probabilistic issues associated with the phenomena of fragmentation, including both historic and recent theories. The introduction is followed by a survey and critical review of predictive methods for fragment size distributions by empirical and physics-based approaches. Special attention is focused on the unique features of dynamic fragmentation in brittle materials. A critical and insightful review is provided of several key ideas, such as the fractal nature of brittle fragmentation and the similarity between hydrodynamic turbulence and brittle fragmentation. Aspects of impact spall processes are explored as a special example of dynamic fragmentation.

This article is not only an authoritative reference, but is also an excellent graduate-level text for a one-semester course on the dynamic fragmentation of solids.

Eglin
May 2008

Yasuyuki (Yuki) Horie
AFRL/RWME

Contents

List of Contributors

John B. Aidun
Multiscale Dynamic Materials
Modeling
Sandia National Laboratories
Albuquerque, NM, USA

Saman Alavi
Department of Chemistry
University of Ottawa
Ottawa, Ontario K1N 6N5
Canada

Ray Engelke
Los Alamos National Laboratory
Los Alamos, NM 87545, USA

Dennis Grady
Applied Research Associates
4300 San Mateo Blvd., NE, A-220
Albuquerque, NM 87110
dgrady@ara.com

Anguang Hu
Department of Chemistry
University of Ottawa
Ottawa, Ontario K1N 6N5
Canada

Stephen A. Sheffield
Los Alamos National Laboratory
Los Alamos, NM 87545
USA

Zhiping Tang
Department of Modern Mechanics
University of Science and Technology
of China
Hefei, Anhui, P.R. China

Tom K. Woo
Department of Chemistry
University of Ottawa
Ottawa, Ontario K1N 6N5
Canada

F. Zhang
Defence R&D Canada–Suffield
PO Box 4000. Medicine Hat
Alberta, T1A 8K6, Canada

1

Condensed-Phase Explosives: Shock Initiation and Detonation Phenomena

S.A. Sheffield and R. Engelke

This article is written for the purpose of acquainting the reader with the concept that condensed-phase explosives "detonate" by virtue of shock wave-induced processes. A detonation wave in its simplest form is a steady reactive wave process, with the front of the reactive wave being a shock wave that takes the material to a high pressure/high temperature state in which the chemical reactions start. Chemical reactions develop as a function of time (on nanosecond time scales) and by means of induced sound waves, support the leading shock wave so it is steady. The models that describe this process from a physical standpoint will be discussed in some detail, along with the fluid flow equations, and the equation of state of both the unreacted explosive and the reaction products.

Similarly, initiation of detonation must at some point involve shock wave processes. The best understood detonation initiation process involves an input shock wave that starts the chemical reactions and eventually develops into a steady detonation wave, usually on the microsecond time scale. This is called a shock-to-detonation transition (SDT). There are major differences in the transition depending on whether the material is "homogeneous" (e.g., liquid or single crystal) or "heterogeneous" (e.g., composed of granules).

Initiation can be started in other ways (e.g., by friction, spark, or flame) which leads to a burning process that sometimes results in a detonation. A burning to detonation process is called a deflagration-to-detonation transition (DDT). This can occur on a hours/minutes/seconds time scale. However, at some point it must involve the development of a shock wave and become a shock initiation. Because of the large difference in timescales, the DDT process is not yet well understood and is not discussed in this chapter.

This work will proceed as follows: (1) introduction and brief history, (2) chemical structures and chemical properties of some condensed-phase explosives, (3) conservation relations for shock processes and equation of state of explosives, (4) steady detonation phenomena in one- and two-dimensions (1-D and 2-D) along with detonation properties of some explosives, (5) shock initiation phenomena of homogeneous and heterogeneous explosives and, finally, (6) summary and possible future developments.

This work is meant to be an introductory discussion of shock waves and their involvement with detonation. More detailed developments can be found in several other sources which will be listed in the appropriate sections and in a "Further Reading" portion of the references.

1.1 Introduction

In this article we concentrate on condensed-phase high explosives (HE) that do work by a shock-induced chemical energy release (detonation). Primarily because of their initial high mass density and the associated high energy density, liquid and solid materials have a major importance in military and blasting applications. Most of the processes to be described are thought to occur in both condensed-phase and gaseous explosives. In this article, we will only discuss condensed-phase materials. Space scales (e.g., the zone of chemical reactions) are relatively small in the condensed phase, making it difficult to measure and understand the processes in detail. Because of this, most of our understanding is phenomenological and is based on studies of one-dimensional (1-D) waves and their interaction with the confinement (i.e., the surrounding material).

Explosives have the property that they exist in an energetically metastable state that can be controllably destabilized. The result of such destabilization can be extreme levels of power generation and extreme physical states in the explosive and its surroundings (e.g., very high pressure). Pressures produced in high-performance explosives are typically in the 300,000 atmosphere range (300 kbar or 30 GPa). The high detonation pressures mean that the explosive and any material in contact with it (e.g., steel) can be treated as *compressible* fluids.

In chemical explosives, the energy storage is usually via the proximity of a fuel and an oxidizer. In high-performance chemical explosives the fuel and oxidizer are usually present in the same molecule, but isolated by chemical bonds, e.g., TNT (2,4,6-trinitrotoluene). Detonation proceeds by a wave process in which the passage of a shock wave triggers the stored energy release and, in turn, some of this energy sustains the shock wave motion.

1.1.1 Relationship of Initiation and Detonation to Shocks

When the term *detonation* is used, most people picture a violent, uncontrollable, chaotic event such as that shown in Fig. 1.1. We will present a picture of a detonation as an orderly event that is governed and rigorously describable in terms of the conservation of mass, momentum, and energy and certain material specific properties of the explosive. When a detonation is viewed experimentally, on the proper time and space scales, the observed shock wave

Fig. 1.1. Photograph of a large piece of condensed-phase high explosive detonating at a firing bunker at Los Alamos National Laboratory. To give an idea of the scale of this picture, the door on the bunker at the bottom of the picture is about 1 m wide.

phenomena are orderly and, when variables are well controlled in the experiment, have some simplicity. The photograph shown in Fig. 1.2 illustrates this; i.e., a detonation wave is progressing from left to right along a cylindrical explosive charge (confined in a metal tube). We will show that this detonation wave is a steady shock wave process (a shock wave that does not change with time or distance traveled) described by compressible fluid mechanics driven by exothermic chemical reactions. Understanding the interplay between the chemistry and shock physics is difficult. The physics is understood reasonably well but understanding of the chemistry is only in its infancy.

In this chapter we discuss the physics of initiation and detonation in condensed-phase high explosives. This will be done from both a theoretical and an experimental standpoint. A reasonably well-developed analytical theory of steady one-dimensional detonation exists. This theory is based on the Euler equations of chemically reactive inviscid compressible flow. Initiation theory is not as well developed, due primarily to a lack of knowledge of how the chemical energy stored in an explosive is released into the reacting flow as a function of the thermodynamic state. Our present knowledge of initiation of detonation is rooted in numerical studies that model experimental measurements using assumed simple energy-release rate forms and material equations of state (EOS).

Fig. 1.2. Photographs of a detonation wave progressing along a cylindrical explosive charge confined by a thin copper tube 36 mm in diameter (outside) with a 2-mm-thick wall. The detonation front is moving from left to right at a speed near 8 km s^{-1}, with the reaction products expanding the copper tube behind the front. The bottom picture was taken 10 µs after the top one. Bright spots on the expanding copper tube are thought to be due to flaws in the copper tube. These pictures are part of a set taken with a high speed rotating mirror framing camera. Photograph courtesy of John Vorthman, Los Alamos National Laboratory.

1.1.2 Brief History: Materials Development

Explosive material technology began when the oxidizer potassium nitrate (KNO_3, saltpeter) was discovered, probably in China or India. Fireworks were reported in China in the twelfth century AD. In England, the first pure KNO_3 was made and mixed with charcoal and sulfur to make a form of black powder in about 1240 AD. A formula for black powder (75.0/15.62/9.38 wt% KNO_3/charcoal/sulfur) was published in Brussels in 1560. This material began to be used in mines in the 1600s. Nitroglycerin (NG), the first molecular explosive, was discovered in 1846 by Sobrero (a professor at the University of Turin in Italy). It was promisingly powerful but dangerously unpredictable. Early use of NG was accompanied by severe accidents; by 1875 Alfred Nobel had learned that by absorbing NG in kieselguhr (a porous siliceous earth) or by making a gelatin with a nitrated cotton material (nitrocellulose), the safety was enhanced. These materials were called *dynamites* and gained almost immediate success in blasting operations. Trinitrotoluene (TNT) was discovered in the late 1800s and became the main explosive used in World War I. It is still widely used today. TNT is relatively insensitive, can be easily melted (melting point 79°C), mixed with other materials (such as aluminum powder), and then cast to a shape. More powerful molecular explosives, such as HMX (cyclotetramethylene tetranitramine), are now of considerable importance for weapons applications. Hundreds of molecular explosives are known, but only relatively few have wide application. This is an active area of research, particularly in the area of high nitrogen-content explosives. Ammonium nitrate (AN) based explosives are of great importance to the blasting industry – in these explosives the fuel and oxidizers are mixed together as opposed to being in the same molecule.

1.1.3 Brief History: Development of Understanding

The physics of detonation was developed early on from observations in gases. The first experimental observations (ca. 1880) of detonation waves (in gases) were made by French workers [9, 77]. Later Chapman [18] and Jouguet [67], independently, gave a theoretical treatment of detonation in gases. In the Chapman–Jouguet (CJ) treatment, a detonation is idealized as a planar mathematical discontinuity that propagates steadily. In the CJ picture, the passage of this discontinuity causes complete release of the stored chemical energy; i.e., there is no spatially resolved chemical-reaction zone. The fluid mechanics required to devise the CJ picture is primarily the mass, momentum, and energy conservation-law jump conditions across a shock wave (i.e., the Rankine–Hugoniot conditions to be discussed in Sect. 1.3). An important result of the CJ treatment is that, given an energetic material, there is a minimum speed at which a steady one-dimensional detonation wave can be propagated in it.

Later, Zeldovich [111], von Neumann [82], and Doering [29] (ZND), independently, refined the CJ model: (1) by assuming that the conservation conditions of mass, momentum, and energy for inviscid flow apply, and (2) by relaxing the assumption of an instantaneous heat release triggered by the shock. They modeled the heat release as a single forward exothermic rate process. Within these assumptions, one finds that steady planar solutions of the flow and shock-jump equations exist. The ZND detonation consists of a steady chemical reaction zone with an attached following flow. The state at the end of steady reaction zone is called the CJ state and it is independent of the chemical-heat-release form. The fluid-mechanical state within the ZND reaction zone is dependent on this form.

There are a number of books which include discussions of detonation theory in more detail than is possible in this chapter. These generally approach the subject from the viewpoint and interests of the particular author(s). Since this subject involves fluid flow, shock waves, chemical reactions, thermodynamics, EOS of the materials involved, and high pressure and high time resolution diagnostics, the discussions can be quite different (see e.g.,

[46,49,65,75,76]). Other valuable sources are Dremin et al. [30], Dremin [31], Cheret [19], and a book of classic papers edited by Johnson and Cheret [66]. We are not exhaustive in the citation of references in this chapter, but rather have tried to present a short, coherent picture of the subject.

1.2 Some Chemical Structures and Chemical Properties of Condensed-Phase HEs

Molecular high explosives have the fuel and oxidizer in the same molecule but separated by chemical bonds. Because of this the early reactions are not controlled by diffusion as would be expected in materials with mechanical mixtures of the fuel and oxidizer. Molecular HEs are in metastable equilibrium, waiting for some energy input before they can begin to react. Some explosives are much easier to initiate than others. We consider a few representative materials here. There are a number of books available that treat the properties of a large number of explosives (e.g., [28, 49, 80]). Explosives are generally classified in accordance with their sensitivity as judged from various experimental tests. Primary explosives are the most sensitive, followed by secondary, and, finally, insensitive explosives. The molecular structures of a few important explosives are shown in Fig. 1.3.

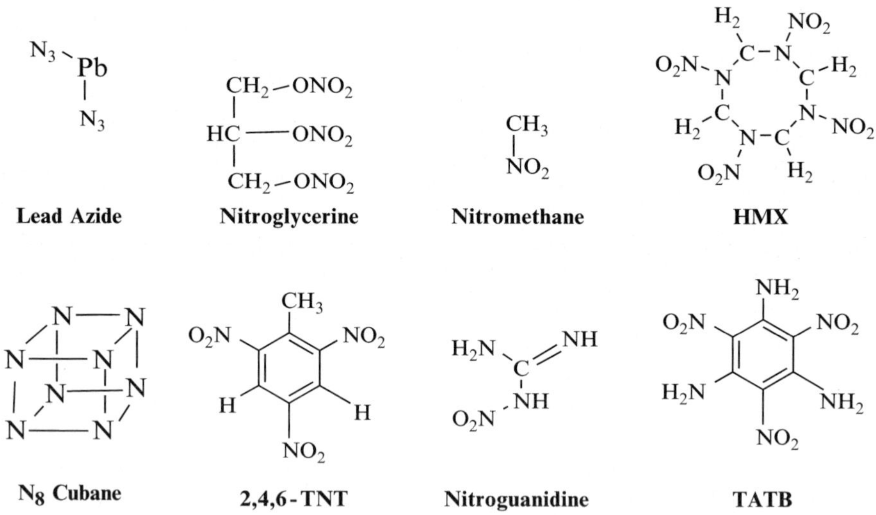

Fig. 1.3. Chemical structure of several molecular high explosives. To be an explosive material, gases must be produced during the chemical reactions. For this reason, molecular explosives have large numbers of nitrogen, oxygen, and hydrogen atoms which make up the molecule. The more moles of gas that are produced in the reaction, the more powerful the explosive.

1.2.1 HE Classifications

Primary explosives can be initiated with rather small inputs of mechanical shock, heat, flame, or spark. Examples of primary explosives are lead azide, lead styphnate, mercury fulminate, and tetrazine. Small amounts of these materials are used in initiators; electrical or mechanical input will reliably initiate them. They, in turn, reliably initiate the output charges, e.g., secondary explosive booster pellets, gun powder, etc. Extreme care is required when handling primary explosives to prevent accidental initiation.

Secondary explosives are considerably less sensitive to the same inputs and can be handled with less concern about accidental initiation. These materials are reliably initiated by moderate to high pressure shock inputs. They are used in high input initiating devices – exploding bridgewire or slapper detonators, output pellets, boosters, and main charges in explosive assemblies. Examples are TNT, HMX, pentaerythritol tetranitrate (PETN), cyclo-1,3,5-trimethylene-2,4,6-trinitramine (RDX), and 2,4,6-trinitrophenyl-methylnitramine (tetryl). Although hundreds of explosives fit in this category, only a relatively few have widespread use.

Insensitive explosives – sometimes referred to as IHEs (insensitive high explosives) – are difficult to initiate and shock inputs for reliable initiation must be quite large, i.e., careful planning is usually required to eliminate failures, particularly when there can be changes in the ambient environment, e.g., low temperature. Examples of materials in this category are 1,3,5-triamino-2,4,6-trinitrobenzene (TATB), and nitroguanidine (NQ). Relatively few HEs fit in this category. An active area of HE synthesis is to find new insensitive materials, preferably organic molecules with high nitrogen content.

It should be noted, some materials are borderline and could be placed in either of two categories, depending on the particular sensitivity property being used as the criterion. For instance, nitroglycerin might be put into the secondary explosive category because it requires a rather high level shock input to initiate it in the absence of heterogeneities (e.g., air bubbles). However, when these are present (which is almost always the case because of NGs relatively high viscosity and its propensity to entrain air) it becomes very shock sensitive, so it is generally classified as a primary explosive.

1.2.2 Properties of Selected HEs

Chemical and physical properties of a selected group of molecular explosives (some from each sensitivity class) are presented in Table 1.1. Most of these are organic molecules containing nitro groups (NO_2). Figure 1.3 shows the chemical structure of several HE molecules. All of these materials produce mostly gases upon reaction. Product gases are typically N_2, CO, CO_2, and H_2O; very little is known about the reaction mechanisms by which the explosive is transformed into reaction products. Also, little is known about how the chemical reactions (e.g., gas production) feeds energy to the detonation front to sustain it.

Table 1.1. Chemical properties of selected molecular explosives[a]

Common name	Chemical name[b]	Chemical formula	Molecular weight	Theoretical density ($g\,cm^{-3}$)	Melting point (°C)	Heat of formation ΔH_f ($kcal\,mol^{-1}$)	Heat of detonation[c] ΔH_{det} ($kcal\,g^{-1}$)
Lead azide	Lead azide	$Pb(N_3)_2$	291.3	4.8	dec.[d]	+112	0.37
NG	Nitroglycerin or glycerol trinitrate	$C_3H_5N_3O_9$	227.1	1.60	13.2	−88.6	1.48
PETN	Pentaerythritol tetranitrate	$C_5H_8N_4O_{12}$	316.2	1.78	140	−128.7	1.51
Tetryl	Trinitro-2,4,6-phenyl-methyl-nitramine	$C_7H_5N_5O_8$	287.0	1.73	130	+4.67	1.45
RDX	Cyclo-1,3,5-trimethylene-2,4,6-trinitramine	$C_3H_6N_6O_6$	222.1	1.81	205[e]	+14.7	1.48
HMX	Cyclotetramethylene-tetranitramine	$C_4H_8N_8O_8$	296.2	1.90	285	+17.9	1.48
HNS	Hexanitrostilbene	$C_{14}H_6N_6O_{12}$	450.3	1.74	316[e]	+18.7	1.36
TNT	2,4,6-trinitrotoluene	$C_7H_5N_3O_6$	227.1	1.65	80.9	−16.0	1.29
NM	Nitromethane	CH_3NO_2	61.0	1.13	−29	−27	1.36
AN	Ammonium nitrate	NH_4NO_3	80.7	1.73	169	−87.3	0.38
TATB	1,3,5-triamino-2,4,6-trinitrobenzene	$C_6H_6N_6O_6$	258.2	1.94	dec.[d]	−36.9	1.08
NQ	Nitroguanidine	$CH_4N_4O_2$	104.1	1.78	257	−22.1	0.88

[a]Data from Meyer [80] and Dobratz and Crawford [28]
[b]Common chemical names are used rather than the names used in Chemical Abstract Index (see Dobratz and Crawford [28])
[c]Heats of detonation are calculated (rather than experimental) and assume H_2O is a gas
[d]Decomposes before melting
[e]Melts with some decomposition

In Table 1.1, the heat of formation is the difference in enthalpy between the molecule and its elements (at their reference state) and the heat of detonation is the enthalpy difference between the explosive molecule and the product molecules that result from the reaction at the standard temperature and pressure state. Comparison of the heat of detonation gives a relative measure of the expected output; higher heats of detonation correspond to more energetic explosives, as is the case for HMX, RDX, and PETN. Data for Table 1.1 were taken from Dobratz and Crawford [28] and Meyer [80].

Research on the synthesis of new explosive materials continues. A current emphasis of this work is in developing explosive molecules that contain more nitrogen atoms. A pure nitrogen containing molecule, N_8 cubane, has been theoretically studied by Engelke and Stine [38] and found to have potential properties that would make it a very important explosive (estimated detonation speed of 14 km s^{-1} and CJ pressure of 130 GPa – much higher than the most powerful explosive used today). However, this molecule has not been synthesized and questions exist about its kinetic stability. Recent theoretical and chemical synthesis work is aimed at understanding this and other high nitrogen compounds.

1.3 Conservation Relations and Equation of State

In order to understand detonation fluid flows, it is necessary to understand the conservation relations across a discontinuity such as those that occur in the initiation and detonation of HEs. These will only be discussed briefly to give the reader an understanding of the form these relationships take in discussions of fluid flow related to detonation. Because an HE starts out as an "unreacted material," it is necessary to have an EOS that describes its behavior in a shock environment. The same is true for the "reaction products" that result from the chemical reactions. A short discussion of some of the more common forms the EOSs take will be presented here. A more in-depth development of EOS forms can be found elsewhere [79].

1.3.1 Conservation Relationships

The fluid flows characteristic of detonation are constrained by the conservation of mass, momentum, and energy. Therefore to discuss such processes quantitatively, we define the mass, momentum, and energy conservation conditions that relate flow quantities across a planar shock wave (or any steady flow or jump discontinuity as shown in Fig. 1.4). These are called the Rankine–Hugoniot jump conditions and can be expressed in the following form, respectively:

$$\rho_o U_s = \rho(U_s - u_p), \tag{1.1}$$

$$-P = \rho(U_s - u_p)^2 - \rho_o {U_s}^2, \tag{1.2}$$

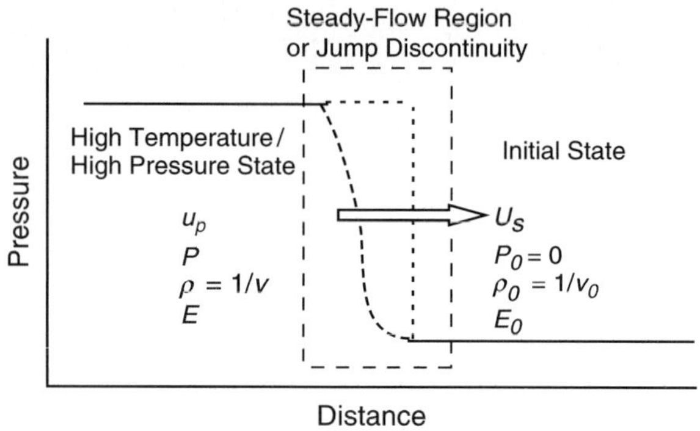

Fig. 1.4. Stations and quantities used in obtaining the conservation conditions in steady flow or across discontinuities. u_p is particle velocity and U_s is wave velocity.

and

$$-P(U_s - u_p) = \rho(U_s - u_p)\left[E + \frac{1}{2}(U_s - u_p)^2\right] - \rho_o U_s \left[E_o + \frac{1}{2}{U_s}^2\right]. \quad (1.3)$$

In these equations, subscript 0 corresponds to a variable's value before passage of the disturbance, while those variables without subscript 0 refer to values immediately after passage. Here, ρ is the mass density, u_p is the mass velocity or particle velocity, U_s is the wave velocity or shock velocity, E is the internal energy per unit mass, and P is the pressure. We have assumed that P_0 is negligible relative to the pressure behind the shock wave; for detonating condensed explosives this is a very good approximation, since the ratio of ambient pressure to detonation pressure is typically of the order of 10^{-5}. Note also that if the process causing the flow is a wave moving at velocity U_s in the lab frame, then the Galilean frame used to obtain (1.1)–(1.3) is the one in which the wave is stationary; i.e., we have transformed from the lab frame to one in which the wave is motionless and matter streams through the plane defined by the wave. For a derivation of these relationships see [100]. An elementary treatment of shock waves in general can be found in [12].

A valuable relationship can be obtained by eliminating u_p between (1.1) and (1.2). One finds that

$$P = K(v_o - v) \quad \text{and} \quad \mathrm{K} \equiv {\rho_o}^2 {\mathrm{U_s}}^2, \quad (1.4)$$

where $v \equiv 1/\rho$ is the volume per unit mass (specific volume) and K is a constant for steady processes.

All 1-D steady flow or flow across jump discontinuities (with initial state $(P_o = 0, v_o)$) *must* have P, v values that satisfy (1.4), to conserve mass and momentum. This line in the P, v plane is called the Rayleigh line (see Fig. 1.5); it will prove useful in defining what detonation processes are possible.

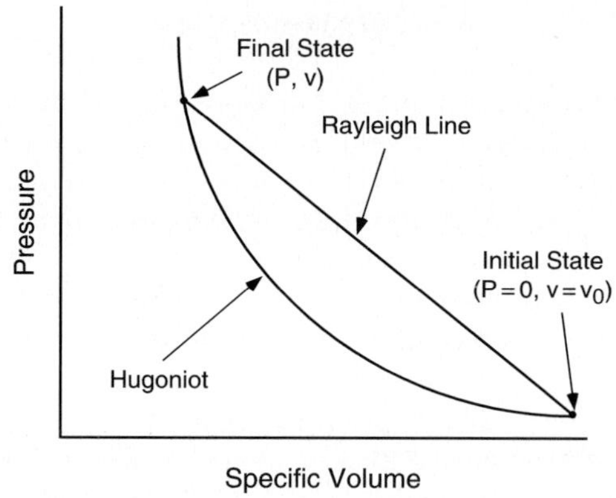

Fig. 1.5. Plot of Hugoniot and Rayleigh line in pressure vs. specific volume plane. A Hugoniot curve is defined as the locus of all possible final states arising from a *single* shock process from an initial state.

If both U_s and u_p are eliminated from (1.3) by use of (1.1) and (1.2), one finds that

$$(E - E_0) - \frac{1}{2}P(v_0 - v) = 0. \tag{1.5}$$

Thus, for 1-D steady processes or for flow across shocks that start from the initial point $(P_0 = 0, v_0, E_0)$, the final state must lie on the curve defined by (1.5), in order to conserve mass, momentum, and energy. Given the dependence of E on P and v for a particular material, (1.5) becomes a relationship between P and v. If the process connecting the initial and final states is a *single* shock process, this P, v relationship is called the principle Hugoniot curve of the material. The Hugoniot curve defines all possible final shock states reachable from the initial state by a *single* shock process.

In the case of a detonating explosive, a minimum of two Hugoniot curves are important, the Hugoniot of the unreacted explosive and the Hugoniot of the products of the reaction. There are other intermediate partially reacted Hugoniots that are present during the reaction process but in this treatment, they are ignored. Since a Hugoniot is a locus of end states achievable by a single shock on the EOS surface of a material, it provides a subset of information about the EOS surface. It is often used as a reference in developing a more complete EOS for a material. A complete description of an EOS surface requires more information. One such development was done by Andrews [1] and Hayes [56] in which they used a Helmholtz EOS form based on the Murnaghan isotherm and other appropriate parameters. This method provides a complete EOS for a material when it is calibrated to that material. Using this form one can obtain the Hugoniot relation, as well as other relationships that describe certain aspects of the EOS surface (e.g., pressure vs. volume, etc.). This form has been used on high explosives as well as other materials in the past [57, 87, 88].

1.3.2 Unreacted Material EOS: Hugoniot

All HE materials produce gases upon reaction, i.e., the products Hugoniot lies above the unreacted Hugoniot in the P, v plane. That is, at a given pressure–temperature state, the products are less dense than the reactants (see Fig. 1.6). This relationship will be illustrated later in the discussion of 1-D detonations. Other organic materials are known to react under the action of shock waves (e.g., anthracene, benzene, carbon disulfide, and acrylonitrile), but the reaction products are more dense than the reactants; they would not be expected to detonate even if they were exothermic [27, 40, 90].

The unreacted Hugoniot of an explosive is important because it defines the state from which the chemical reactions in a detonation start (this point is often called the von Neumann spike or chemical peak; see Sect. 1.4.1). In other words, the HE is shocked to a high pressure/high temperature state on the unreacted Hugoniot, then the reactions start. The fact that the initial reactions occur rapidly (nanosecond or tens of nanoseconds timescale) at this condition makes it difficult to measure this state accurately. What is usually done is to measure several unreacted Hugoniot states at much lower input pressures and then extrapolate to the higher pressures.

The Hugoniot is usually assumed to be of the (experimentally based) linear form

$$U_s = a + bu_p, \tag{1.6}$$

where a and b are constants determined from experimental data.

Some HEs have had considerable efforts made to measure the unreacted Hugoniot, while others have had little or no experiments done for this purpose. Two of the most widely studied materials are an HMX-based explosive and a TATB-based explosive, primarily because of their use in nuclear weapons. The HMX-based explosive most widely studied is PBX 9501, a plastic bonded

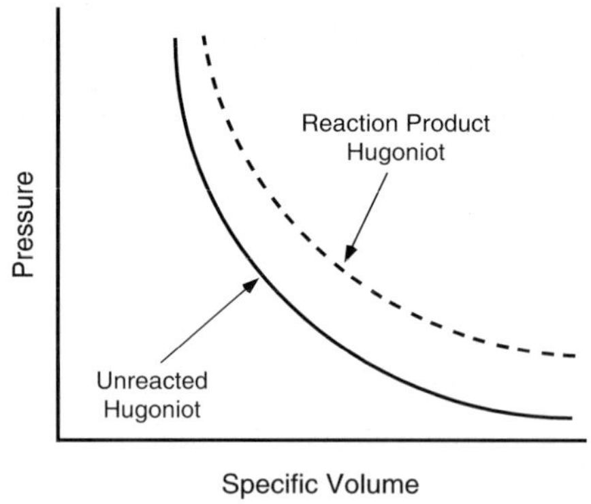

Fig. 1.6. Plot of unreacted and reaction products Hugoniots showing their relative position in P–v space; e.g., the products Hugoniot lies above the unreacted Hugoniot. This is expected because the reactions produce high-density gases.

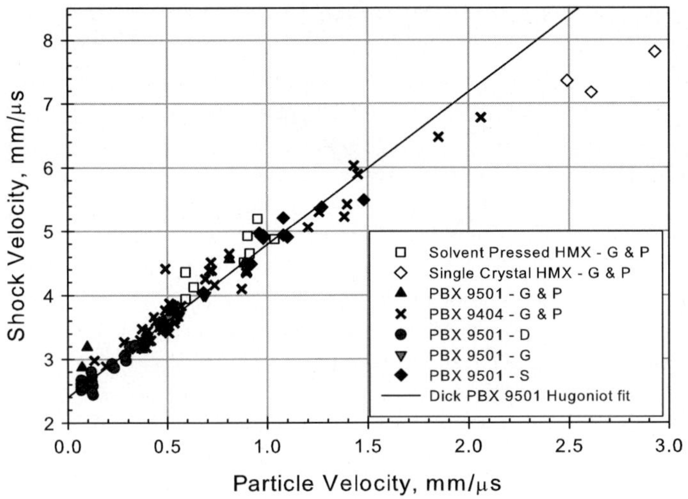

Fig. 1.7. Plot of PBX 9501 Hugoniot Data from a number of sources including G & P [49], D [26], G [51], and S [94]. The line is a fit to some of the data obtained by Dick et al. [26] and applies up to a particle velocity of 1 mm (μs)$^{-1}$. PBX 9404 data has also been included because its composition is very much like PBX 9501.

explosive consisting of 95 wt% HMX, 2.5 wt% estane, and 2.5 wt% bisdinitropropyl acetal:bisdinitropropyl formal eutectic. Some of the Hugoniot data available for PBX 9501 are shown in Fig. 1.7. Much of the data are at inputs up to 10 GPa, considerably below the von Neumann spike point. The line shown is the fit given by Dick et al. [26] and is based on the lower pressure data. Notice that the highest pressure points do not lie on the line defined by the low pressure data, particularly those for single crystal HMX. Recent studies have been aimed at explaining this observation; it has been conjectured that a phase transition occurs somewhere above 10 GPa based on static diamond anvil cell measurements [110].

The TATB-based HE which has been fairly thoroughly studied is a plastic bonded explosive called PBX 9502. It consists of 95 wt% TATB and 5 wt% Kel-F 800 plastic. Its unreacted Hugoniot is shown in Fig. 1.8 [24,53]. Notice there is a slope discontinuity or softening in the Hugoniot at about 8 GPa ($u_p \approx 0.8$ mm (μs)$^{-1}$) so that the lower pressure Hugoniot data does not extrapolate to the high pressure data. The linear fits for each of the regions are shown in the figure. The softening will considerably effect the state at the von Neumann spike point. The authors believe that the reason for this discontinuity is due to an endothermic crosslinking reaction between the benzene rings of two TATB molecules; such a reaction has been clearly observed in shocked solid TNT via mass-spectroscopic experiments [41].

Unreacted Hugoniot data are usually obtained from shock initiation experiments; such experiments will be discussed below. During the early stages

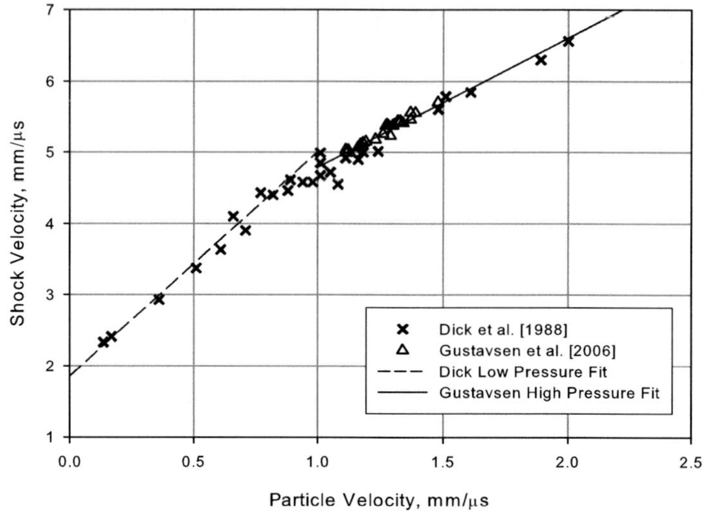

Fig. 1.8. Plot of PBX 9502 Hugoniot Data from Dick et al. [24] and Gustavsen et al. [53]. Density for the PBX 9502 in this plot were between 1.88 and 1.89 g cm^{-3}. The dashed line is the fit of Dick et al. [24] to the low pressure data (below a particle velocity of 0.8 mm (μs)$^{-1}$). The solid line is a fit by Gustavsen et al. [53] to the data at higher pressures.

of these experiments very little reaction takes place so the states measured are essentially unreacted. Table 1.2 gives estimates for the unreacted Hugoniots of a few explosives. Notice that the densities are not at the theoretical maximum densities and that the unreacted Hugoniot depends strongly on the initial density of the material. Both solid and liquid TNT are shown to indicate that there is a large difference in the Hugoniots depending on the material's phase. The constant "a" is often thought of as the zero pressure sound speed. However, linear liquid explosive Hugoniots do not go through the zero pressure sound speed. This can be rectified by using the "universal liquid Hugoniot" [109] which is

$$\frac{U_s}{c_0} = 1.37 - 0.37 \exp^{-(2u_p/c_0)} + 1.62 \frac{u_p}{c_0}, \tag{1.7}$$

where c_0 is the room condition sound speed for the liquid. This form has been used to accurately match the Hugoniot for NM and other liquids. In the case of NM it can be used to develop unreacted Hugoniots for the liquid at different initial temperatures.

Other methods have been used to produce unreacted Hugoniot and EOS information. Static high pressure diamond anvil cell experiments have produced data on HMX to very high pressures [110]. Front surface impact experiments, in which the HE is mounted on the front of a projectile and

Table 1.2. Unreacted Hugoniot estimates for several HEs based on linear fits

Common name[a]	Chemical formula	Density ($g\,cm^{-3}$)	Constant "a" ($mm\,(\mu s)^{-1}$)	Constant "b"	Reference[b]
AN (s)	NH_4NO_3	0.86	0.84	1.42	D&C
AN (s)	NH_4NO_3	1.73	2.20	1.96	D&C
HMX (s)	$C_4H_8N_8O_8$	1.891	2.901	2.058	D&C
PBX9501 (s)	95/2.5/2.5 wt%	1.833	2.501	2.261	G&P
PBX9501 (s)	95/2.5/2.5 wt% HMX/Estane/ BDNPA&BDNPF	1.844	2.40	2.39	D1
HNS (s)	$C_{14}H_6N_6O_{12}$	1.57	1.00	3.21	D&C
NM (l)	CH_3NO_2	1.13	2.0	1.38	D&C
NQ (s)	$CH_4N_4O_2$	1.69	3.048	1.725	D&C
PETN (s)	$C_5H_8N_4O_{12}$	0.82	0.47	1.73	D&C
PETN (s)	$C_5H_8N_4O_{12}$	1.75	2.26	2.32	G&P
XTX8003 (e)	80/20 wt% PETN/Sylgard	1.53	1.59	3.24	G&P
RDX (s)	$C_3H_6N_6O_6$	1.0	0.4	2.0	D&C
RDX (s)	$C_3H_6N_6O_6$	1.80	2.87	1.61	D&C
PBX9407 (s)	94/6 wt% RDX/Exon 461	1.60	1.328	1.993	G&P
Comp B-3 (s)	60/40 wt% RDX/TNT	1.70	3.03	1.73	D&C
TATB (s)	$C_6H_6N_6O_6$	1.876	1.663	2.827	G&P
PBX9502 (s)	95/5 wt%	1.896	3.263	1.678	G&P
PBX9502 (s)	95/5 wt%	1.89	1.857	3.15	D2
PBX9502 (s)	95/5 wt% TATB/Kel-F 800	1.89	2.97	1.81	G
LX-17 (s)	92.5/7.5 wt% TATB/Kel-F 800	1.90	2.33	2.32	D&C
Tetryl (s)	$C_7H_5N_5O_8$	1.4	1.61	1.97	G&P
Tetryl (s)	$C_7H_5N_5O_8$	1.7	2.48	1.42	G&P
TNT (s)	$C_7H_5N_3O_6$	1.635	2.1	2.34	G&P
Octol (s)	25/75 wt% TNT/HMX	1.80	3.01	1.72	D&C
TNT (l)	$C_7H_5N_3O_6$	1.47	2.14	1.57	D&C

Data may apply over only certain input ranges
[a](s) denotes a solid; (l) a liquid; (e) extrudable
[b]D&C denotes Dobratz and Crawford [28]
G&P denotes Gibbs and Popolato [49]
D1 denotes Dick et al. [26]
D2 denotes Dick et al. [24]; low pressure fit
G denotes Gustavsen et al. [53]; high pressure fit

then impacted on a high impedance window have been used to provide high time resolution unreacted Hugoniot information [94]. The Sandia National Laboratories Z-pinch magnetic-compression machine ([54], [55], and [86]) has been used to produce isentropic loading of HE samples [3, 4, 52, 63]; this provides information about an isentrope on the EOS surface. Also other forms for the unreacted explosive EOS developed by Andrews [1], Hayes [56], Davis [21] and others are available; see Menikoff [79].

An experimental method that provides several unreacted Hugoniot points in a single experiment is that of multiple magnetic particle-velocity measurements on a shock initiating explosive as the wave grows to a detonation. The jump in each individual gauge measurement is assumed to be unreacted and since the shock velocity is measured during the buildup process with "shock tracker" gauges (see Sect. 1.5.3), the jump in particle velocity at the front can be paired with the shock velocity at that position to produce an unreacted Hugoniot point from each gauge. Using this method, up to ten data pairs can be obtained in a single experiment. Information from several multiple-magnetic-gauge shock-initiation experiments on PBX 9502 is shown in Fig. 1.9 [53]. The multiple magnetic-gauge method will be discussed in more detail in Sect. 1.5.3.

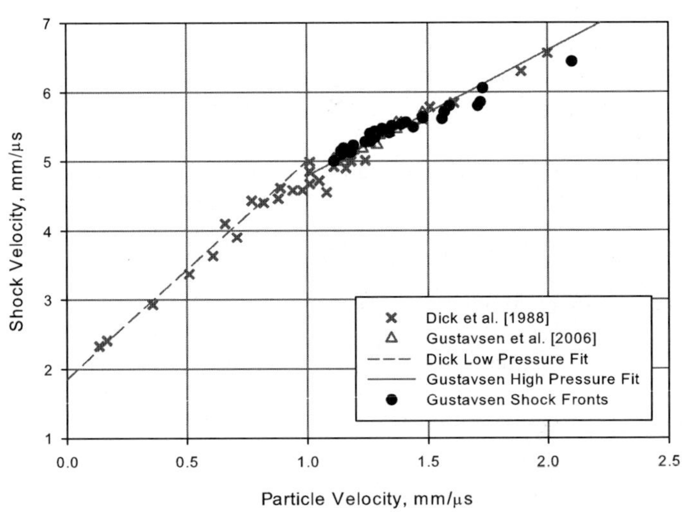

Fig. 1.9. Plot of PBX 9502 Hugoniot data from magnetic particle velocity measurements from Gustavsen et al. [53]. In these multiple gauge experiments, each gauge record is assumed to be a shock so that an unreacted Hugoniot point can be obtained from it by using the shock velocity obtained from the experiment. Data from three experiments are shown producing a total of 30 new Hugoniot points. The backdrop for this figure is Fig. 1.8.

1.3.3 Reaction Products Hugoniot and EOS

All explosives produce gases upon reaction., i.e., the products Hugoniot lies above the unreacted Hugoniot (see Fig. 1.6). Obtaining information about the reaction products Hugoniot is quite challenging and an ongoing area of research. Let us assume that at the end of the reaction zone (the reaction zone will be discussed in Sect. 1.4.1), the products do not continue to react and are at a high temperature/high pressure state (called the Chapman–Jouguet or CJ state) in which they can do work on the surroundings. It is this assumption of no further reaction that leads to several of the methods used to estimate the chemical reaction-zone length as discussed below. A number of years ago, it was realized that a "cylinder test" could be used for this purpose [69]. A cylinder test consists of a carefully prepared copper cylinder that is filled with pressed explosives – see Fig. 1.10 [16, 60]. Electrical pin switches are placed along the external cylinder wall to determine the detonation speed. The expansion of the copper cylinder wall produced by the detonating explosive is measured with a streak camera as a function of time to determine the work the expanding gas explosive products are doing on the copper.

Estimates of the reaction products Hugoniot are usually made by using a 2-D reactive hydrodynamics computer code to model the cylinder test; the parameters of an assumed EOS form are varied until agreement with the measured copper wall speed as a function of time is obtained. These parameters are then used for the particular explosive in other situations. This method was pioneered at Lawrence Livermore National Laboratory by Kury et al. [69] and Lee et al. [72] using the JWL EOS described in Lee et al. [71]. It is

$$P = A\left\{1 - \frac{\omega}{R_1 V}\right\} \exp^{-R_1 V} + B\left\{1 - \frac{\omega}{R_2 V}\right\} \exp^{-R_2 V} + \frac{\omega E}{V} \tag{1.8}$$

and the pressure (P_s) on the isentrope as a function of volume is

$$P_s = A \exp^{-R_1 V} + B \exp^{-R_2 V} + C V^{\{-\omega+1\}}, \tag{1.9}$$

where A, B, and C are constants in GPa pressure units, R_1, R_2 and ω are pure numbers, $V = v/v_0$ is the volume of the detonation products divided by the volume of the undetonated HE, P and P_s are pressure in GPa, and E is the detonation energy per unit volume in GPa-m^3 m^{-3}.

Derivation of JWL parameters has been carried out for a number of explosives by various workers. Dobratz and Crawford [28] have compiled the information from these studies; a number of representative explosives from this reference are given in Table 1.3.

Other methods are being studied to obtain information about the reaction products EOS including a variant of the cylinder test called the "sandwich test" [61] in which a slab of explosive is detonated along a line by a line wave generator and the detonating slab drives a thin metal layer on both sides. This type of test was originally conceived at Lawrence Livermore National

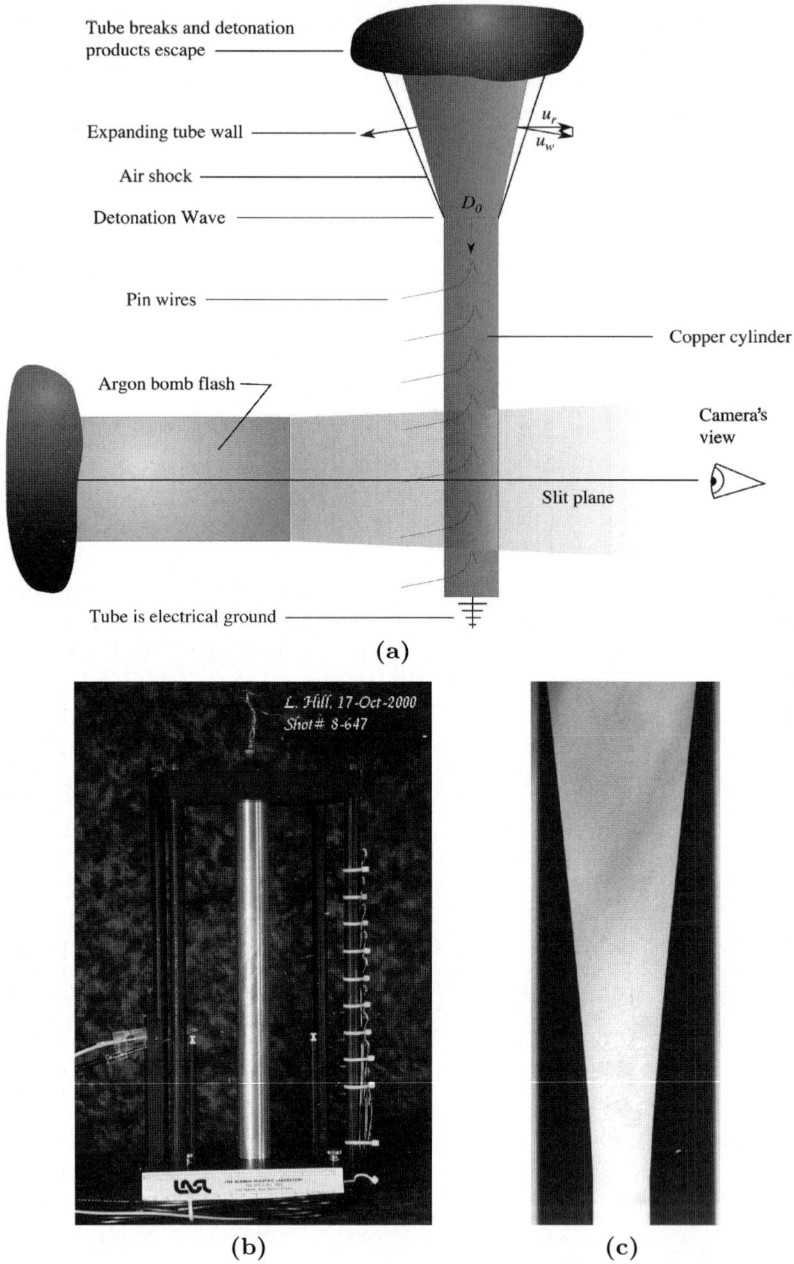

Fig. 1.10. Cylinder test setup by Catanach et al. [16]. (**a**) is a schematic of the cylinder test showing the various aspect of the test including the copper cylinder, the pin wires to measure detonation speed, and an argon bomb lighted profile of the expanding cylinder taken by a streak camera. (**b**) is a picture of one of the tests. (**c**) is a streak camera record of the expanding copper cylinder vs. time (time increases in vertical direction). This figure was provided by L. G. Hill, Los Alamos National Laboratory [60].

Table 1.3. Reaction products parameters for several high explosives – JWL form[a]

Common name[b]	Density ($g\,cm^{-3}$)	A Mbar	B Mbar	C Mbar	R_1	R_2	ω
Comp A-3(s)[c]	1.65	6.113	0.1065	0.0108	4.4	1.2	0.32
Comp B(s)[d]	1.717	5.242	0.077678	0.01082	4.20	1.100	0.34
HMX(s)	1.891	7.783	0.07071	0.00643	4.20	1.00	0.30
HNS(s)	1.65	4.631	0.08873	0.01349	4.55	1.35	0.35
HNS(s)	1.40	3.665	0.06750	0.01163	4.8	1.4	0.32
HNS(s)	1.00	1.627	0.1082	0.00658	5.4	1.8	0.25
LX-10-1(s)[e]	1.865	8.807	0.1836	0.01296	4.62	1.32	0.38
LX-17(s)[f]	1.90	4.46	0.01339	0.01306	3.85	1.03	0.46
NM(l)	1.128	2.092	0.05689	0.00770	4.40	1.20	0.30
PBX 9404(s)[g]	1.84	8.524	0.1802	0.01207	4.6	1.30	0.38
PBX 9407(s)[h]	1.60	5.73187	0.14639	0.01200	4.6	1.40	0.32
PBX 9501(s)[i]	1.84	8.524	0.1802	0.01207	4.55	1.30	0.38
PBX 9502(s)[j]	1.895	4.603	0.09544	0.01343	4.0	1.70	0.48
PETN(s)	1.77	6.170	0.16926	0.00699	4.40	1.320	0.25
PETN(s)	1.50	6.253	0.2329	0.01152	5.25	1.60	0.28
PETN(s)	1.26	5.731	0.2016	0.01267	6.00	1.80	0.28
Tetryl(s)	1.73	5.868	0.10671	0.00774	4.40	1.20	0.28
TNT(s)	1.63	3.712	0.03231	0.01045	4.15	0.95	0.30

[a]Data are from Dobratz and Crawford [28]
[b](s) denotes a solid; (l) a liquid
[c]Comp A-3 composition 91/9 wt% RDX/Wax
[d]Comp B composition 63/36/1 wt% RDX/TNT/Wax
[e]LX-10-1 composition 94.5/5.5 wt% HMX/VitonA
[f]LX-17 composition 92.5/7.5 wt% TATB/Kel-F 800
[g]PBX 9404 composition 94/3/3 wt% HMX/Nitrocellulose/Plasticizer
[h]PBX 9407 composition 94/6 wt% RDX/Exon 461
[i]PBX 9501 composition 95/2.5/2.5 wt% HMX/Estane/BDNPA-F
[j]PBX 9502 composition 95/5 wt% TATB/Kel-F 800

Laboratory [71]. This test setup is shown in Fig. 1.11. This type experiment has the advantage of not stretching the driven metal walls and, therefore, larger expansions can be observed. This type test is also modeled using a 2-D computer code to estimate EOS parameters.

Other EOS forms besides the JWL are being studied to try to upgrade the fit to match various types of initiation experiments, including the multiple magnetic-gauge experiments.

1.4 Detonation Phenomena

In this section we present the classical theory of 1-D detonation, implications of 2-D steady flow, followed by a discussion of some experiments used to obtain steady detonation data. Detonation parameters of some of the more

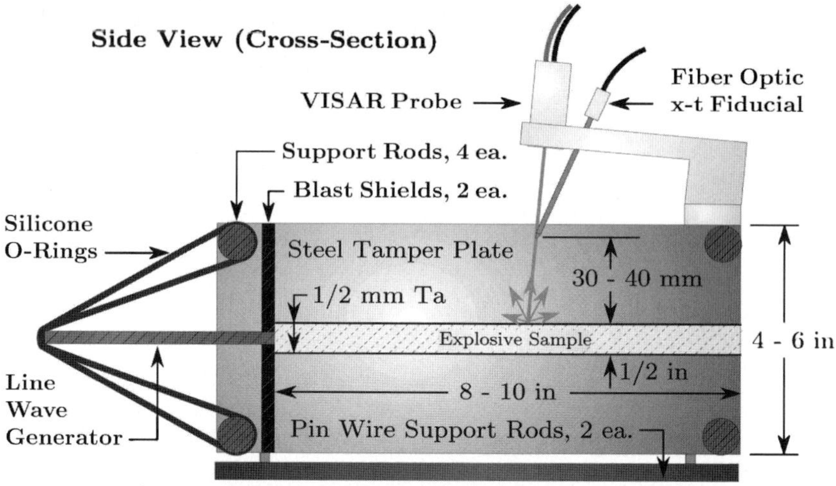

(a)

(b)

Fig. 1.11. Sandwich test as designed and performed by Hill [61]. (**a**) shows a cross-section view of the test through the center of the assembly shown in (**b**) of this figure with the centerline from the top to the bottom. (**b**) is a picture of a sandwich test. An explosive line-wave generator initiates the explosive sample along the entire end. The detonating explosive then pushes the top and bottom tantalum plates up and down, respectively. Pins are arranged down the side to measure the detonation speed. A framing camera records the movement of the upper plate, as does a VISAR velocity interferometer system, and a fiber optic pin that measures the arrival of the plate at its end. This test measures the plate pushing ability of the detonation reaction products. This figure was provided by L. G. Hill, Los Alamos National Laboratory.

common high explosives are given along with some measurement methods used to obtain these parameters. A short discussion of observations leading to ideas of three-dimensional (3-D) flow in detonations is included.

1.4.1 1-D Steady Detonation

The CJ theory of planar steady detonation is the simplest case, therefore it will be considered first. In this case there is no resolved reaction zone. Next we give a discussion of ZND theory; this theory includes a finite chemical reaction zone. Finally a discussion is given of the flow following the end of the reaction zone – called a "Taylor" wave flow [99].

Chapman–Jouguet Theory

We now derive the Chapman–Jouguet (CJ) theory of steady planar detonation. Equations (1.4) and (1.5) are used to do this. In order to close the set of equations, we need to specify a relationship between the explosive's internal energy per unit mass (E), its pressure (P), and the volume per unit mass (v); i.e., an equation of state (EOS). A simple such relationship, commonly used, is

$$E(P, v, \lambda) = \frac{Pv}{\gamma - 1} - q\lambda. \tag{1.10}$$

In (1.10), γ is a constant characteristic of the material, q is the chemical heat stored in the material per unit mass, and λ is a variable that measures the proportion of the chemical energy that has been released into the flow; $\lambda = 0$ and $\lambda = 1$ correspond to no and complete chemical energy release, respectively. Note that in this type of formulation no knowledge is required concerning the actual individual chemical processes occurring or their dependence on thermodynamic state. The $Pv/(\gamma - 1)$ part of (1.10) is called a "γ-law" equation of state. For condensed-phase explosives, a γ value of 3 gives a reasonable description of many observed detonation phenomena. This EOS form is particularly useful since the detonation properties of interest can be obtained analytically.

Use of (1.10) in (1.5) yields

$$\left[\frac{Pv}{\gamma - 1} - q\lambda\right] - \left[\frac{P_0 v_0}{\gamma - 1} - q\lambda_0\right] = \frac{1}{2}P(v - v_0). \tag{1.11}$$

For condensed-phase explosives, it is a very good approximation to set $P_0 = 0$ in (1.11), because of the very high pressures (P) which they generate. Furthermore, within the assumption of zero chemical reaction-zone length, $\lambda = 0$ before shock wave passage and $\lambda = 1$ immediately afterward. With these assumptions, (1.11) becomes

$$\left[\frac{Pv}{\gamma - 1} - q\right] - \frac{1}{2}P(v - v_0) = 0. \tag{1.12}$$

Note that except for the v_0 term, (1.12) only depends on properties of the fully reacted material. Elimination of P from (1.12) using (1.4) yields a quadratic equation for the specific volumes possible at the detonation shock

$$v^2 - Bv + C = 0, \tag{1.13}$$

where $B = 2\gamma v_0/(\gamma+1)$ and $C = (\gamma-1)\left[1 + 2q/D^2\right] v_0^2/(\gamma+1)$. The discriminant $(B^2 - 4C)$ determines the properties of the roots of (1.13). At this point, the value of "D" in the constant C is free. For B^2 greater than, equal to, or less than $4C$, there are, respectively, two, one, or zero real roots of (1.13). The case of a single real root (i.e., a unique solution) corresponds to

$$D_{CJ}^2 = 2q\left(\gamma^2 - 1\right). \tag{1.14}$$

This is the Chapman–Jouguet solution. For D values smaller than that given by (1.14), there are no real solutions consistent with the conservation laws and the assumption of zero chemical reaction-zone length.

Using (1.14) in (1.13) and (1.4) shows that $v_{CJ} = \gamma v_0/(\gamma+1)$ and $P_{CJ} = \rho_0 D_{CJ}^2/(\gamma+1)$, where $\rho_0 = 1/v_0$ and use of the v_{CJ} result in (1.1) gives $u_{CJ} = D_{CJ}/(\gamma+1)$.

It is of interest of compute various Chapman–Jouguet values for a typical high performance explosive. If we use typical experimental values of $v_0 = 1/(1.8)\,\mathrm{cm^3\,g^{-1}}$ and $D = 8.0\,\mathrm{km\,s^{-1}}$ with $\gamma = 3$, one finds that $P_{CJ} = 290$ kbar and $v_{CJ}/v_0 = 3/4$. Note that the yield strength of common steels is ca. 5 kbar and that the volume per unit mass of the explosive is reduced by ca. 25% by the detonation shock. For these detonation parameters, one megawatt of power is generated by a 1.9 cm^2 area on the detonation front.

Another important property of the CJ solution that can be derived is that the flow at the CJ plane is exactly sonic. That is, a sound signal generated at that plane travels exactly at the speed of the shock wave. This fact will be proven at the end of the Taylor wave section (see below). This property isolates (i.e., protects) the detonation shock wave from events in the flow following the detonation and, therefore, partially accounts for the detonation shock's stability.

1.4.2 Zeldovich–von Neumann–Doering Theory

Zeldovich [111], von Neumann [82], and Doering [29] (ZND) advanced a theory of steady planar detonation beyond the CJ theory by finding a solution of the flow equations with a resolved chemical reaction zone. We will only give a qualitative discussion of their results and refer the reader to other sources for a more rigorous treatment [39, 46].

ZND assumed that the detonation process consists of a shock wave that takes the unreacted explosive from its initial state $(P, v, \lambda) = (0, v_0, 0)$ to a "spike" point $(P_s, v_s, 0)$ (see Fig. 1.12). Here we are using a notation similar to that used in the previous section. As implied by Fig. 1.12, in traversing

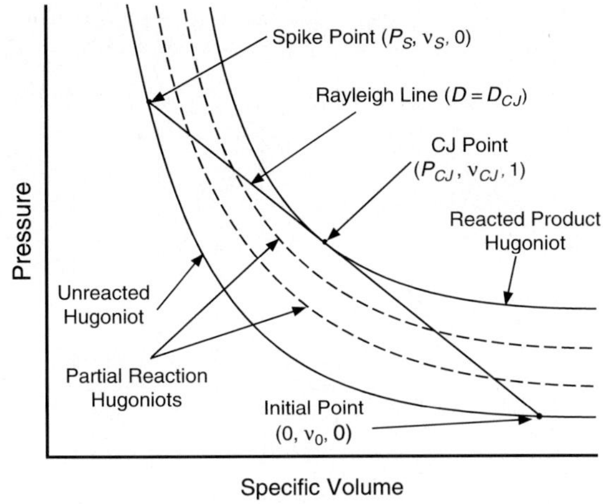

Fig. 1.12. Pressure vs. specific-volume plane construction to show the relationship between the unreacted and reaction products Hugoniots and the various states on the detonation wave profile.

the ZND reaction zone, one proceeds down the Rayleigh line from the spike point $(P_s, v_s, 0)$ to the CJ point $(P_{CJ}, v_{CJ}, 1)$ crossing all the partial reaction Hugoniots in the process. The most striking feature of this model is that as the chemical energy is released into the reaction zone, the pressure *falls.* There is direct experimental evidence that this is true [89]. The detonation speed is determined from the common value of the Rayleigh line and the *fully reacted* Hugoniot; again one finds $D_{CJ}^2 = 2q\left(\gamma^2 - 1\right)$. That is, the detonation shock wave and the end of the ZND reaction zone move with a speed identical to that of the CJ model. The end of the ZND reaction zone is a sonic point which shields the chemical reaction zone from perturbations in the flow behind the reaction zone. Within this model, one rejects detonation solutions with $D > D_{CJ}$ because in this case the flow is subsonic at the point of complete reaction and, therefore, is not protected by a sonic point from the following flow. Solutions with $D < D_{CJ}$ are impossible because there is no path down the Rayleigh line to reach the point of complete reaction.

A sketch of a 1-D detonation wave propagating from left to right is shown in Fig. 1.13. The spike point (called the von Neumann spike or the chemical peak), which is a result of the initial shock taking the material from the unreacted state to a point on the unreacted Hugoniot determined by the detonation speed (i.e., the Rayleigh line), is shown. This is an unreacted state at a very high temperature and pressure that starts the chemical reactions. Behind the shock is a region of reaction which is a subsonic flow so that the chemistry can support the initial shock by sonic disturbances. Chemical reactions go on in this region leading to the gaseous products of the detonation. At the end of the reaction zone is the CJ point or sonic point. This is followed by the Taylor wave or rarefaction region behind the detonation wave. Flow

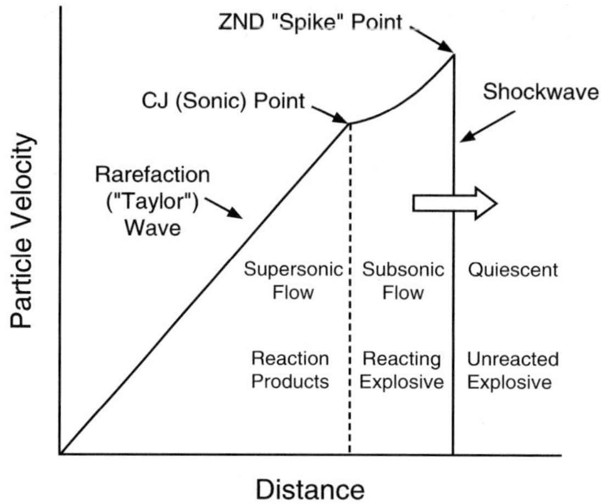

Fig. 1.13. Sketch in the particle velocity vs. distance plane of a detonation wave traveling from left to right showing the leading shock wave, von Neumann spike, chemical reaction zone, and Taylor wave regions.

in this region is supersonic and so perturbations rising there (other than new shock waves) cannot reach the sonic point. Therefore, after transients have damped out, the 1-D detonation wave propagates steadily. The shape of the pressure profile in the reaction zone depends on the detailed chemical kinetic processes taking place there. Although this is a description of a 1-D flow, detonation wave measurements indicate that it is descriptive of some 2-D detonation flows also. More discussion about measurements of detonation waves is given in Sects. 1.4.6 and 1.5.3.

The initial shock and very early time (picoseconds) flow in a detonation wave undoubtedly involve complicated processes. What these processes are exactly is unknown and is being currently studied by very fast laser-based research. A review chapter with many references on this work is contained in Moore et al. [81]. Tarver [97, 98] has presented a hypothetical picture of the processes involved in the approach to equilibrium which he calls the "non-equilibrium ZND model" (see Fig. 1.14). This model is based on postulated ideas of what processes are likely to be involved; none of the early time phenomena has been experimentally observed yet. This remains a challenge to the detonation experimentalists.

1.4.3 Taylor Wave

So far we have examined the flow in a detonation from the "spike" point to the sonic point at the CJ plane. In both the CJ and the ZND models it is assumed that the detonating material is inviscid; i.e., transport processes (viscosity, diffusion, etc.) can be neglected. In such a fluid the only other

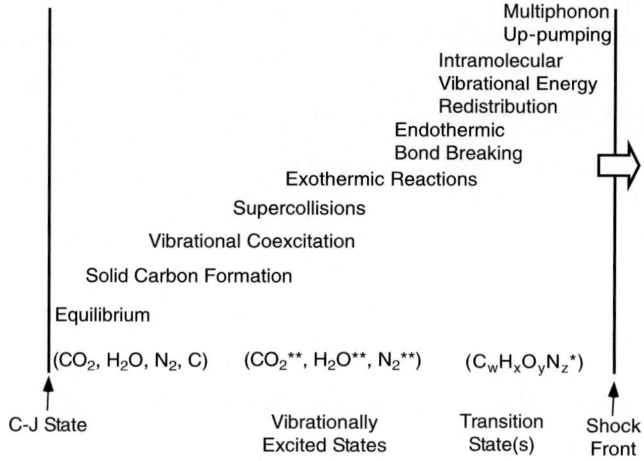

Fig. 1.14. Hypothetical processes involved in the non-equilibrium ZND detonation reaction zone as envisioned by Tarver [97]. The processes labeled are thought to be important in the area of the reaction zone as shown although the extent of each area is unknown.

sources of entropy production (i.e., irreversibility) are chemical reactions or shock waves. In the simple picture we are treating, the only shock in the system is the detonation shock and at the CJ plane the chemistry is complete. Therefore, the flow behind the CJ plane is isentropic. The isentropes for a γ-law EOS are given by

$$P = C\rho^{\gamma}, \tag{1.15}$$

where C is a constant.

The mechanical equations governing the fluid flow behind the CJ plane, under these circumstances are the equations governing mass and momentum conservation, i.e.,

$$\frac{\partial \rho}{\partial t} + u\frac{\partial \rho}{\partial x} + \rho\frac{\partial u}{\partial x} = 0, \tag{1.16}$$

and

$$\frac{\partial u}{\partial t} + u\frac{\partial u}{\partial x} + \gamma C\rho^{\gamma-2}\frac{\partial \rho}{\partial x} = 0, \tag{1.17}$$

where (1.15) has been used to re-express the pressure term in the momentum equation (1.17).

Taylor [99] recognized that since there are no pertinent space or time scales in the part of the flow behind the CJ plane that it should be self-similar. He introduced the self-similarity variable $z \equiv x/t$ into (1.16) and (1.17) to express this observation. Use of this variable into the two partial differential equations reduces them to the ordinary differential equations

$$(u - z)\frac{d\rho}{dz} = -\rho\frac{du}{dz}, \tag{1.18}$$

and

$$(u - z)\frac{du}{dz} = -\gamma C \rho^{\gamma-2}\frac{d\rho}{dz}. \tag{1.19}$$

Division of (1.19) by (1.18) produces the equation

$$\left[\gamma C \rho^{\gamma-3}\right]^{1/2} d\rho = du. \tag{1.20}$$

This equation integrates to yield the Taylor wave flow behind the reaction zone, which is

$$u = \frac{D_{CJ}}{\gamma - 1}\left[\frac{2\gamma}{\gamma+1}\left(\frac{\rho}{\rho_{CJ}}\right)^{\frac{\gamma-1}{2}} - 1\right]. \tag{1.21}$$

In obtaining (1.21), we have evaluated the integration constant, resulting from the integration of (1.20), at the CJ point. That is, on the CJ isentrope from which the Taylor wave expansion begins and have used the CJ values $P_{CJ} = \frac{\rho_0 D_{CJ}^2}{\gamma+1}$, $u_{CJ} = \frac{D_{CJ}}{\gamma+1}$, and $\frac{\rho_{CJ}}{\rho_0} = \frac{\gamma+1}{\gamma}$ to simplify the expression.

Equation (1.21) can be used in (1.19) to determine the mass density dependence on space and time in the Taylor wave which is

$$\rho(x,t) = \rho_{CJ}\left[\frac{(\gamma-1)\frac{x}{t} + D}{\gamma D}\right]^{\frac{2}{\gamma-1}}. \tag{1.22}$$

Use of (1.22) in (1.21) gives the particle speed dependence on x and t, which is

$$u(x,t) = \frac{2\frac{x}{t} - D}{\gamma+1}. \tag{1.23}$$

Figure 1.13 shows the $u(x,t)$ Taylor wave solution at a particular time. Note that if the back boundary of the explosive is a free surface (i.e., the density and pressure are zero there), then (1.22) shows that this surface will move at a speed of $x/t = -D/(\gamma-1)$. For a γ=3 material, the free expansion speed is half the detonation speed (i.e., $-D/2$).

Finally, using (1.15) we can show that the flow at the CJ point is sonic. The sound speed in a material is given by $c^2 = \left(\frac{\partial P}{\partial \rho}\right)_S$, where the subscript "S" indicates that the partial derivative is to be taken along the appropriate isentrope – here the CJ isentrope, so

$$c_{CJ} = \sqrt{\gamma C_{CJ}\rho_{CJ}^{(\gamma-1)}} = \frac{\gamma D_{CJ}}{(\gamma+1)}, \tag{1.24}$$

where the form of the result has been simplified by using results obtained above for ρ_{CJ} and P_{CJ}. Now since $u_{CJ} = D_{CJ}/(\gamma+1)$, we find by using (1.24) that

$$u_{CJ} + c_{CJ} = D_{CJ},$$

i.e., the CJ point is a sonic point.

1.4.4 2-D Steady Detonation

As noted above (within the ZND model) the speed of the shock wave in a steady 1-D detonation is independent of the chemical reaction-zone structure. Therefore, in this case, measurement of the 1-D detonation speed provides no information on the reaction-zone structure (e.g., its length). However, if one introduces a second space dimension into the experiment, this shortcoming is corrected. Suppose, for example, that one constructs a long right circular cylinder of an explosive (i.e., with cylinder length much greater than the cylinder diameter) and detonates the cylinder at one end. This type of experimental configuration is called a "rate stick" (see Fig. 1.15).

For a sufficiently large diameter cylinder, a 2-D wave will proceed through the cylinder and, after an initial transient, will propagate steadily. The speed of such a wave is dependent on the diameter of the cylinder. Optical experiments show that the center of such a steadily propagating wave leads the wave position at the cylinder's edge [37], i.e., the shock wave is curved. The pressure jump across the curved shock decreases as one moves from the center of the charge to its edge. This can be seen from the fact (since we are assuming an inviscid flow) that the appropriate shock-jump conditions across the curved shock require that the quantity being calculated is the value projected onto the local normal to the shock surface (see Fig. 1.16). This implies, e.g., that the pressure behind the shock wave decreases as one moves from the center of the charge to its edge. Since the chemical-heat release will certainly depend on the local thermodynamic state, this implies that the chemical heat will be released behind a steady 2-D shock in a way that is dependent on the geometric form of the detonation shock wave. Furthermore, this dependence can be explored by firing charges of various diameters – since this expands the range of wave shapes and speeds produced by the detonation waves and, hence, the range of thermodynamic states explored by the experiments.

It is important to note that a fundamental property of the 2-D geometry is that if one decreases the diameter of the cylinder being detonated below a certain value, it is impossible to propagate a steady detonation wave at all. This smallest cylinder diameter is called the "failure diameter" ($\equiv d_f$) of the material. If the explosive cylinder is confined within a second (inert) material, d_f depends on the properties of the confining material.

An important characteristic of an explosive is its "diameter-effect" curve. Such a curve is generated by measuring the detonation speeds in a sequence of rate sticks with varying diameters. The standard way to display the resultant data is to plot the speeds on the ordinate and the reciprocal charge radii on the abscissa (see Fig. 1.17). Note that a diameter-effect curve has finite length because all explosives have a failure diameter. This plane is useful because the $1/R \to 0$ intercept ($R \to \infty$) corresponds to a 1-D detonation. Thus, extrapolation to $1/R = 0$ gives an estimate of the planewave detonation speed.

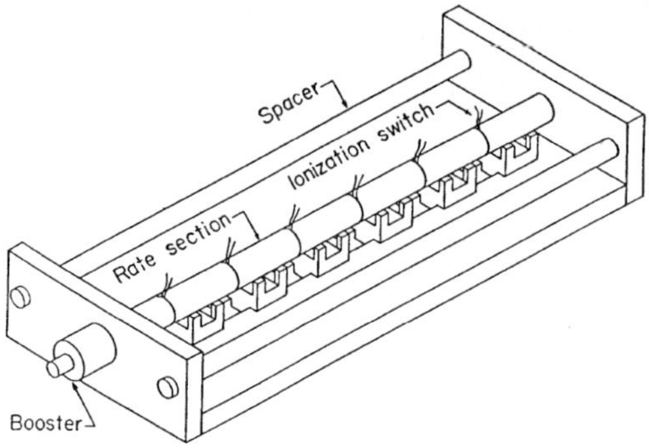

(a)

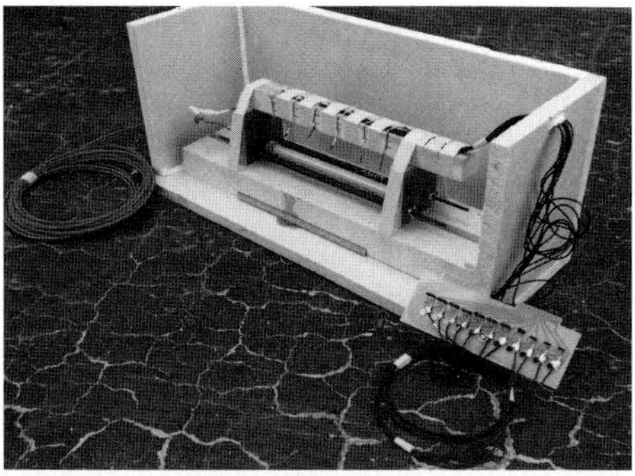

(b)

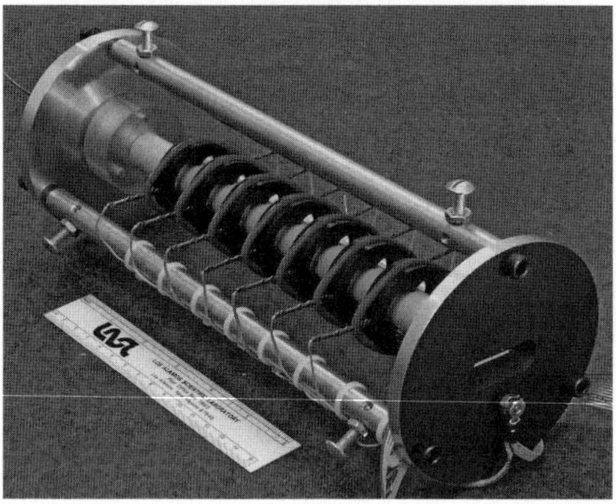

(c)

Fig. 1.15. Schematic of a rate stick showing the pieces of explosive with electrical switch wires ("pin switches") between each segment, shown in (**a**). The pieces are squeezed to a length equal to the sum of the individual pellet lengths. A picture of a rate stick ready to be fired is shown in (**b**). The detonator cable is on the left and the electrical pin-switch circuit is on the right. A modern rate stick designed by Hill et al. [59,62] which can be used at different initial temperatures and gives access for a detonation front curvature measurement is shown in (**c**).

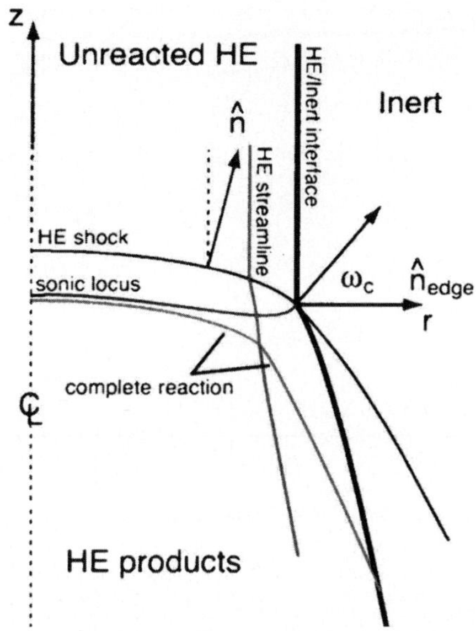

Fig. 1.16. Schematic of the right half of a 2-D detonation front showing the shock curvature, reaction zone, sonic locus, and the completion of reaction. This figure shows the component of the shock velocity vector normal to the shock surface. This component is called D_n. There is a second component of the detonation speed that is tangential to the shock surface; this component is not shown. This component corresponds to a perfect slip flow parallel to the shock surface. The D_n component determines the "jump conditions" across the shock (Illustration from John Bdzil, Los Alamos National Laboratory).

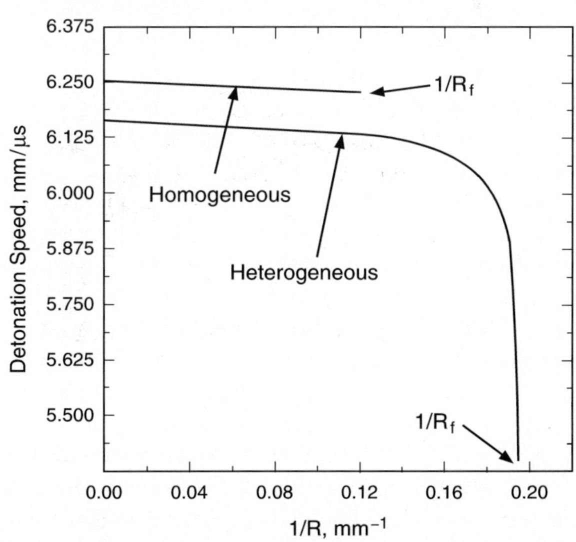

Fig. 1.17. Diameter-effect curves for homogeneous and heterogeneous high explosives. Steady 2-D detonation speed is plotted vs. the reciprocal of charge radius [15].

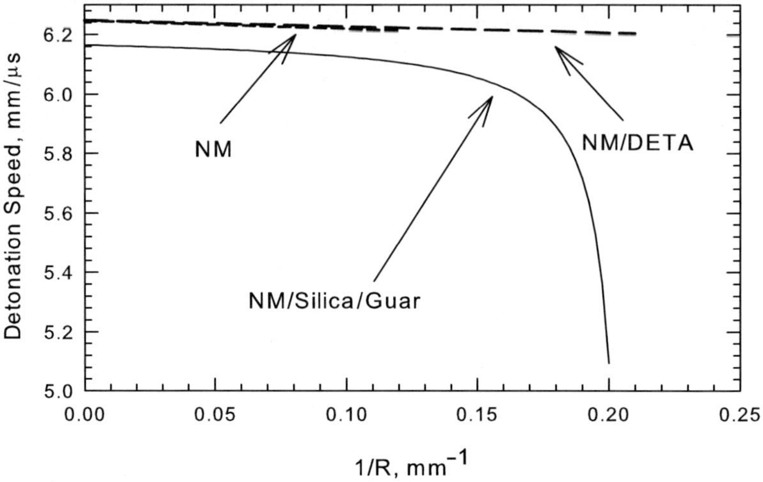

Fig. 1.18. Diameter-effect curves for three types of nitromethane based explosives: neat nitromethane (NM), NM with 0.03 wt% DETA (diethylene triamine), and gelled NM with 6 wt% of silica beads added to make it a heterogeneous explosive [35, 36].

Two types of diameter-effect curves are shown on Fig. 1.17. The straight line curve is characteristic of explosives not containing "hot spots" (e.g., liquid or single crystal). The curve with pronounced downward concavity is typical of explosives containing "hot spots" (e.g., porous granular materials).

As noted above, diameter-effect curves contain reaction-zone property information. The three curves shown on Fig. 1.18 illustrate this explicitly. In the research done to produce these curves, a liquid homogeneous explosive (nitromethane – NM – CH_3NO_2) was chosen as the canonical material and a diameter-effect curve was generated for it.

A second material was then constructed from the NM by adding a very small amount (0.03 wt%) of a chemical sensitizer and the diameter-effect curve of this material was measured [36]. This material, still being homogeneous, produced the same type of curve as NM (see the curve labeled NM/DETA). The effect of the sensitizer being to extend the curve (i.e., to reduce the material's failure diameter) and to reduce the curve's slope relative to neat NM. A third type of explosive was produced by adding a small amount of gelling agent to the NM along with ca. 6 wt% of very small silica beads; the gelling agent suspends the beads in the material [35]. This procedure produces the curve labeled NM/Silica/Guar on the figure. This curve is profoundly different from the other two curves because the beads undergo shock interactions with the detonation wave and, hence, modify the chemical-heat-release function of the material. Analysis of the data [36] obtained from these three cases (see Fig. 1.19) shows that the chemical sensitizer reduces the 1-D reaction-zone length by ca. 20%, while the silica beads roughly double its length. The silica beads are found to produce a chemical-heat-release function with

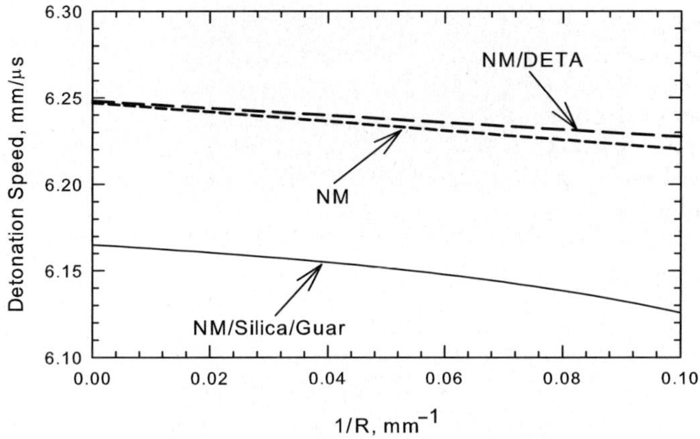

Fig. 1.19. Blowup of the diameter-effect curves in Fig. 1.18 showing the slopes of each of the curves. The ratio of the slopes is a measure of the relative reaction-zone length of these closely related materials [36].

greatly reduced thermodynamic-state dependence [37]. These are examples of how diameter-effect curves can be used to obtain information on reaction-zone properties.

This relationship between diameter-effect curve slopes to reaction-zone length ratios has been recently used to determine the reaction-zone lengths of various NM/DETA mixtures [42, 43] and also of the reaction-zone length of deuterated NM [44] relative to that of neat NM. In all cases, the slopes of these closely related materials order in the way expected. The experimental measurement of absolute reaction-zone lengths (and not just their relative length) has yet to be fully accomplished and is an area on ongoing research.

1.4.5 Detonation Shock Dynamics

There is more information present in a set of rate stick experiments than is contained in the resultant diameter-effect curve. This further information can be obtained if one measures the geometrical form of the shock wave's front [59]. Such measurements are usually made with a streak camera using reflected light from either the polished surface of the explosive or from a mirror placed in contact with the explosive's surface. The detonation wave destroys the reflectivity when it contacts the surface and, consequently, allows one to record the shock shape (see Fig. 1.20). As noted above only the component of the shock velocity normal to the shock front determines the state behind the shock front. That is, the fluid mechanical (and hence the thermodynamic) state behind a position of high shock curvature is more moderate than that behind a location on the shock which is less curved. Therefore, there is a continuum of shock (thermodynamic) states experienced in moving from the center line of a rate stick to its edge. Given a hydrodynamic-chemical kinetic

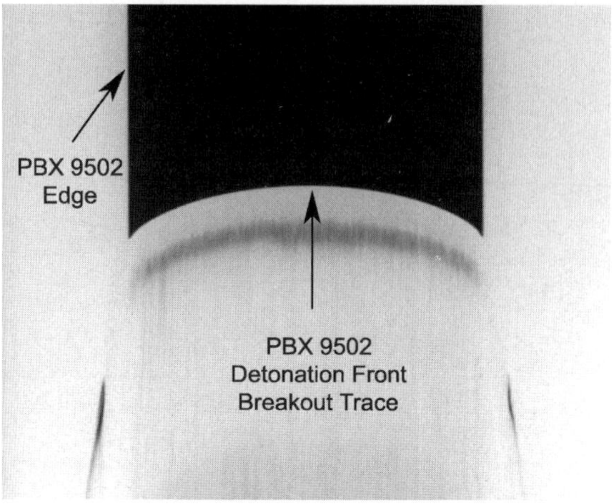

Fig. 1.20. PBX 9502 detonation front curvature measured by recording, with a streak camera, the front break out from the end of a rate stick [59, 62]. This trace came from measuring the break out trace from the rate stick shown in Fig. 1.15c. Photo from L. G. Hill of Los Alamos National Laboratory [59].

theory of the flow in a rate stick, one can view a set of measured shock shapes and detonation speeds as an inverse problem for defining the chemical kinetics driving the rate stick flow.

Bdzil and his coworkers [6, 8] have utilized this observation to generate a theoretical treatment for propagating a shock wave's evolution even for *nonsteady* detonation flows in complex explosive geometries. Their treatment is a generalization of Widom's theory [108] for shock shape evolution in chemically inert materials (called by Widom "shock dynamics"). Paralleling this, Bdzil termed his method "detonation shock dynamics".

In its simplest form this theory employs the relationship between the detonation speed normal to the shock (D_n) at any point to the shock curvature (κ) at that point. Bdzil and his coworkers derived differential equations that, given an experimentally obtained D_n vs. κ relationship, show how to evolve the time dependent shock wave shape for detonations propagating in complex explosive geometries. The D_n vs. κ function is determined from a set of shock curvatures and detonation wave speeds obtained in *steady-state* rate stick experiments. A somewhat thorny numerical analysis problem is accurately obtaining κ from the experimental shock shapes since the experimental data has to be numerically differentiated twice.

A central assumption of the theory is that the D_n vs. κ relationship is a universal function that controls the local shock evolution. Hill et al. [59] have experimentally shown that this assumption is valid (for the explosive PBX 9502) in steady-state rate stick flows for curvatures within ca. 2% of a rate stick's edge. Beyond this point the shock curvatures becomes large (i.e., $\kappa > 0.25\,\text{mm}^{-1}$) and the D_n vs. κ functions deviate for different diameter rate sticks.

The primary power of the D_n vs. κ method is that is allows an improved ability to accurately calculate the time dependent shock surface evolution in complex explosive geometries with greatly reduced numerical computing resources. Furthermore, it produces marked improvement of predicted shock shapes over earlier methods used to propagate detonations in such situations (e.g., Huygen's construction).

1.4.6 Reaction-Zone Measurements

The main features of the ZND model are borne out by experiment, e.g., direct measurement of the particle velocity vs. time in detonating explosives show a particle-velocity discontinuity (shock front) followed by a zone of decreasing particle velocity (see Fig. 1.21). X-ray radiography of the explosive products has been used to show that the flow behind the shock is sonic [34]. This work was done by tracking strong rarefaction waves introduced into the reaction products.

The profile shown in Fig. 1.21 was measured using a laser-based VISAR interferometer system [5] to transduce the interface particle velocity history of a detonation wave in NM interacting with a polymethyl methacrylate (PMMA) window which had a less than 1 μm thick aluminum mirror deposited on it. The shock impedance of the PMMA is between that of the unreacted NM and its reaction products, making it the least perturbing window possible so that the profiles shown are very close to what would be the in-situ profile in the NM. The time resolution of the measurement is about 2 ns.

Laser velocity interferometer measurements are the method of choice for measuring detonation wave profiles as the detonation wave interacts with a

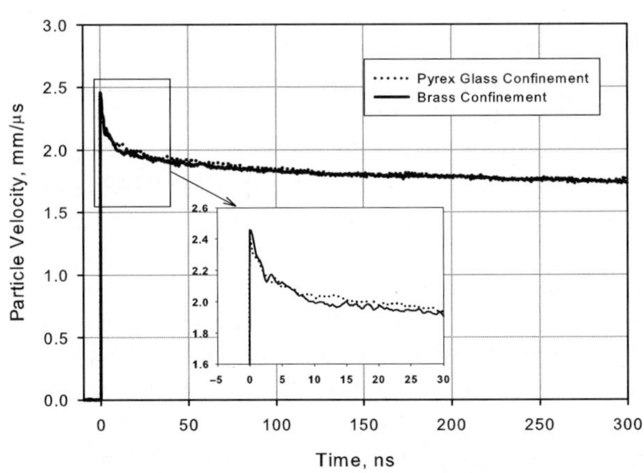

Fig. 1.21. Experimentally measured interface-particle velocity history showing the ZND reaction-zone structure in detonating nitromethane. The CJ point is estimated to be at about 15 ns after shock front passage. Two experiments are shown, one with a Pyrex-glass tube confinement and one with a brass-tube confinement, indicating that there is very little difference due to the different confining materials [93].

calibrated window material. These systems operate on the fact that laser light is Doppler shifted by the moving interface surface. The Doppler-shifted light is routed through an optics system that beats the light from one time against light collected from a slightly later time (time difference 100 ps to about 1 ns, depending on the optical system setup). This produces fringes, the number of which is directly proportional to the velocity change of the surface. The optical system can be adjusted to produce fringe constants between about 0.1 and 1.8 mm (μs)$^{-1}$/fringe (newer systems can go much higher and lower than this, depending on the optics design). The types of interferometers used are VISAR [5], Fabry–Perot [33], and ORVIS [11]. More recently ORVIS has been setup to look at a line rather than just a single point; this development will be very important in helping to understand the reaction zones of heterogeneous explosives [101,102].

Measurements of this kind have been made using various types of interferometer systems (with a few ns resolution) on PETN [95], TATB [89,96], HMX [50], TNT [70,89], and a few other materials. However, it is still difficult to tell where the CJ point is located because, contrary to what is shown in Fig. 1.13, there is no abrupt slope change in the particle velocity vs. distance relationship at the end of the reaction zone.

If the window impedance is quite different from the unreacted and reaction products Hugoniot, perturbations in the measurement occur. The degree of perturbation depending on the shock impedance mismatch. For example, if a lithium fluoride window is used, a shock is reflected back into the reaction-zone probably causing it to decrease in length; for a window of lower impedance than the high explosive (e.g., water) a rarefaction wave is reflected and one would expect a slowing down of the chemistry causing a consequential increase in the reaction-zone length. With a free surface, the reactions would be expected to be frozen at the state where the rarefaction reduces the pressure to zero. This technique has been used in Los Alamos National Laboratory "detonation mass spectrometer" experiments to sample various layers in a chemically reacting material [10,40].

A number of thin metal plate-push studies have been done in the past to try to determine the reaction-zone length in various detonating explosives [22,32]. These reaction-zone measurements are suspect because of the shape of the detonation wave (including the reaction zone and the Taylor wave) and the interactions that occur in the metal plates which result from the wave reflecting as a rarefaction wave from the plate free surface. These shock interactions cause spall of the metal plates unless they are backed by a window. This leads to errors in the reaction-zone length estimates. An example of a recent observation of spall phenomena using a new diagnostic, proton radiography, is shown on Fig. 1.22. This radiograph is of a copper plate pushed by detonating PBX 9501 and shows numerous spall layers in the plate as it moves. Because of these difficulties, reaction-zone lengths and times from these early studies should only be considered as approximate.

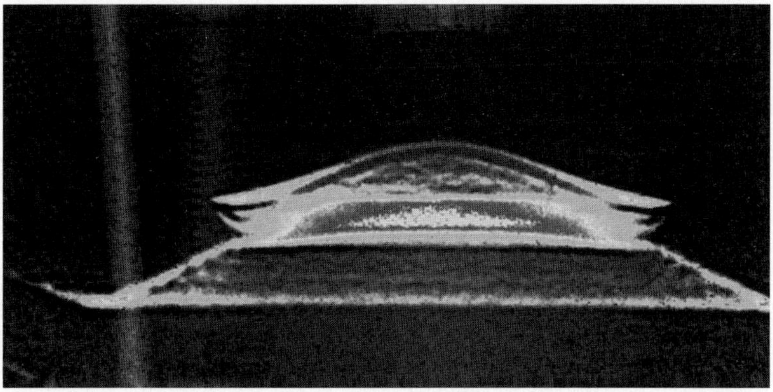

Fig. 1.22. Proton radiograph of a spalling copper plate which has been pushed by detonating PBX 9501. This work was done at the LANSCE accelerator at the Los Alamos National Laboratory. A detonator initiates a cylinder of PBX 9501 in contact with a copper plate. The detonation wave propagates through the PBX 9501 and pushes the copper plate. The plate spalls in a number of places because of the interaction of the rarefaction wave from the copper free surface with the reaction zone and Taylor wave which follow the front of the detonation shock [64].

1.4.7 Corner Turning

The ability of a detonation wave progressing through a high explosive to turn a corner is important for certain applications. The classic configuration for studying corner turning ability is a cylindrical charge that has an abrupt diameter change to a larger diameter. As the detonation wave encounters the larger diameter, rarefaction waves are generated at the corners and move toward the center of the charge. It is possible for the rarefaction waves to penetrate the reaction zone in such a fashion that the detonation wave is extinguished. Also, as the wave turns the corner, some amount of the HE can be left unreacted or only partially reacted giving rise to what are called "dead zones".

Several tests have been developed to measure corner turning properties. One is called the "mushroom" test and is shown in Fig. 1.23. In this test a small cylindrical charge is initiated by a detonator and it, in turn, initiates a hemisphere of explosive. The breakout of the detonation wave is monitored using a streak camera which looks at the charge head on and also records the light from mirrors on each side of the hemisphere so that the breakout on the sides can be recorded. From the streak record, a breakout pattern for the material is determined as a function of angle around the hemisphere. Breakout data from one experiment is shown in Fig. 1.24.

If the explosive does not do well at corner turning, "dead zones" can develop. These are areas where the explosive does not detonate even at late times. The presence of dead zones in corner turning experiments can be established by using X-ray or proton radiography to take a picture of the detonating

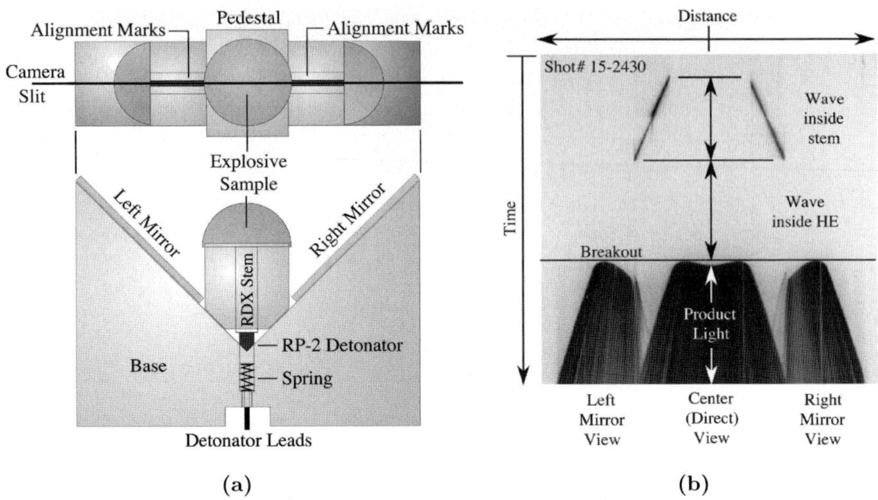

Fig. 1.23. Mushroom test to measure the corner turning ability of an explosive [58]. (**a**) is a schematic showing the detonator, RDX stem to initiate the sample, and the hemisphere of explosive being tested. Mirrors are placed on both sides of the sample to measure the breakout on the side. (**b**) is a streak recording of the breakout traces from the mushroom test showing the center breakout as well as both sides. Schematic and streak records were provided by L. G. Hill of Los Alamos National Laboratory.

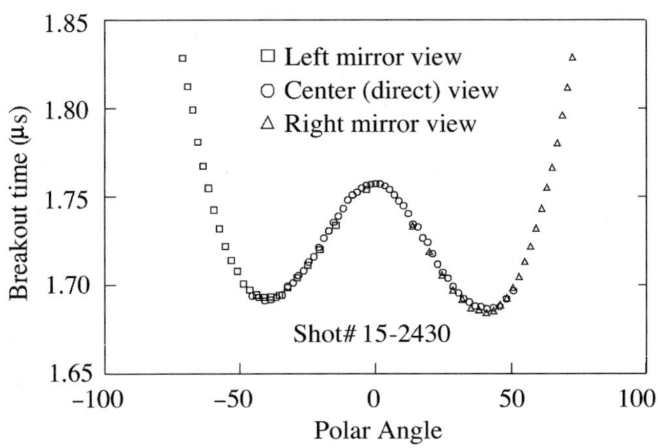

Fig. 1.24. Data reduction from a PBX 9502 mushroom test showing the high explosive's ability to turn corners [58]. The data are the breakout times for the various angles from the centerline of the sample. The parts of the data obtained from the center, right, and left traces (see Fig. 1.23b) are shown as different symbols. Samples with better corner turning ability spread to larger angles at earlier times. This test has been able to assess the effect of density and particle size on the corner turning ability of PBX 9502. This figure was provided by L. G. Hill of Los Alamos National Laboratory.

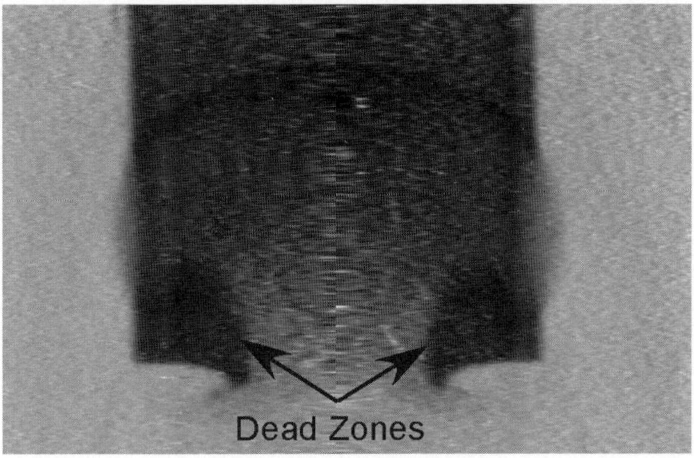

Fig. 1.25. Proton radiograph of a PBX 9502 experiment. A 12-mm diameter PBX 9502 donor cylinder is in contact with a 50-mm diameter PBX 9502 acceptor cylinder so the detonation wave moves through the donor and into the acceptor. The detonation wave must spread (corner turn) in the acceptor. The proton radiograph was taken after the detonation front has moved into the corner turning region. The detonation front is visible as a front moving upward. The dead zones are visible as dark regions [45].

explosive at various times. Proton radiography is the best diagnostic but it involves very large machines (e.g., LANSCE accelerator at Los Alamos National Laboratory) and difficult setups; therefore, a large number of these experiments have not been done. A proton radiograph is shown in Fig. 1.25. The dead zones show up as different density materials. It has not yet been possible to determine the extent of material reacted from these experiments. This research is ongoing from both an experimental as well as a theoretical standpoint (see [23]).

1.4.8 Detonation Properties of Selected Explosives

Some detonation properties of explosives are relatively easy to measure. These include detonation speed, diameter-effect curves, failure diameter, detonation wavefront shape, and corner turning ability. Well developed measurement techniques are used to make these measurements and provide information that is quite reliable. For example, if rate stick measurements are done carefully, accuracies of 1–2 $\mathrm{m\,s^{-1}}$ out of 8,000 $\mathrm{m\,s^{-1}}$ are possible [15].

Other detonation properties are much more difficult to measure (e.g., reaction-zone length). Measuring the CJ state conditions are difficult because of our inability to determine the location of the sonic plane. Probably the best measurement that has been made of a CJ state is that of PBX 9501 which came out of "overdriven" experiments to help determine the reaction products

Hugoniot [48]. "Overdriven" means the detonation is driven at a higher speed than the steady 1-D value. This was done by using explosively driven flyer plates to impact PBX 9501 at pressures considerably above the CJ state and then measuring the sound speed in the products using an optical method. In such a flow there is no sonic plane. The result of this measurement was a CJ state pressure of 34.8 ± 0.3 GPa. This compares to previous estimates ranging from 33 to 37 GPa. Similar experiments on PBX 9502 (a TATB based HE with a much longer reaction zone) have proven much more difficult because of the effects of the long reaction zone on the measurements.

Table 1.4 is a tabulation of some detonation properties for several explosives. Densities are given for each entry because the values are density dependent. Detonation speed is the most easily and most accurately measured detonation parameter and there is a considerable amount of data available, including initial density and temperature dependence. High detonation speeds are associated with high outputs (higher CJ state properties); this is apparent from the table entries. As indicated earlier, the CJ values are less reliable.

Von Neumann spike and reaction-zone data are only estimated for a few materials because of the measurement difficulties. Better measurements are now being made by laser velocity interferometry and in-situ gauge techniques. TATB has one of the longer reaction zones at about 2 mm and NM has one of the shorter ones at about 50 μm.

Explosives are often mixed with polymeric materials to make them easier to process, e.g., to press, mold, machine; such materials are called plastic bonded explosives (PBX). Some examples of PBXs are PBX 9404, PBX 9501, and PBX 9502. Parameters of other explosives, not considered here, can be found in Dobratz and Crawford [28] and Meyer [80]. TNT is often mixed with more sensitive or higher performance explosives to provide castable mixtures (examples are cyclotol and pentolite). Aluminum particles have also been added to provide high pressure at late times. PETN has been mixed with silicon rubber to form an extrudable explosive called XTX; this material has been used to make explosive plane-wave lenses and line-wave generators. There are other mixtures that have been developed to use as moldable explosives for various purposes.

Many properties depend on initial temperature, e.g., nitroglycerine has a failure diameter of 3.9 mm at an initial temperature of −20°C and 1.1 mm at 70°C.

1.4.9 3-D Detonation

We have discussed detonation in 1-D and 2-D. When the third space dimension is added, new interesting wave phenomena appear. These effects have been extensively studied both experimentally and theoretically in gaseous explosives. This work has been possible principally because the wave structure in gases is visible and recordable in experiments. Experiments can be designed

Table 1.4. Detonation properties of selected high explosives[a]

Material[b]	Composition	Theoretical density	Detonation speed[c]	Det. speed as function of density	Estimated CJ pressure[c]	Estimated Spike pressure[c]	Reaction Zone Length[c]	Failure Diameter[c]
		$(\mathrm{g\,cm^{-3}})$	$(\mathrm{mm\,(\mu s)^{-1}})$	$(\mathrm{mm\,(\mu s)^{-1}})$	(GPa)	(GPa)	(mm)	(mm)
Lead Azide(s)	—	4.80	5.3 (4.60)	—	—	—	—	0.4 – 0.6 (3.14)
Nitroglycerine(l)	—	1.60	7.7	—	25.3	—	0.2	2.2 [d]
PETN(s)	—	1.78	8.26 (1.76)	$3.19+3.7(\rho-0.37)$	30 (1.67)	≈37(1.76)[g]	—	$8.38<d_c<19$ (1.77)
XTX 8003(e)	80/20 wt%[e]	1.56	7.3 (1.53)	—	17 (1.55)	≈20(1.53)[g]	—	0.36 (1.53)[f]
Tetryl(s)	—	1.73	7.85 (1.71)	—	26 (1.71)	≈50(1.7)[g]	—	—
RDX(s)	—	1.82	8.7 (1.77)	$2.56+3.47\rho$	34 (1.77)	—	2.9 (1.80)	5.2 (0.9)
HMX(s)	—	1.905	9.11 (1.89)	—	39 (1.89)	≈52(1.89)[g]	—	—
PBX 9501(s)	95/5 wt%[e]	1.86	8.83 (1.84)	—	—	≈454(1.84)[g]	—	<1.52 (1.83)
PBX 9404(s)	94/6 wt%[e]	1.87	8.8 (1.84)	$2.176+3.6\rho$	37 (1.84)	—	—	<1.18 (1.846)
TNT(c)	—	1.65	6.93 (1.64)	$1.873+3.187\rho$	21 (1.63)	≈24(1.64)[g]	0.3 (1.63)	14.5 (1.62)
Cyclotol(c)	75/25 wt%[e]	1.77	8.3 (1.76)	—	31 (1.75)	—	—	6.0 (1.74)
Pentolite(c)	50/50 wt%[e]	1.71	7.53 (1.70)	—	25. (1.70)	—	—	6.7
Nitromethane(l)	—	1.13	6.35	—	13	20	≈0.05[h]	16.2[d]
Ammonium Nitrate(s)	—	1.725	5.27 (1.30)	—	—	—	—	12.7 (0.95)
TATB(s)	—	1.938	7.76 (1.88)	$0.343+3.94\rho$	29 (1.88)	≈32(1.88)[g]	—	$6.35<d_c<9.53$ (1.6–1.7)
PBX 9502(s)	95/5 wt%[e]	1.942	7.71 (1.90)	—	—	≈38(1.89)[g]	3.3 (1.895)	$8.<d_c<10.$ (1.894)
Nitroguanidine(s)	—	1.775	7.93 (1.62)	$1.44+4.015\rho$	—	≈42(1.69)[g]	—	$1.27<d_c<1.43$ (1.52)

[a]Data for this table obtained from Meyer et al. [80], Dobratz and Crawford [28], Gibbs and Popolato [49]
[b](s) denotes solid; (l) liquid; (e) extrudable/moldable; (c) castable
[c]Densities are given in parentheses after each property value in $\mathrm{g\,cm^{-3}}$
[d]Confined in Pyrex
[e]XTX 8003–PETN/Sylgard 182; PBX 9501 and 9404 –HMX/Plasticizer; Cyclotol–RDX/TNT; Pentolite–PETN/TNT; PBX 9502–TATB/Kel-F 800
[f]Confined in polycarbonate
[g]Estimated Spike Pressure calculated using the detonation speed from this table and the linear unreacted Hugoniot data from Table 1.2. Numbers should be considered rough estimates
[h]See Fig. 1.21

in which a detonation cell structure is easily observed. The exact nature and reason for the appearance of the cells is still an ongoing research area [84].

It is assumed that this cell phenomena (also called cross-wave structure) happens in condensed-phase explosives although the scale is much smaller. There have been several studies made to measure cells in condensed phase HEs. These observations have been made mostly in liquid explosives. A typical experiment uses NM diluted with acetone; this makes the cell structure visible in head-on framing camera pictures. Most experiments have used acetone loadings of 20 wt% or more [30, 46, 78, 103]. This would be expected to have a dramatic effect on the chemistry taking place. The only evidence that neat NM has these features on a very small scale comes from light reflection measurements [30]. These structures would be expected to affect reaction-zone measurements such as those shown for NM in Fig. 1.21 but it appears not to be the case. This is an indication that the structure is so small that it does not affect the few hundred μm diameter of a VISAR measurement (due to averaging) or that they are not present. New instrumentation such as line ORVIS interferometers should help determine if these structures are present in neat NM.

1.5 Shock Initiation Phenomena; Shock-to-Detonation Transition

Initiation of explosives can happen in a number of ways both planned and unplanned. The planned events are usually that of using the explosive for a designed purpose or experimental studies for the purpose of understanding a particular explosive material. Special care is taken to design and understand the initiation process to the point that the desired outcome is achieved.

Unplanned events occur when explosives are subjected to inputs (for example, high temperatures, mechanical impacts, etc.) that can cause the material to explode (run away chemical reaction) or possibly even detonate. These processes usually involve rapid burning (deflagration) and under special circumstances (such as heavy confinement) can result in a detonation. This process is called deflagration-to-detonation transition (DDT). Only in a few situations is this understood from a physics standpoint and, in those cases, it involves the formation of a shock wave leading to a shock-to-detonation transition (SDT). It is the SDT process we discuss in detail in this section. We will not discuss DDT initiation.

Since detonation is a self-sustaining, shock-generated chemical reaction, at some point the initiation process must lead to the formation of a shock. The most direct initiation method is to mechanically introduce a shock, e.g., with a thrown plate. If the resultant pressure input is sufficiently large, the shock will grow to detonation. Shock inputs obviously have a particular duration, depending on how they are generated. If the shock duration is long compared to the initiation time, it has no relevance. Durations of the same order as the initiation time or shorter can become important because pressure relief from the rear may quench the initiation reactions. Qualitative differences occur in SDT of homogeneous and heterogeneous materials, principally due to the hot spot phenomena generated in the latter case. “Hot spots” are high pressure/temperature regions produced locally by shock wave interaction with physical irregularities.

Most SDT experiments are designed to be 1-D so that the input is well understood and controlled.

Shock initiation depends on an HEs state, mass density, particle-size distribution, nature of the several constituent materials (if it is a mixture), initial temperature, etc. Solid HEs are usually pressed to high densities and the particle size and nature of the constituents may be changed in this operation. We will discuss both homogeneous and heterogeneous SDT in this section. This will be followed by experimental methods used to measure shock initiation properties and then a listing of some HEs initiation properties.

1.5.1 1-D Homogeneous HE SDT

Homogeneous high explosives are typically liquids or single crystals where there are no physical imperfections (e.g., bubbles or voids) that can cause perturbations in the input shock and the flow behind it. Homogeneous materials viewed with the macroscopic probes characteristic of detonation physics experiments appear uniform.

Homogeneous SDT has been studied in detail by Campbell et al. [13] and Chaiken [17], leading to the classical homogeneous initiation model shown in the time–distance diagram of Fig. 1.26. The model involves the following processes: the explosive is shocked and, after an induction period τ (which depends on the initial shock pressure), a thermal explosion occurs at the explosive/driver interface. This explosion produces a superdetonation (detonation in the explosive precompressed to higher mass density by the initial shock) that runs forward and eventually overtakes the initial shock. After the superdetonation overtakes the input shock, the detonation wave decays over time to a steady detonation.

The simplest theory predicts the induction time (time to thermal explosion) using Arrhenius type kinetics as

$$\tau = \frac{C_v R T^2}{AQE} e^{\frac{E}{RT}}, \tag{1.25}$$

where C_v is the constant volume specific heat, R is the gas constant, A is a collision factor, Q is the heat release, T is the shock temperature, and E is the Arrhenius activation energy. A number of assumptions, including no heat transfer, lead to this simplified form of a more general form developed by Frank–Kamenetskii [47]. Because of the exponential dependence of the

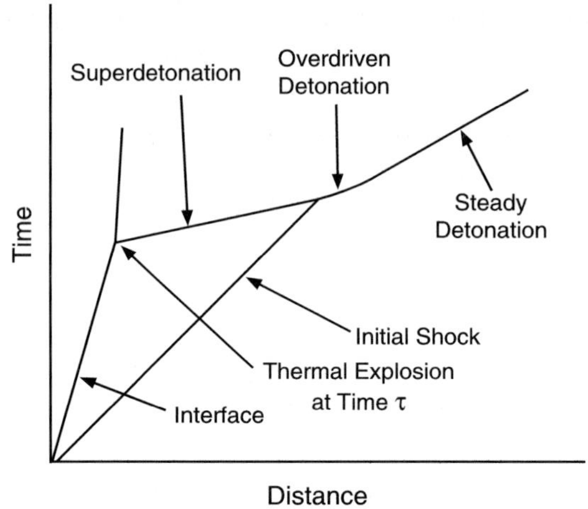

Fig. 1.26. Time–distance diagram for a homogeneous initiation process according to Campbell et al. [13] and Chaiken [17] in which a thermal explosion occurs at the driver-explosive interface, where the explosive has been shocked the longest, and a superdetonation (running into precompressed material) immediately develops. The superdetonation speed is higher than the normal detonation speed and overtakes the original shock, producing an overdriven detonation in the undisturbed material. Eventually, the overdriven detonation decays to a steady detonation.

Arrhenius rate on E, this relation indicates the induction time is a very sensitive function of the temperature (provided $E \gg RT$); i.e., a small increase in temperature yields a large increase in reactivity and a consequent large decrease in τ.

It seems reasonable to suspect that a superdetonation does not form immediately after the thermal explosion as depicted in Fig. 1.26 but rather that some type of extended reactive-wave buildup process occurs. Recent in-situ particle velocity measurements of this growth process in nitromethane have shown a developing reactive wave over relatively long time and distance scales. When these measurements are transformed into the time–distance plane, the trajectories of the various waves become those shown in Fig. 1.27. This modified homogeneous initiation model consists of the following processes: (1) the explosive is initially shocked, raising the pressure and temperature to the point that, after an induction time, a chemical reaction starts; (2) the reaction produces waves that coalesce and strengthen; (3) a reactive compression or shock wave is formed within the material; (4) the reactive shock builds in strength which may reach a steady superdetonation condition (detonation in a precompressed material); (5) finally this reactive shock overtakes the original shock wave producing an overdriven condition that eventually settles down to a steady detonation [91]. Examples of this behavior in several materials will be presented later in this section.

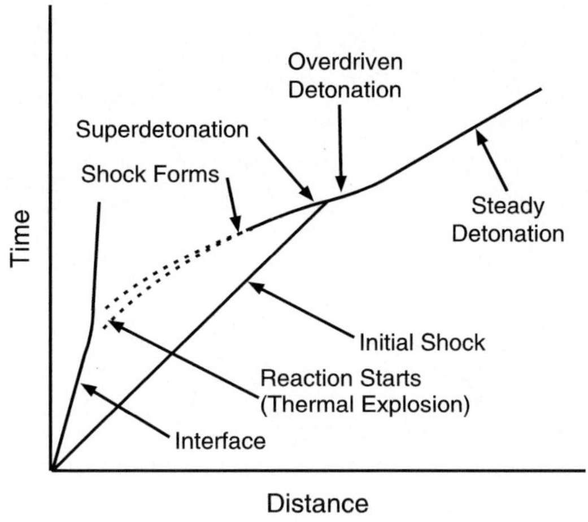

Fig. 1.27. Modified time–distance diagram for a homogeneous explosive SDT initiation process. It accounts for the processes observed in in-situ particle velocity gauging experiments on NM which indicate a buildup over significant time (distance) to the superdetonation [91].

1.5.2 1-D Heterogeneous HE SDT

Heterogeneous explosives are usually pressed, cast, or extruded into the shapes or parts desired. Here, heterogeneous means a material that contains any kind of imperfections that can cause fluid-mechanical irregularities ("hot spots") when a shock or detonation wave passes over them. Such hot spots cause associated space/time fluctuations in the thermodynamic fields (e.g., the pressure or temperature fields). These thermodynamic variations affect the local chemical heat-release rate. When averaged over a sufficiently large space scale, such variations convoluted with the underlying chemical rate(s) produce an average heat-release rate that is a combination of chemistry and mechanics.

Examples of conditions that cause hot spots are: (1) voids collapsing by impact of one side on the other, (2) shock wave propagation through irregular particles that cause complex shock interactions, (3) shock wave interactions between particles and voids that cause jetting, (4) plastic flow involving crystal breakage and shearing, and (5) shock impedance mismatch between components of the explosive that cause stagnation points and shock reflections and interactions. Based on this, one could expect differences in initiation behavior to occur due to, for example, the particle size of material used to fabricate the explosive and this is the case. Some of these have been modeled using wave propagation computer codes (e.g., see [75, 83]). Because of these hot spots, heterogeneous initiation is qualitatively different from that in homogeneous materials.

Heterogeneous SDT was also studied in detail by Campbell et al. [14], leading to much of the basic understanding that exists today. They showed

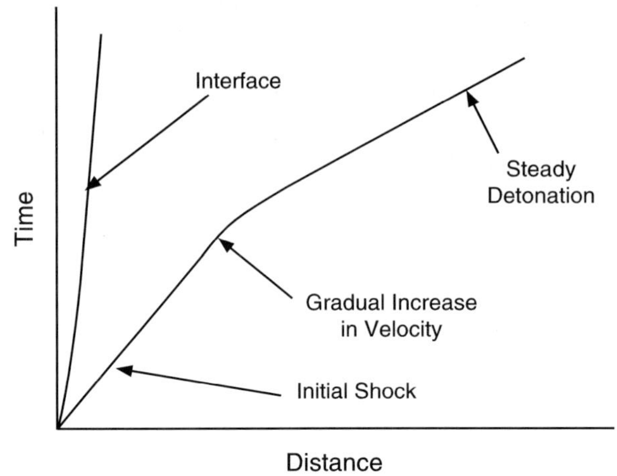

Fig. 1.28. Time–distance diagram for a heterogeneous explosive SDT initiation process. Much of the growth is in or near the front causing an increase in wave speed up to detonation.

that wave growth occurs at the front as well as behind it. This leads to a more smooth growth of the initiating shock to a detonation (see Fig. 1.28), in contrast to the abrupt changes observed in the homogeneous case (see Figs. 1.26 and 1.27). In this case, no overshoot in the detonation speed is observed, leading to the idea that the transition to detonation occurs at or near the shock front.

Little is known about the exact mechanisms that produce the shock wave growth other than that the hot spots develop at the inhomogeneities, stimulating chemical energy release locally and making the explosive initiate at much lower pressure inputs than would otherwise be the case. For example, pure liquid nitromethane can be made to initiate with a sustained shock input of about 8–9 GPa, while the same material with a large number of carborundum heterogeneities will initiate at 2–3 GPa [14]. Many experimental studies have looked at hot spot phenomena, but the size and nature of the inhomogeneities have not been sufficiently experimentally controlled to allow detailed understanding.

Because the hot spots produce reaction locally, heterogeneous explosives have a much less sensitive state dependent chemical heat release rate than homogeneous materials. The dependence on the bulk material pressure and temperature (all important in the homogeneous case) is secondary.

1.5.3 Experimental Methods For Making Shock Initiation Measurements

For many years, the method of choice for making shock initiation measurements was to use explosively driven inputs to the sample which was machined in the form of a wedge. The explosive driver system was either in direct contact with the wedge or a plate was thrown at the wedge by the driver system. In the last few years, gas-gun driven plates have been used because the guns are available and the input to the wedge is a well characterized flat-topped wave. Also, multiple magnetic gauges and manganin pressure gauges have been developed to the point that in-situ measurements of the growth to detonation are possible. We will discuss both explosively driven method and the gas gun driven method for making these measurements in this section.

1.5.4 Explosively Driven Wedge Experiments

Explosively driven wedge experiments have provides a great deal of SDT data on various explosives. Campbell et al. [13, 14] used this technique in their experiments that led to our basic understanding of SDT. This method yields distance- or time-to-detonation data from sustained shock input experiments. A precisely machined explosive wedge is mounted on a plane wave explosive driver/attenuator system that is usually 8–12 in. in diameter (see Fig. 1.29a). The wedge is illuminated by a light source (usually explosively-compressed argon) and a streak camera is used to monitor the light changes as a function of time along the wedge surface as the shock interacts with the surface; such a trace is shown in Fig. 1.29b. The point at which the distance–time streak record has a large change in slope is the distance- or time-to detonation for that particular pressure input. This is clearly shown in Fig. 1.29c in which the slope (shock speed) is plotted vs. distance of shock travel into the wedge. By doing several experiments, in which the input shock pressure is varied, a set of pressure vs. distance- or time-to-detonation is compiled. This is then plotted on a "Pop-plot" (see below).

A careful set of explosively driven wedge experiments was done on PBX 9502 at several different temperatures by Dallman and Wackerle [20] (see Fig. 1.29). It was possible to do these experiments because the heating and cooling could be done remotely at an a firing site that had this capability. Run distance-to-detonation vs. input pressure are shown in Fig. 1.30 for three temperature for PBX 9502 in a log–log plot. The data shows that PBX 9502 (TATB based) becomes as shock sensitive as PBX 9404 (HMX based) when the temperature of the PBX 9502 is increased to 250°C. This plot is called a "Pop-Plot" after Ramsay and Popolato [85] who first plotted shock initiation data in this way (in a log–log plot).

Most of the initiation sensitivity data that is available in the literature comes from explosively driven wedge tests; only recently have gas gun experiments become important as will be discussed below. The Pop-Plots for a number of other explosives are shown on Fig. 1.31 to give some idea of the relative sensitivity of the various HE materials. No liquids are shown on this plot because they are homogeneous explosives and are mapped in a different Pop-Plot plane. The densities for each of the materials is given in the figure caption because shock initiation sensitivity depends strongly on density. It is

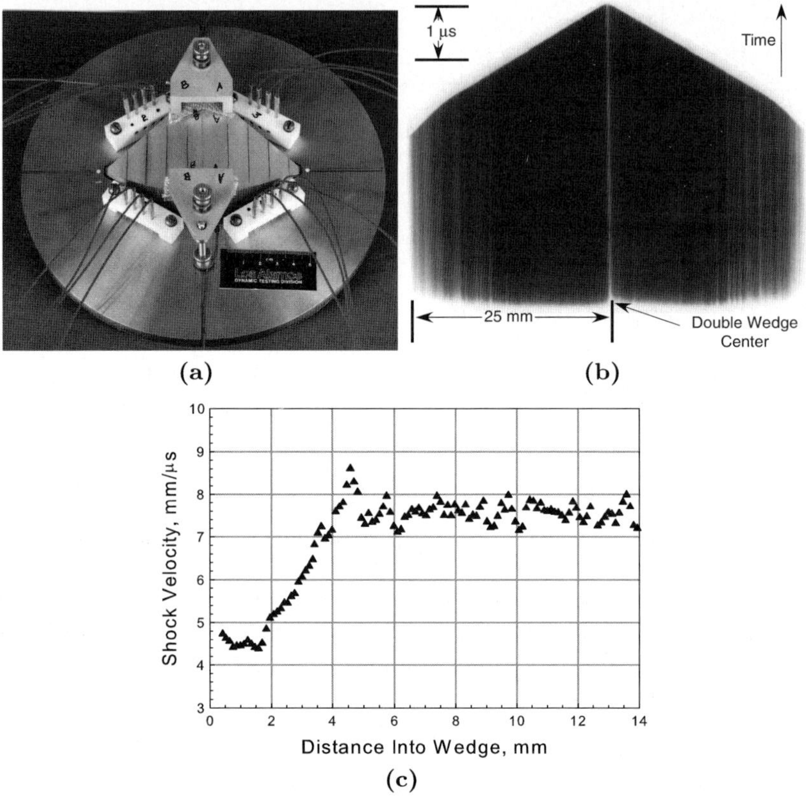

Fig. 1.29. (**a**) The configuration for an explosively driven wedge experiment. The 16 wires are for electrical pins to measure the input to the explosive wedge. (**b**) A streak-camera record showing shock contact with the wedge surface. Shock contact reduces the reflectivity and therefore the light intensity. The dark area in the streak record photograph is the region of high reflectivity. (**c**) The shock speed plotted vs. distance into the wedge obtained from the space–time trajectory of the shock contact with the wedge surface. The distance-to-detonation point is at about 4 mm. Pictures and data are from Dallman and Wackerle [20].

interesting to note that low density PETN is the most sensitive HE plotted with a line to the far left in the figure. Three different densities of PETN are shown and, as expected, sensitivity decreases at density increases. TATB and nitroguanidine (NQ) are the most insensitive and are on the far right. It is interesting to note that pressed TNT is considerably more sensitive than cast TNT at nearly the same density. This is due to the number of hot spots in the pressed material being much greater. The cast TNT line falls essentially on top of the PBX 9502 line (which has been dashed so the two lines can be distinguished).

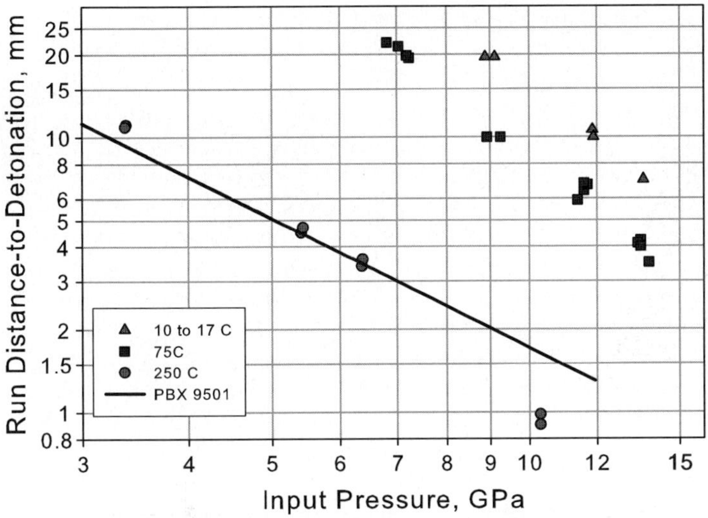

Fig. 1.30. Shock initiation data in the form of a Pop-Plot of Input Pressure vs. Distance-to-detonation for PBX 9502 at ambient, 75°C, and 250°C. A line for ambient temperature PBX 9501 is also shown, indicating that PBX 9502 at 250°C is about the same sensitivity as ambient temperature PBX 9501. Data for PBX 9502 are from Dallman and Wackerle [20]. Line for PBX 9501 is from Gibbs and Popolato [49].

1.5.5 Multiple Gauge Measurements

In the last two decades it has become possible to do in-situ gauging on explosive materials to measure the SDT initiation from a mechanical standpoint. Both manganin gauges [104, 105] and magnetic particle velocity gauges have been used [92,106,107] for this purpose. These methods provide unprecedented insight into the buildup from the shock to the detonation. They give direct information on wave profiles in the flow and contain implicit information on how the chemical heat release is occurring. Because the gauges are embedded in the explosive, there is concern about the perturbations they might cause. However, these gauges are being made as thin as possible and out of materials with a shock impedance very close to that of the explosive; this minimizes the perturbations.

A number of studies have been done at LLNL using manganin gauges to measure the pressure profiles of several explosives during the shock initiation process, including TATB based materials at temperatures up to 250°C. The manganin gauges are somewhat thicker than the magnetic gauges used at LANL so they are more perturbing. These manganin gauges have been put in at up to *seven* positions in an explosive sample [104], providing information about the pressure profile as it builds during shock initiation.

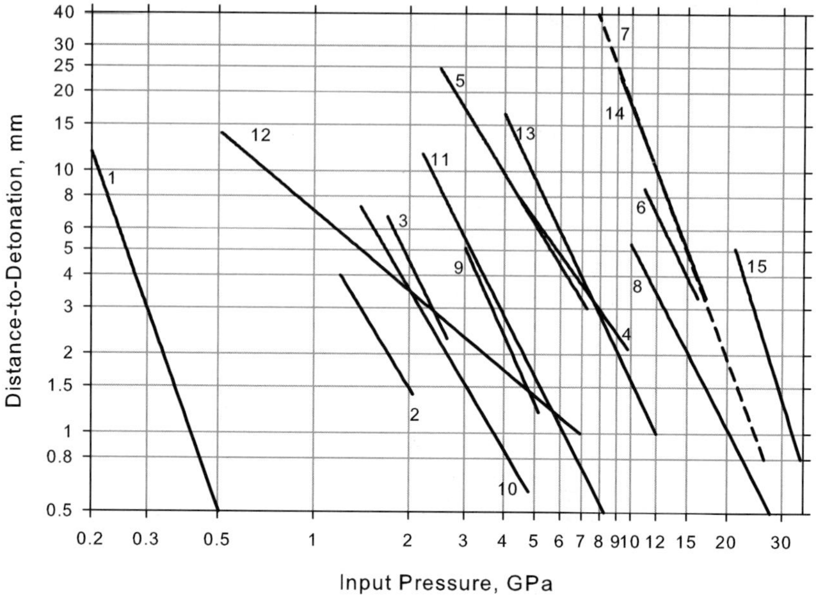

Fig. 1.31. Shock initiation data in the form of a Pop-Plot of Input Pressure vs. Distance-to-detonation for several different HE materials. Explosives (with densities in $\mathrm{g\,cm^{-3}}$) are (1) PETN (ρ_0=1.0); (2) PETN (ρ_0=1.6); (3) PETN (ρ_0=1.75); (4) HMX (ρ_0=1.89); (5) PBX 9501 (95/2.5/2/5 wt% HMX/Estane/(BDNPA/BDNPF), ρ_0=1.844); (6) TATB (ρ_0=1.876); (7) PBX 9502 (95/5 wt% TATB/Kel-F 800, ρ_0=1.889); (8) TATB Superfine (ρ_0=1.81); (9) XTX 8003 (80/20 wt% PETN/Sylgard 182, ρ_0=1.53); (10) PBX 9407 (94/6 wt% RDX/Exon 461, ρ_0=1.60); (11) Tetryl (ρ_0=1.70); (12) Tetryl (ρ_0=1.40); (13) Pressed TNT (ρ_0=1.63); (14) Cast TNT (ρ_0=1.635); (15) NQ (ρ_0=1.688). The length of the lines is an indication of the applicable region of the data in input pressure space. Data were taken from Gibbs and Popolato [49] and Dobratz and Crawford [28].

We will concentrate on the multiple magnetic particle velocity gauge method used at LANL because it is becoming a very important means of providing reactive wave profiles for modeling SDT data for both homogeneous and heterogeneous explosives. Figure 1.32 shows how such an experiment is put together for a solid explosive sample. A gas-gun driven projectile provides the shock input. A gauge membrane 60 μm thick is epoxied between two precisely machined pieces of explosive. It is important to make sure the gauge ends (where the measurement takes place) are parallel with the top of the sample. The gauge is usually installed at an angle of 30° with respect to the front of the sample. A small cut is taken off the top of the glued assembly to ensure a smooth flat surface at the projectile–target impact face. The target is carefully placed in the magnetic field so the gauge ends are perpendicular to the magnetic field lines. After impact, the shock moves past each gauge and a voltage signal proportional to the gauge length, magnetic field

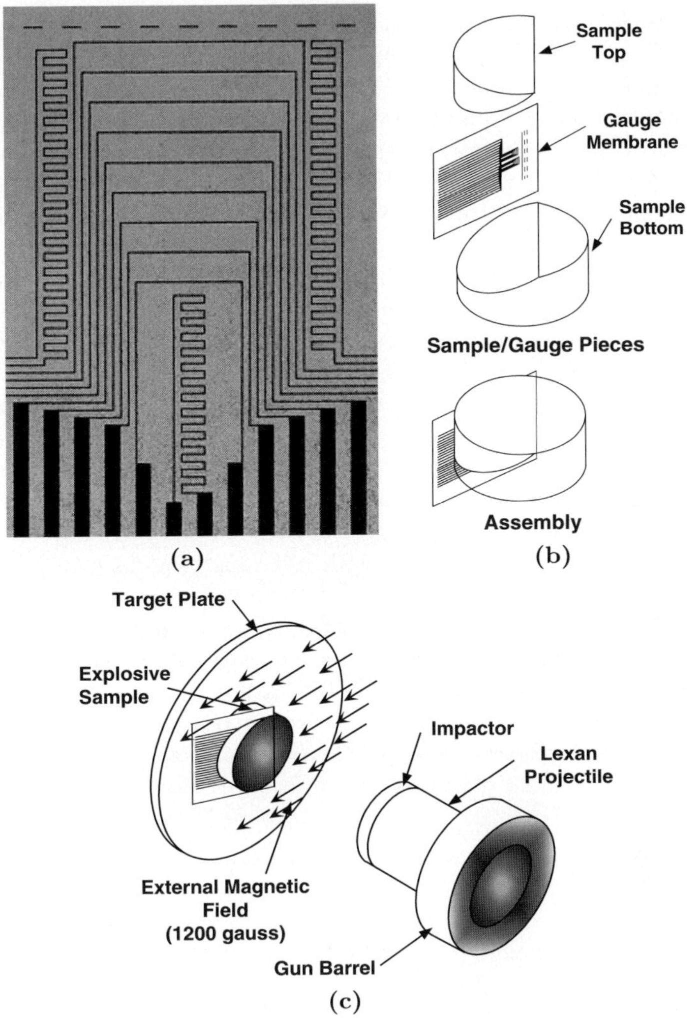

Fig. 1.32. Various aspects of a magnetic particle velocity gauge experiment using an explosive sample: (**a**) shows the gauge pattern – nine particle velocity gauges and three "shock tracker" gauges, (**b**) shows how the gauge is installed in the sample, and (**c**) shows how the target is situated in the gas gun. The technique was originally developed by Vorthman [106] and Vorthman et al. [107] and has been innovated over the years to the method presently used at LANL [92].

strength, and the gauge velocity vs. time is produced; this voltage is recorded by a digitizer.

An important aspect of the multiple magnetic-gauge method is the amount of experimental information that comes from a single experiment. The initial gauges provide an unreacted Hugoniot point for the material. All the gauges provide in-situ measurements of the particle velocity as a function of time at several Lagrangian positions, up to and including detonation (depending on the experimental parameters). If the SDT experiment is such that the explosive sample reaches a detonation condition during the measurements, the position and time detonation is reached can be determined providing a Pop-Plot point for the explosive. The "shock trackers," three in each experiment, track the shock front vs. time, providing information for a distance–time diagram to determine the Pop-Plot point and the shock wave front speed as a function of time and position. This allows several unreacted Hugoniot points to be determined for each experiment as shown in Fig. 1.9 discussed earlier in this chapter. An estimate (few percent accuracy) of the detonation speed is also possible from the distance–time diagram. Obviously this is a large amount of information from a single experiment, much more than can be obtained from an explosively driven wedge experiment. For this reason, the magnetic-gauge experiments have largely replaced the wedge experiments at LANL. This method has proven accurate enough to distinguish samples of PBX 9501 with density differences of $0.005\,\mathrm{g\,cm^{-3}}$ by Gustavsen et al. [51] and between samples of PBX 9502 with different particle sizes [53].

These kind of experiments have been performed on both solid and liquid explosives producing SDT data on heterogeneous and homogeneous explosive materials, respectively. As indicated in the early part of this section, there are dramatic differences between the initiation characteristics of these two types of materials because of the hot spots developed at the heterogeneities in the heterogeneous explosives. We will present multiple gauge waveforms from experiments that will emphasize these differences below.

1.5.6 Multiple Magnetic-Gauge Measurements: Homogeneous Explosives

In order to do experiments on homogeneous explosives, it was necessary to modify the sample shown in Fig. 1.32 so that the gauge would be suspended in a cavity which could be filled with a liquid explosive and then sealed up. This was done by making a polymethyl methacrylate (PMMA) cell that was coated on the inside with epoxy to keep the liquid explosive from attacking the PMMA, especially when using nitromethane-based liquid explosives. The early experiments using this technique resulted in the modification to the homogeneous initiation model as shown in Fig. 1.27. This experiment was done on nitromethane sensitized with a chemical (diethylene triamine (DETA)). Recent experiments have been done on neat nitromethane, producing waveforms that corroborate the modified homogeneous initiation model. An example is shown in Fig. 1.33.

It is apparent from this figure that the modified homogeneous initiation model applies. If the shock tracker data are used, along with the gauge arrival time data, an distance–time plot can be constructed (see Fig. 1.34).

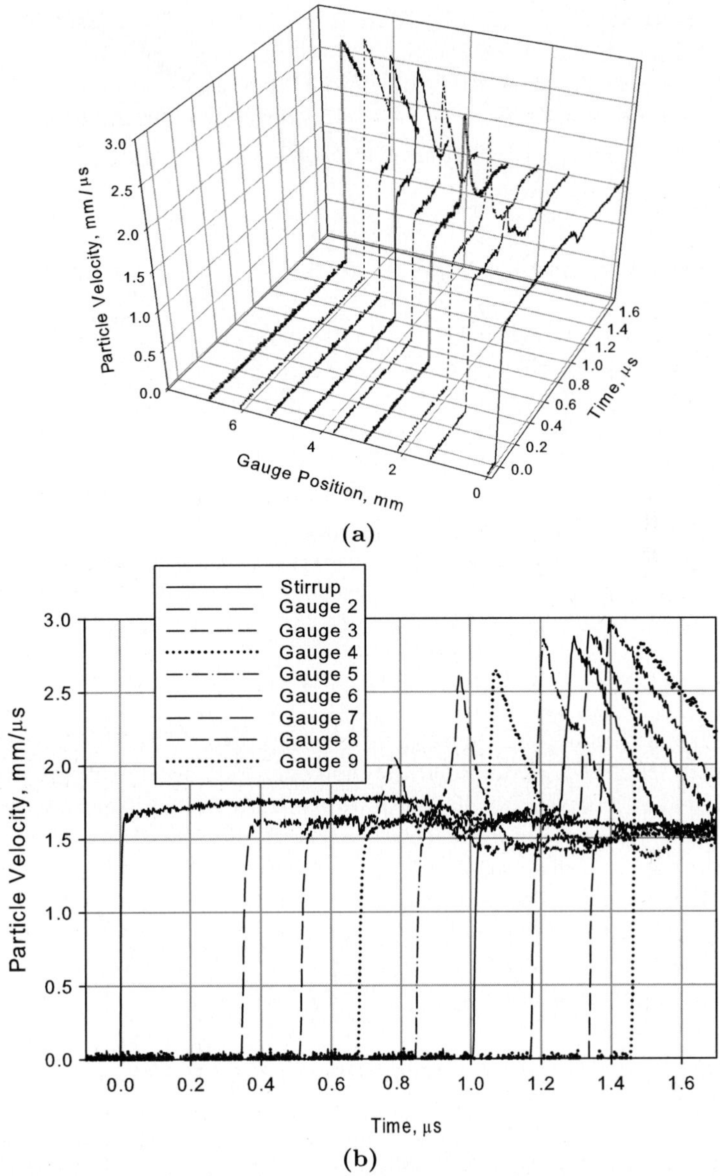

Fig. 1.33. Homogeneous shock initiation data for neat nitromethane from a multiple magnetic-gauge experiment. (**a**) is a 3-D plot showing each waveform with the reactive wave growing into a steady superdetonation by gauge 5 which overtakes the initial shock at gauge 8. This can be seen more easily by looking at the 2-D plot in (**b**). Also shown is data from a "stirrup" gauge that is a single gauge located at the input to the nitromethane. The gas-gun-driven projectile imparted a sustained shock into the nitromethane of 9.1 GPa on this experiment.

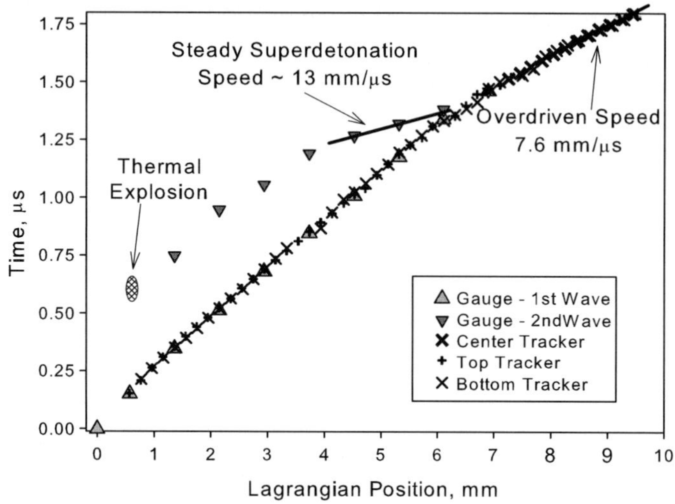

Fig. 1.34. Time–distance plot of the data from Fig. 1.33 obtained from shock tracker data and gauge arrival time data. It is thought that the thermal explosion that starts the reaction occurs where it is shown, slightly into the nitromethane.

This diagram is obviously very similar to that shown in Fig. 1.27, indicating that this experiment agrees with the modified homogeneous initiation model. Several liquid explosives have been studied using this gauging technique with the data indicating the homogeneous model applies to all of them. The liquid explosives that have been studies include neat nitromethane, DETA sensitized nitromethane (mixtures containing DETA wt% from 0.2 to 5.), deuterated nitromethane, isopropyl nitrate, and FEFO (bis-(2-fluoro-2,2-dinitroethyl)formal). Based on this data, the homogeneous initiation model appears to be universal for all liquid explosives. There is some evidence that it may also apply to single crystal solid explosives. However, orientation effects and initiation anomalies have been observed [25].

1.5.7 Multiple Magnetic-Gauge Measurements: Heterogeneous Explosives

Heterogeneous explosives have voids, grain boundaries, etc., which perturb the flow so that hot spots are produced. These hot spots are energy concentrations that greatly increase the chemical reaction rate near them. Because of this the initiation and detonation behavior is qualitatively different from that of homogeneous explosives. Experiments in which multiple magnetic gauges are used to make in-situ measurements of the flow field substantiate this difference. Instead of a thermal explosion driven reaction, reaction starts immediately after the passage of a shock at the hot spots and builds as a function of time. In materials with a large number of heterogeneities, the wavefront grows directly to a detonation. As the number of heterogeneities is decreased, there is some growth in the front and some behind the front, with the reactive wave behind the front catching up to the front as it turns into a detonation. There is very little overshoot in the detonation speed as the transition to a detonation occurs, in contrast to the large overshoot in velocity and the relatively long time and space required to settle down to a steady detonation in homogeneous explosives.

The heterogeneous HEs that have been studied most using in the magnetic-gauge technique are high density PBX 9501 and PBX 9502 weapon explosives. Neither of these is heterogeneous enough that the major part of the growth is in the wavefront. However, there is a definite difference between PBX 9501 and PBX 9502, with the latter appearing to be more heterogeneous, probably because of the pores that occur in the TATB grains that make up the pressings. Particle velocity waveforms from a PBX 9501 experiment are shown in Fig. 1.35a. These waveforms show some growth at the front and a relatively large reactive hump after the front that is overtaking it. This can be contrasted to waveforms obtained from a PBX 9502 experiment in which there is more growth in the front and less of a reactive hump, as shown in Fig. 1.35b. This is thought to be the result of porosity in the TATB grains which make up the pressed sample. However, this is yet to be substantiated.

What is apparent from Fig. 1.35 is that a difference in heterogeneity of an HE sample can be measured using this gauging technique. In a study of various PBX 9501 pressings at slightly different pressed densities, it was possible to measure differences between samples in which the density varied by only $0.005\,\mathrm{g\,cm^{-3}}$ [51]. This method has yet to be exploited in a study in which controlled heterogeneities are put in an explosive sample; the authors feel that a continuum of wavefront profiles ranging from completely heterogeneous (growth only in the front) to purely homogeneous would be possible if this were done.

A time distance diagram has been plotted for the experiment shown in Fig. 1.35a by using the shock tracker and gauge data. The arrival time at each gauge, along with the arrival time at each shock tracker, is shown in Fig. 1.36. This diagram looks very much like Fig. 1.28 in which there is a relatively slow change in the slope (speed) followed by a more rapid slope change as the transition to a detonation occurs. There is no abrupt change as the transition takes place nor is there a relatively long overdriven detonation region as occurs in a homogeneous initiation.

It would be expected that a more heterogeneous HE (one of lower density with more voids) would have more growth in the front with little or no hump behind it. This type of experiment would represent purely heterogeneous behavior. Both PBX 9501 and PBX 9502 have both growth in the front and growth behind the front, although the amount of growth is different for the two materials. The initiation in these materials can be described as having both some heterogeneous and some homogeneous nature with the heterogeneous initiation process dominating by a considerable amount.

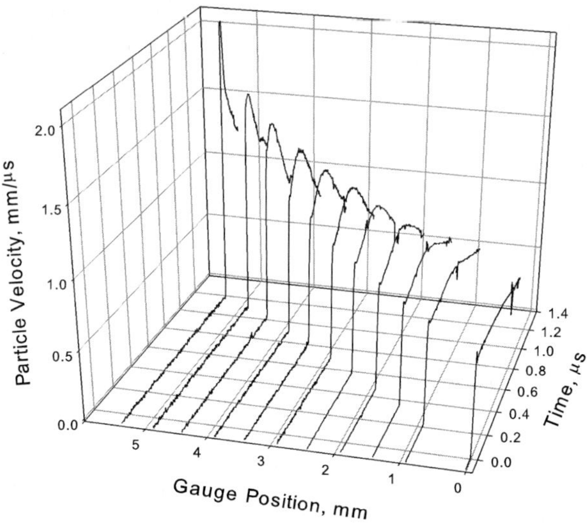

(a)

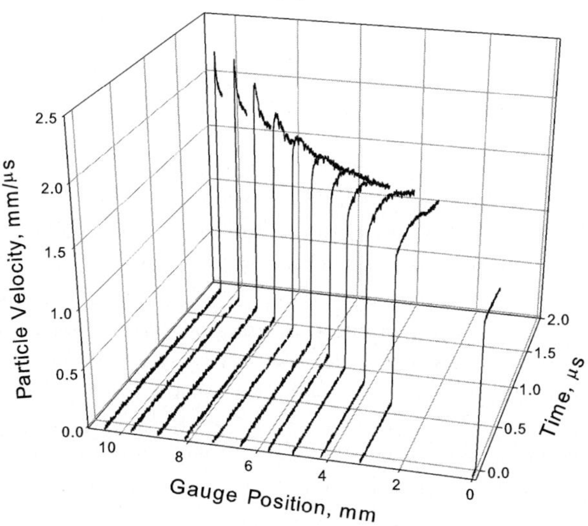

(b)

Fig. 1.35. Particle velocity vs. time plots of the magnetic-gauge data obtained from two experiments. (**a**) is a PBX 9501 (HMX based) experiment in which the shock input pressure to the HE was 3.4 GPa. It turned over to a detonation at about 5.5 mm into the sample. (**b**) is a PBX 9502 (TATB based) experiment in which the input shock pressure to the HE was 11.6 GPa. It turned over to a detonation at about 10 mm into the sample. There are profiles from a stirrup gauge at the impact interface and 10 particle velocity gauges for each experiment.

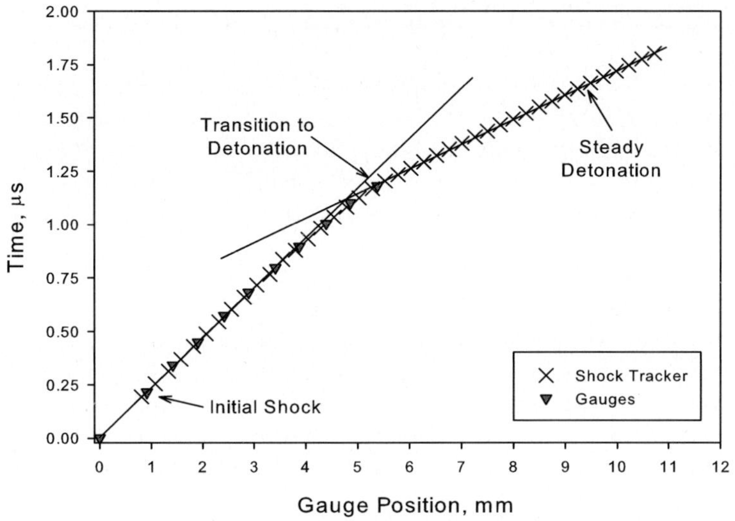

Fig. 1.36. Time vs. distance plot for the PBX 9501 experiment shown in Fig. 1.35a. Both the shock tracker data and the gauge arrival time data are plotted. Two lines have been drawn in, one at the initial shock slope and the other at the detonation speed slope, to provide a reference as to how the shock front changes with time.

From these two previous figures, it is clear that there is a large amount of information generated in each multiple magnetic-gauge experiment which is providing valuable new information about initiation in heterogeneous HE materials. The particle velocity waveforms are proving to be valuable for initiation modelers to compare their calculations to as they go about developing reaction rate models for the HEs of interest. Because of this, there is a need for experiments on additional materials of interest.

1.5.8 Numerical Modeling of Initiation

Modeling the shock initiation of HEs has received considerable attention for many years. One of the early models that was used for a number of years is called "Forest Fire" (see [74–76] for the details). This model was based on Pop-plot data for a particular explosive with a dependence of detonation delay time on shock pressure. It was used to model both homogeneous and heterogeneous HE initiation. When Lagrangian gauge measurements began to be made in the flow with Manganin and magnetic particle velocity gauges, a new model was indicated. The Ignition and Growth Model was introduced by Lee and Tarver [73] which was intended to include formalisms that described both ignition of the HE and growth to detonation. This model employs a relatively large number of adjustable parameters, some of which are nearly the same for most explosives. This model has been used successfully since its introduction by a large number of researchers to model HE initiation processes in a variety of different HE experiments (see [70, 95, 96]). It is still the model of choice today (see [68]).

These models all assume an unreacted HE goes to HE products with a single reaction rate and that the intermediate states are simply mixtures of unreacted material or products. Clearly this is a simplified modeling approach because the real situation involves many chemical reactions and many associated rates. However, it is not yet possible to measure and understand these chemical reactions and rates so this is what can be done with the present state of the experimental art.

Similar problems are associated with modeling detonation fronts as they propagate in one- or two-dimensions. The assumptions are still those of an unreacted material going to reaction products with a single reaction rate. Even when the reaction zone is made up of a large number of zones in the calculations, there are numerical instabilities and other problems that limit the accuracy of the calculations. Some of these problems have been addressed by [2, 7, 8].

Bdzil et al. [7] make the following statement, "Whatever improved modeling paradigms are developed for the continuum response of heterogeneous explosives, we expect that, for the foreseeable future, models will have to be calibrated to experiments if they are to make the accurate predictions of detonation propagation necessary for weapons simulations." This indicates the necessity of making better experimental measurements and enhancing our understanding of the chemistry involved in HE initiation and detonation so that the models can be continuously upgraded in the future.

1.6 Summary and Future Developments

Developments in condensed-phase explosive technology in the last 150 years have led to substantial increases in understanding of the initiation and detonation process. For a simple chemical reaction rate, compressible fluid mechanics shows that a planar steady detonation is a chemical-reaction-supported shock. The shock takes the unreacted material to a high pressure (the von Neumann spike); this is followed by decreasing pressure in a chemical reaction zone in which the explosive is transformed into gaseous reaction products. The chemical reaction zone terminates at a sonic point (the CJ state). In-situ and interface velocity measurements of the reaction zone validate the 1-D theory. Spectroscopic measurements of the reaction-zone chemistry are in their infancy but are a current area of research.

Explosives are classified according to sensitivity as primary (the most sensitive), secondary (which are intermediate), and insensitive explosives (the least sensitive). Sensitivity is measured by a number of well-defined test methods in which sensitivity to shock, impact, heat, and electrical inputs are measured. Initiation and 2-D steady detonation processes depend on the physical nature of the explosive, i.e., whether it is homogeneous or heterogeneous.

Behavior in the two cases is qualitatively different. Homogeneous explosives are typically liquids or single crystals; their initiation and detonation behavior depend on the bulk temperature and pressure generated by an input shock. Heterogeneous explosives have voids, grain boundaries, etc., which perturb the flow so that hot spots are produced; these hot spots are energy concentrations where the chemical reaction rate is greatly increased. These regions control the initiation and detonation behavior of heterogeneous explosives. Explosives are used in gun propellants, rocket propellants, and pyrotechnics; these materials are designed to burn (deflagrate) rather than detonate. Under some accident conditions, these explosives can burn too rapidly or even detonate with great consequent damage to the surroundings.

Important areas of explosive research have been neglected in our exposition. Examples of this are: (1) the study of the hydrodynamic stability of detonations to small perturbations and (2) studies concerning numerical solution of the partial differential equations governing multi-dimensional time-dependent initiation and detonation. Discussion of these and other important topics can be found in the "Further Reading" list.

We have shown that, in contrast to the intuitive feeling that detonations are disorderly, irreproducible, and chaotic processes, they can be quite orderly and are governed by the conservation equations of compressible fluid mechanics along with material constitutive relations.

Material has been discussed showing that there are explosives with widely varying characteristics. These include, e.g., explosives with different sensitivities to accidental insult, level of power and pressure generation during detonation. Because of this variation, one can choose an explosive appropriate to the application. Further work is being done to increase this variety of choice. An important example of this is the study of insensitive explosives for use in nuclear weapons. The goal being to produce a situation where it is virtually impossible to produce detonation of the explosive, except by design.

1.7 Glossary

C-J (Chapman–Jouguet) Model: The simplest model of planar steady detonation.
DDT (Deflagration-to-Detonation Transition): The mechanical compaction and reaction build-up processes an explosive goes through to make the transition from deflagration to detonation.
Detonation: A shock wave driven by a rapid release of chemical energy.
Diameter-Effect Curve: The relation between steady 2-D detonation speed and the lateral charge size.
EOS (Equation of State): The functional relationship between internal energy, pressure, and volume for a material in thermodynamic equilibrium.
Failure Diameter: The diameter of a cylindrical explosive charge below which one cannot propagate a steady detonation.

Heterogeneous Explosive: A material in which the release of chemical heat is controlled by local high temperature and pressure regions near flow irregularities ("hot spots").
Homogeneous Explosive: A material in which the release of chemical heat is controlled by the (uniform) bulk temperature and pressure.
Hot Spots: Localized regions in a shocked explosive where energy is concentrated by the fluid flow in the vicinity of physical imperfections.
Hugoniot: The locus of points reachable by a single shock process from an initial state.
Initiation: The process by which an initial input into an explosive develops into a detonation wave.
Isentrope: The locus of points originating from an initial state along which the system entropy is constant.
Overdriven Detonation: An overdriven detonation means the detonation is driven at a higher speed than the steady 1-D value.
Proton Radiography: A radiographic technique that uses the scattering of high energy protons by a material as a mass density probe.
Rankine–Hugoniot Conditions: The three mathematical conditions that guarantee conservation of mass, momentum, and energy across flow discontinuities (e.g., shocks); they also apply between any two locations in 1-D steady flow.
SDT (Shock-to-Detonation Transition): The reaction build-up process an explosive goes through to make the transition from shock to detonation.
Sensitivity: A ranking of the ease with which a detonation can be induced in an explosive by various types of insults.
Sonic Point: A position in the fluid flow where sound wave motion changes from subsonic to supersonic or vice versa.
Superdetonation: A detonation which occurs in a explosive precompressed to a higher mass density by an initial shock.
Taylor Wave: The pressure relief wave (rarefaction) following a planar steady detonation in most circumstances.
Thermal Explosion: A rapid chemical reaction that occurs when the temperature in an explosive gets above a threshold value.
von Neumann Spike: The leading (high pressure) point in the ZND model of planar detonations (sometimes referred to as the chemical peak).
ZND (Zeldovich, von Neumann, Doering) Model: The simplest model of planar steady detonation with a resolved chemical reaction zone.

Acknowledgments

We thank Larry Hill, Rick Gustavsen, Dave Holtcamp, Eric Ferm, Dave Moore, Jay Dallman, and John Vorthman of LANL and Craig Tarver of LLNL for pictures, schematics, etc. that were used in this chapter.

References

1. Andrews, D.J.: Equation of state of the alpha and epsilon phases of iron. Ph.D. Thesis, Washington State University, Pullman, WA (1970)
2. Aslam, T.D.: Shock compression of condensed matter – 2005, American Institute of Physics Publication CP845, p. 931 (2006)
3. Baer, M.R., Hall, C.A., Gustavsen, R.L., Hooks, D.E., Sheffield, S.A.: Shock compression of condensed matter – 2005, American Institute of Physics. In: Furnish, M.D., Elert, M., Russell, T.P., White, C.T. (eds.) AIP Conference Proceedings CP845, New York, p. 1307 (2005)
4. Baer, M.R., Hall, C.A., Gustavsen, R.L., Hooks, D.E., Sheffield, S.A.: J. Appl. Phys. **101**, 034906 (2007)
5. Barker, L.M., Hollenbach, L.E.: Rev. Sci. Instrum. **36**, 1617 (1965)
6. Bdzil, J.B., Stewart, D.S.: Phys. Fluids **1**, 1261 (1989)
7. Bdzil, J.B., Aslam, T.D., Henninger, R., Quirk, J.J.: Los Alamos Science, Los Alamos National Lab. Report Los Alamos Science, No. 28, p. 96 (2003)
8. Bdzil, J.B., Stewart, D.S.: Annu. Rev. Fluid Mech. **39**, 263 (2007)
9. Berthelot, M., Vielle, P.: C. R. Hebd. Sceances Acad. Sci. **93**, 18 (1881)
10. Blais, N.C., Fry, H.A., Greiner, N.R.: Rev. Sci. Instrum. **64**, 174 (1993)
11. Bloomquist, D.D., Sheffield, S.A.: J. Appl. Phys. **54**, 1717 (1983)
12. Boslough, M.B., Asay, J.R.: Basic principles of shock compression. In: Asay, J.R., Shahinpoor, M. (eds.) High Pressure Shock Compression of Solids. Springer-Verlag, New York (1992)
13. Campbell, A.W., Davis, W.C., Travis, J.R.: Phys. Fluids **4**, 498 (1961)
14. Campbell, A.W., Davis, W.C., Ramsay, J.B., Travis, J.R.: Phys. Fluids **4**, 511 (1961)
15. Campbell, A.W., Engelke, R.: The diameter effect in high-density heterogeneous explosives. Sixth Symposium (International) on Detonation Proceedings, Office of Naval Research Report ACR-221, Arlington, VA, p. 642 (1976)
16. Catanach, R.A., Hill, L.G., Harry, H.H., Aragon, E.C., Murk, D.M.: Cylinder Test Specification, Los Alamos National Lab. Report LA-13643-MS (1999)
17. Chaiken, R.F.: J. Chem. Phys. **33**, 760 (1960)
18. Chapman, D.L.: Phil. Mag. **47**, 90 (1899)
19. Cheret, R.: Detonation of Condensed Explosives. Springer-Verlag, New York (1993)
20. Dallman, J.C., Wackerle, J.: Proceedings of the Tenth International Detonation Symposium, Office of Naval Research ONR 33395-12, Arlington, VA, p. 130 (1993)
21. Davis, W.C.: Comb. Flame **120**, 399 (1999)
22. Deal, W.E.: J. Chem. Phys. **27**, 796 (1957)
23. deOliveira, G., Kapila, A.K., Schwendeman, D.W., Bdzil, J.B., Henshaw, W.D., Tarver, C.M.: Thirteenth International Detonation, Office of Naval Research Report ONR 351-07-01, Arlington, VA, p. 13 (2006)
24. Dick, J.J., Forest, C.A., Ramsay, J.B., Seitz, W.L.: J. Appl. Phys. **63**, 4881 (1988)
25. Dick, J.J., Mulford, R.N., Spencer, W.J., Pettit, D.R., Garcia, E., Shaw, D.C.: J. Appl. Phys. **70**, 3572 (1991)

26. Dick, J.J., Martinez, A.R., Hixson, R.S.: Proceedings of the Eleventh International Symposium on Detonation, Office of Naval Research ONR 33300-5, Arlington, VA, p. 317 (1998); see also Dick, J.J., Martinez, A.R., Hixson, R.S.: Planer Impact Response of PBX 9501 and Its Components Below 2 GPa, Los Alamos National Laboratory Report LA-13426-MS (1998)
27. Dick, R.D.: J. Chem. Phys. **52**, 6021 (1970)
28. Dobratz, B.M., Crawford, P.C.: LLNL Explosive Handbook, UCRL-52997, Lawrence Livermore Natl. Lab., Livermore, CA (1985)
29. Doering, W.: Ann. Phys. **43**, 421 (1943)
30. Dremin, A.N., Savrov, S.D., Trofimov, V.S., Shvedov, K.K.: Detonation Waves in Condensed Media, Translation from Russian, National Technical Information Service AD-751417, Springfield, VA (1970)
31. Dremin, A.N.: Toward Detonation Theory. Springer-Verlag, New York (1999)
32. Duff, R.E., Houston, E.: J. Chem. Phys. **23**, 1268 (1955)
33. Duran, M., Laharrague, P., Lalle, P., Le Bihan, A., Morvan, J., Pujols, H.: Rev. Sci. Instrum. **48**, 275 (1977)
34. Engelke, R., Davis, W.C.: Bull. Am. Phys. Soc. **18**, 1483 (1973)
35. Engelke, R.: Phys. Fluids **22**, 1623 (1979)
36. Engelke, R.: Phys. Fluids **23**, 875 (1980)
37. Engelke, R., Bdzil, J.B.: Phys. Fluids **26**, 1210 (1983)
38. Engelke, R., Stine, J.R.: J. Phys. Chem. **96**, 5689 (1990)
39. Engelke, R., Sheffield, S.A.: Explosives Encyclopedia of Applied Physics, vol. 6. CCH Publishers, New York (1993)
40. Engelke, R., Blais, N.C.: J. Chem. Phys. **101**, 10961 (1994)
41. Engelke, R., Blais, N.C., Sheffield, S.A., Sander, R.K.: J. Phys. Chem. A **105**, 6955 (2001)
42. Engelke, R., Sheffield, S.A., Stacy, H.L.: Phys. Fluids **16**, 4143 (2004)
43. Engelke, R., Sheffield, S.A., Stacy, H.L., Quintana, J.P.: Phys. Fluids **17**, 1 (2005)
44. Engelke, R., Sheffield, S.A., Stacy, H.L.: J. Phys. Chem. A **110**, 7744 (2006)
45. Ferm, E.N., Morris, C.L., Quintana, J.P., Pazuchanic, P., Stacy, H., Zumbro, J.D., Hogan, G., King, N.: Shock Compression of Condensed Matter – 2001, American Institute of Physics. In: Furnish, M.D., Thadhani, N.N., Horie, Y. (eds.) AIP Conference Proceedings 620, New York, p. 966 (2001)
46. Fickett, W., Davis, W.C.: Detonation. University of California Press, Berkeley (1979)
47. Frank-Kamenetskii, D.A.: Diffusion and Heat Exchange in Chemical Kinetics. Princeton University Press, Princeton, NJ (1955)
48. Fritz, J.N., Hixson, R.S., Shaw, M.S., Morris, C.E., McQueen, R.G.: J. Appl. Phys. **80**, 6129 (1996)
49. Gibbs, T.R., Popolato, A.: LASL Explosive Property Data. University of California Press, Berkeley (1980)
50. Gustavsen, R.L., Sheffield, S.A., Alcon, R.R.: Shock compression of condensed matter – 1997, American Institute of Physics. In: Schmidt, S.C., Dandekar, D.P., Forbes, J.W. (eds.) AIP Conference Proceedings 429, New York, p. 739 (1997)
51. Gustavsen, R.L., Sheffield, S.A., Alcon, R.R., Hill, L.G.: Shock Initiation of New and Aged PBX 9501 Measured with Embedded Electromagnetic Particle Velocity Gauges, Los Alamos National Lab. Report LA-13634-MS (1999)

52. Gustavsen, R.L., Dattelbaum, D.M., Orler, E.B., Hooks, D.E., Alcon, R.R., Sheffield, S.A., Hall, C.E., Baer, M.R.: Shock compression of condensed matter – 2005, American Institute of Physics. In: Furnish, M.D., Elert, M., Russell, T.P., White, C.T. (eds.) AIP Conference Proceedings 845, New York, p. 149 (2005)
53. Gustavsen, R.L., Sheffield, S.A., Alcon, R.R.: J. Appl. Phys. **99**, 114907 (2006)
54. Hall, C.A.: Phys. Plasmas **7**, 2069 (2007)
55. Hall, C.A., Asay, J.R., Knudson, M.D., Stygar, W.A., Spielman, R.B., Pointon, T.D., Reisman, D.B., Toor, A., Cauble, R.C.: Rev. Sci. Instrum. **72**, 3587 (2001)
56. Hayes, D.B.: Experimental determination of phase transformation rates in shocked potassium chloride. Ph.D. Thesis, Washington State University, Pullman, WA (1973)
57. Hayes, D.B.: A $P^n t$ detonation criterion from thermal explosion theory, Sixth Symposium (International) on Detonation Proceedings, Office of Naval Research Report ACR-221, Arlington, VA, p.76 (1976)
58. Hill, L.G., Seitz, W.L., Forest, C.A., Harry, H.H.: Shock compression of condensed matter – 1997, American Institute of Physics. In: Schmidt, S.C., Dandekar, D.P., Forbes, J.W. (eds.) AIP Conference Proceedings 429, New York, p. 751 (1997)
59. Hill, L.G., Bdzil, J.B., Aslam, T.D.: Proceeding of the Eleventh International Symposium on Detonation, Office of Naval Research ONR 33300-5, Arlington, VA, p. 1029 (1998)
60. Hill, L.G., Catanach, R.A.: *W-76 and PBX-9501 Cylinder Tests*, Los Alamos National Laboratory Report LA-13442, Los Alamos, NM (1998)
61. Hill, L.G.: Shock compression of condensed matter – 2001, American Institute of Physics. In: Furnish, M.D., Thadhani, N.N., Horie, Y. (eds.) AIP Conference Proceedings 620, New York, p. 149 (2001)
62. Hill, L.G., Bdzil, J.B., Davis, W.C., Critchfield, R.R.: Thirteenth International Symposium on Detonation, Office of Naval Research Report ONR 351-07-01, Arlington, VA, p. 331 (2006)
63. Hooks, D.E., Hayes, D.B., Hare, D.E., Reisman, D.B., Vandersall, K.S., Forbes, J.W., Hall, C.A.: J. Appl. Phys. **99**, 124901 (2006)
64. Holtkamp, D.B. et al.: Shock compression of condensed matter – 2003, American Institute of Physics. In: Furnish, M.D., Gupta, Y.M., Forbes, J.W. (eds.) AIP Conference Proceedings 706, New York, p. 477 (2003)
65. Johansson, C.H., Persson, P.A.: Detonics of High Explosives. Academic Press, New York (1970)
66. Johnson, J.N., Cheret, R.: Classic Papers in Shock Compression Science. Springer-Verlag, New York (1998)
67. Jouguet, E.: J. Math. Pure Appl **6**, 347 (1905)
68. Kapila, A.K., Schwendeman, D.W., Bdzil, J.B., Henshaw, W.D.: Combustion Theory and Modeling **11**, 781 (2007)
69. Kury, J.W., Hornig, H.C., Lee, E.L., McDonnel, J.L., Ornelias, D.L., Finger, M., Strange, F.M., Wilkins, M.L.: Fourth Symposium (International) on Detonation – 1965, Office of Naval Research ACR-126, Washington D.C., p. 3 (1965)
70. Kury, J.W., Breithaupt, R.D., Tarver, C.M.: Shock Waves **9**, 227 (1997)
71. Lee, E.L., Hornig, H.C., Kury, J.W.: Adiabatic Expansion of High Explosive Detonation Products, Lawrence Livermore Laboratory Report UCRL-50422, Livermore, CA (1968)

72. Lee, E.L., Finger, M., Collins, W.: JWL Equation of State Coefficients for High Explosives, Lawrence Livermore Laboratory Report UCID-16189, Livermore, CA (1973)
73. Lee, E.L., Tarver, C.M.: Phys. Fluids **23**, 2362 (1980)
74. Mader, C.L., Forest, C.A.: Two-Dimensional Homogeneous and Heterogeneous Detonation Wave Propagation, Los Alamos Scientific Laboratory Report LA-6259 (1976)
75. Mader, C.L.: Numerical Modeling of Detonation. University of CA Press, Berkeley (1979)
76. Mader, C.L.: Numerical Modeling of Explosives and Propellants, 2nd edn. CRC Press, Boca Raton, FL (1998)
77. Mallard, E., Le Chatelier, H.: C. R. Hebd. Sceances Acad. Sci. **93**, 145 (1881)
78. Mallory, H.D.: J. Appl. Phys. **38**, 5302 (1967)
79. Menikoff, R.: Empirical equations of state for solids. Chapter 4 in Shock Wave Science and Technology Reference Library, Vol. **2** Solids I, ed. Y. Horie. Springer-Verlag, New York (2007)
80. Meyer, R., Kohler, J., Homburg, A.: Explosives, 5th edn. Wiley-VCH Verlag GmbH, Weinheim (2002)
81. Moore, D.S., McGrane, S.D., Funk, D.J.: Chapter 5: ultrashort laser shock dynamics. Chapter in Shock Wave Handbook: Shock Waves in Solids. Springer-Velag, New York, p. 177 (2006)
82. von Neumann, J.: Progress report on the theory of detonation waves, OSRD Report No. 549 (1942). In: John von Neumann Collected Works, vol. 6. Pergamon Press, New York, p. 203 (1963)
83. Nunziato, J.W., Kipp, M.E., Setchell, R.E., Walsh, E.K.: Shock Initiation in Heterogeneous Explosives, Sandia National Laboratories Report SAND81-2173 (1982)
84. Pintgen, F., Eckett, C.A., Austin, J.M., Shepherd, J.E.: Combust. Flame **133**, 211 (2003)
85. Ramsay, J.B., Popolato, A.: Proceedings of the Fourth Symposium (International) on Detonation, Office of Naval Research ACR-126, Washington D.C., p. 233 (1965)
86. Reisman, D.B., Toor, A., Cauble, R.C., Hall, C.A., Asay, J.R., Knudson, M.D., Furnish, M.D.: J. Appl. Phys. **89**, 1625 (2001)
87. Sheffield, S.A., Mitchell, D.E., Hayes, D.B.: The equation of state and chemical kinetics for hexanitrostilbene (HNS) explosive. Sixth Symposium (International) on Detonation Proceedings, Office of Naval Research Report ACR-221, Arlington, VA, p. 748 (1976)
88. Sheffield, S.A., Duvall, G.E.: J. Phys. Chem. **79**, 1981 (1983)
89. Sheffield, S.A., Bloomquist, D.D., Tarver, C.M.: J. Chem. Phys. **80**, 3831 (1984)
90. Sheffield, S.A.: J. Chem. Phys. **81** 3048 (1984)
91. Sheffield, S.A., Engelke, R., Alcon, R.R.: Proceedings of the Ninth Symposium (International) on Detonation, Office of Naval Research OCNR 113291-7, Arlington, VA, p. 39 (1989)
92. Sheffield, S.A., Gustavsen, R.L., Alcon, R.R.: Shock compression of condensed matter – 1999. In: Furnish, M.D., Chhabildas, L.C., Hixson, R.S. (eds.) American Institute of Physics, AIP Conference Proceedings 505, New York, p. 1043 (1999)

93. Sheffield, S.A., Engelke, R., Alcon, R.R., Gustavsen, R.L., Robbins, D.L., Stahl, D.B., Stacy, H.L., Whitehead, M.C.: Twelfth International Detonation Symposium, Office of Naval Research Report ONR 333-05-2, p. 159 (2002)
94. Sheffield, S.A., Gustavsen, R.L., Alcon, R.R., Robbins, D.L., Stahl, D.B.: Shock compression of condensed matter – 2003. In: Furnish, M.D., Gupta, Y.M., Forbes, J.W. (eds.) American Institute of Physics, AIP Conference Proceedings 706, p. 1033 (2003)
95. Tarver, C.M., Breithaupt, R.D., Kury, J.W.: J. Appl. Phys. **81**, 7193 (1997)
96. Tarver, C.M., Kury, J.W., Breithaupt, R.D.: J. Appl. Phys. **82**, 3771 (1997)
97. Tarver, C.M.: Shock compression of condensed matter – 1999. In: Furnish, M.D., Chhabildas, L.C., Hixson, R.S. (eds.) American Institute of Physics, AIP Conference Proceedings 505, New York, p. 873 (1999)
98. Tarver, C.M.: Shock compression of condensed matter – 2001. In: Furnish, M.D., Thadhani, N.N., Horie Y. (eds.) American Institute of Physics, AIP Conference Proceedings 620, New York, p. 42 (2001)
99. Taylor, G.I.: Proc. R. Soc. A **CC**, 235 (1950)
100. Thompson, P.A.: Compressible Fluid Dynamics. McGraw-Hill, New York (1972)
101. Trott, W.M., Asay, J.R.: Shock compression of condensed matter – 1997. In: Schmidt, S.C., Dandekar, D.P., Forbes, J.W. (eds.) American Institute of Physics, AIP Conference Proceedings 429, New York, p. 837 (1997)
102. Trott, W.M., Castaneda, J.N., O'Hare, J.J., Baer, M.R., Chhabildas, L.C., Knudson, M.D., Davis, J.P., Asay, J.R.: Dispersive Velocity Measurements in Heterogeneous Materials, Sandia National Laboratories Report SAND2000-3082 (2000)
103. Urtiew, P.A., Kusubov, A.S., Duff, R.E.: Combust. Flame **14**, 117 (1970)
104. Urtiew, P.A., Cook, T.M., Maienschein, J.L., Tarver, C.M.: Proceedings of the Tenth International Detonation Symposium, Office of Naval Research ONR 33395-12, Arlington, VA, p. 139 (1993)
105. Vantine, H., Chan, J., Erickson, L., Janzen, J., Weingart, R.: Rev. Sci. Instrum. **51**, 116 (1980)
106. Vorthman, J.E.: Shock compression of condensed matter – 1981, American Institute of Physics. In: Nellis, W.J., Seaman, L., Graham, R. A.(eds.) AIP Conference Proceedings 78, New York, p. 680 (1981)
107. Vorthman, J.E., Andrews, G., Wackerle, J.: Proceedings of the Eighth Detonation (International) on Symposium, Office of Naval Research NSWC MP-86-194, Arlington, VA, p. 951 (1986)
108. Widom, G.B.: Linear and Nonlinear Waves. Wiley, New York (1974)
109. Woolfolk, R.W., Cowperthwaite, M., Shaw, R.: Thermochim. Acta **5**, 409 (1973)
110. Yoo, C.S., Cynn, H.: J. Chem. Phys. **111**, 10229 (1999)
111. Zeldovich, Ya.B.: Sh. Eksp. Theo. Fiz. **10**, 542 (1940) (English Translation: NACA TM 1261, 1960)

Further Reading

112. Engelke, R., Sheffield, S.A.: Chapter 7 – initiation and propagation of detonation in condensed-phase high explosives. High-Pressure Shock Compression of Solids III, p. 171. Springer-Verlag, New York (1998)

113. Lewis, B., von Elbe, G.: Combustion, Flames, and Explosions of Gases. Academic Press, Orlando (1987)
114. Encyclopedia of Explosives and Related Items, in ten volumes, Picatinny Arsenal/Large Caliber Weapons Systems Lab., Dover, NJ (1960–1983)
115. Proceedings of the First to Thirteenth International Symposia on Detonation, sponsored by the Office Naval Research, Washington D.C. (1951–2006)
116. Zeldovich, Ya.B., Raizer, Y.P.: Physics of Shock Waves and High-Temperature Hydrodynamic Phenomena. Academic Press, New York (1968)
117. Zukas, J.A., Walters, W.P.: Explosive Effects and Applications. Springer, New York (1997)

2

First Principles Molecular Simulation of Energetic Materials at High Pressures

F. Zhang, S. Alavi, A. Hu, and T.K. Woo

2.1 Introduction

The availability of relatively inexpensive and powerful computing technology has profoundly changed the way in which modern scientific research is conducted in numerous fields. At a rather trivial but highly relevant practical level, the rapid increase in computer power has considerably sped up the pace of theoretical schemes and approaches for simulating bulk material at the atomic level. The results of the simulations have been invaluable in the guidance of experiments and for providing insight into system behavior, particularly under extreme conditions of temperature and pressure. In this respect, a major role has been played by molecular dynamics (MD), which since the pioneering efforts of the 1960s [1, 2] has developed into a mature and active discipline that has been used as a means of simulating and understanding the properties of real systems. More recently, major progress has been achieved in the development of ab initio or "first principles" MD, in which the potential energy and interatomic forces are derived from accurate quantum mechanical electronic structure calculations that are performed as the simulation proceeds [3–5]. This has greatly improved the predictive power of the simulation and opens the way for the reliable simulation of processes in which chemical bonds are formed and broken. The formation and breaking of chemical bonds is simulated with great difficulty in MD based on classical mechanics and empirically derived reactive potentials [6–9].

The development of ab initio MD, along with the rapid development of computer power, has inspired its application to studies of energetic materials and detonation processes. There are several difficult problems currently encountered in the field of molecular condensed matter detonation. Is the initiation of detonation in existing molecular explosives controlled by the thermal decomposition via shock temperature or by initial molecular collisions occurring within the shock front? [6, 10] Could the initial molecular collision and subsequent bond energy release be made the main mechanisms for the initiation and propagation of detonation in a new generation of molecular explosives, whose detonation velocity could be several times that of current explosives? Would pressure dissociation or momentum-induced decomposition at low temperatures be another possible mechanism for detonation to generate high-speed solid particles products? Experimental support to answer these questions must be derived under extreme conditions of pressure and temperature from observations inside the shock front, thereby requiring measurements on the time scales of 10^{-2}–10^{0} ps. Molecular simulation has offered an alternative avenue towards gaining insights into these challenging problems and guidance to experimental research with energetic materials and detonation processes. High-energy-density materials must feature metastability and a large energy content that mostly originates in transformations of the molecular, atomic and electronic structures. Successful synthesis of these materials and control of the energy release from the structural bonds strongly rely on the understanding of the chemical processes and physics of structural transformations at the atomic level.

In the following we review computations that have recently been carried out in our group on high energy density materials. Quantum mechanics-based, density functional theory (DFT) [11] methods are used to provide atomic-level insight into the physical and chemical properties and processes which occur in these materials. Compared to classical molecular scale simulation methods that are based on empirically derived potentials, the simulations presented here allow for the simulation of complicated covalent bond breaking and formation, such as in the case of molecular decomposition.

The cases discussed include the reaction of nitromethane under binary collision conditions, in the absence and presence of spectator molecules which participate in secondary collisions in the system. Additionally, the reaction of liquid nitromethane under high pressure conditions is studied. Our work on the solid-state phases of covalent nitrogen is also summarized. There are several areas where computations can contribute in these studies. First, direct information about optimized crystallographic lattice parameters and the geometrical parameters of systems can be determined. This can be done for a variety of uncompressed lattices or when external isotropic or anisotropic compression is applied to the crystal. Besides the structural, elastic, and phonon-modes parameters, other important energetic and electronic properties can be evaluated. Among the list of electronic properties some representative examples are the band structure and the total or partial density of states. Furthermore, additional insight can be obtained from population analyses of charge distribution, bond order, and electron and spin density maps.

2.1.1 Introduction to First Principles Molecular Simulation

When studying processes at the molecular level, the most fundamental property of interest is the molecular potential energy surface or simply the potential energy surface (PES). The potential energy surface of a system relates the total energy of a molecular system with the geometric arrangement of

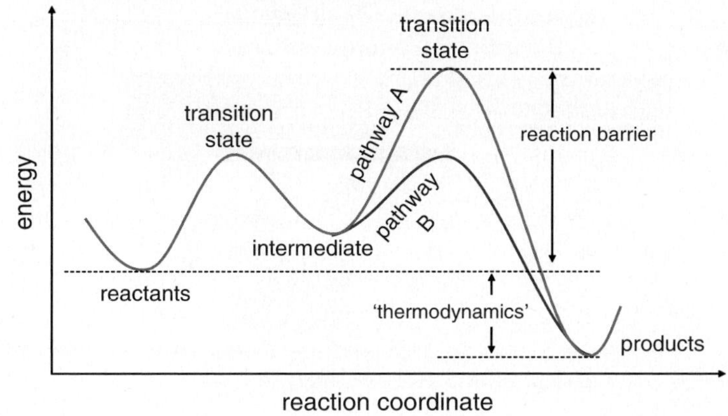

Fig. 2.1. Generic potential energy surface for a chemical reaction. The reaction coordinate represents a combination of nuclear coordinates which links the reactant molecules on the potential energy surface to the products

the atoms that make up the system. Since each atomic nucleus has three spatial degrees of freedom associated (x, y, z in Cartesian coordinate space) the potential energy surface is a 3N dimensional hypersurface where N is the number of atoms that make up the system. A simple one-dimensional potential energy surface of a generic chemical reaction is depicted in Fig. 2.1.

In principle, if the potential energy surface for a given reaction can be calculated, important quantities such as the reaction enthalpy, which is related to the energy content of a material, the reaction barriers, the reaction mechanism, and the structure of a material, among other quantities, can be determined. Thus, a detailed understanding of a reactive process can be achieved via computer modeling of the potential energy surface that in turn provides an avenue for rationally studying reactions and predicting new products at different temperature and high pressure conditions.

A wide array of chemical simulation methods exist and have been used for decades, as outlined in Fig. 2.2. For processes which do not involve the formation or breaking of chemical bonds, the motion of reactants on the potential energy surface can be adequately described by classical potentials. These are also referred to as empirical potentials, molecular mechanics methods, force field methods and ball and spring potentials. For these processes, the inter- and intramolecular forces between the atoms can be described loosely as forces among classical springs (harmonic or anharmonic). The force constants among these "springs" are determined empirically by comparing the predictions of the simulations of thermodynamic properties with experimental results, or higher level ab initio calculations. Thus for processes such as melting, evaporation, and non-reactive shockwaves, classical potentials may be used. Classical potentials have the advantage that they are relatively inexpensive to calculate and, as a result, simulations with thousands to millions of atoms can be

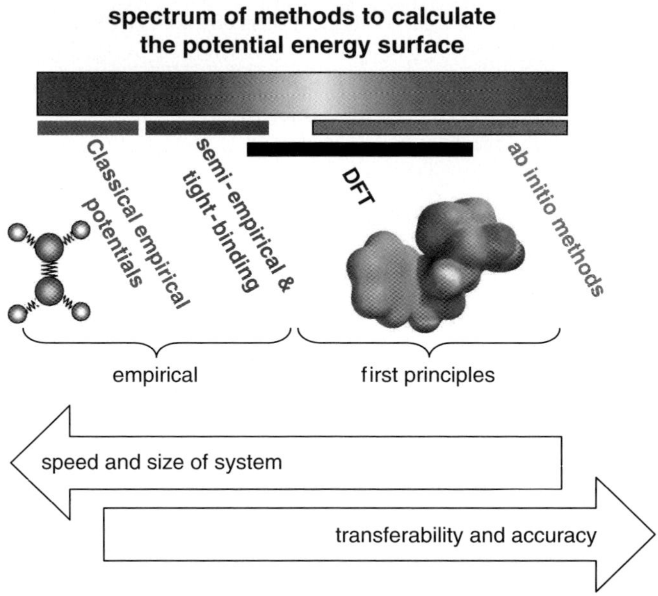

Fig. 2.2. Spectrum of methods to calculate the potential energy surface. Moving toward the left, methods tend to be faster, allowing for more atoms to be simulated. Moving toward the right, methods are more transferable, and often more accurate

performed. The main disadvantages of classical potentials are that they can not be generally used to simulate bond breaking/forming processes, and they are limited in their transferability. The latter caveat means that the empirical constants in the potentials are chosen to reproduce a certain set of properties of a limited given class of materials, and as a result these potentials may be of limited value when simulating materials and properties different from the training set. For simulating processes which involve the formation and breaking of covalent bonds, the use of classical potentials is limited in value, and in these cases, quantum mechanical methods are most often used. We note that there are some 'empirical' potentials that do allow for covalent bond breaking and formation [12–14], but these are still very limited to the class of materials for which the potentials are parameterized.

With quantum mechanical based methods, there is an attempt to solve for the Schrödinger equation or equivalent equations, which provide the electronic structure of the system either in the form of a wave function or electron density. The detailed electronic structure allows one to evaluate the energy of the system, the forces exerted on the nuclear centers, and therefore the all-important potential energy surface. There exists a hierarchy of quantum mechanics methods used in molecular simulation, ranging from fast, empirical methods to more computationally demanding 'first principles' methods. The choice of method to be used in a simulation generally involves striking a balance between accuracy and computational cost and typically, these two features of a method run contrary to one another. That is, there is usually a high computational price to pay for high accuracy.

In this work, we will restrict ourselves to simulations using 'first principles' quantum mechanical methods. First principles methods sometimes referred to as ab initio methods, use little or no empirical data to construct the electronic structure. As a result, they have a high degree of transferability since they are not constrained by a particular training set. This makes first principles quantum mechanical methods well suited to simulating materials under extreme conditions where available experimental data can be limited or non-existent. The first principles methods of choice for simulating materials and molecules are density functional theory (DFT) methods. DFT strikes a favorable balance between accuracy and computational efficiency that has led to its widespread use in the molecular and material sciences. DFT calculations can routinely be performed on systems containing hundreds or thousands of atoms with modern computers.

2.1.2 Density Functional Theory

The basic premise for DFT is that all ground state properties of a chemical system are uniquely determined by the total electron density rather than the full multielectron wavefunction. This is known as the first Hohenberg–Kohn theorem [15]. This theorem implies that the electronic energy of a molecular system is a functional of the electron density. A functional is simply a function of a function, for example, the value of a variable I has a functional relationship with integrable functions $f(x)$, shown as $I[f]$ for the following case, $I\left[f\right]=\int_0^\infty f(x)dx$. The value of I depends on the function $f(x)$. In this case, the energy, $\mathrm{E}[\rho]$, is a functional of the electron density, $\rho(\mathbf{r})$, which is in turn a function of the three spatial coordinates ($\mathbf{r}$ or x, y, z in Cartesian space). From the electronic energy of a system, the total energy of the system can be easily determined for a given geometric arrangement of the atoms that make up the system, and, therefore, the potential energy surface can be evaluated. The electronic energy functional is typically divided into components that account for the electron kinetic energy, electron–nuclear interaction potential energy and the electron–electron interaction potential energy:

$$E[\rho] = \underbrace{T[\rho]}_{\substack{\text{kinetic energy}\\ \text{of the electrons}}} + \underbrace{E_{\mathrm{Ne}}[\rho]}_{\substack{\text{nuclear}-\text{electron}\\ \text{interaction energy}}} + \underbrace{E_{\mathrm{ee}}[\rho]}_{\substack{\text{electron}-\text{electron}\\ \text{interaction energy}}}, \tag{2.1}$$

where ρ is the electron density and the square brackets indicate a functional dependence in which the variable ρ is a function of other variables, in this case the spatial coordinates.

In 1965, Kohn and Sham recognized that the utility of (2.1) was severely limited by the lack of a kinetic energy functional, $T[\rho]$ [16]. In fact, previous attempts to use DFT with approximate kinetic energy functionals [17, 18], predicted that chemical bonding should not occur at all [19]. To tackle this situation, Kohn and Sham ingeniously introduced a 'fictitious' non-interacting reference system (analogous to Hartree–Fock orbitals) built from a set of single electron wave functions, or Kohn–Sham orbitals, $\{\psi_i(\mathbf{r})\}$, that sum to give the real electron density ρ,

$$\rho(\mathbf{r}) = \sum_i |\psi_i(\mathbf{r})|^2 \tag{2.2}$$

where the summation index, i, runs over occupied orbitals (loosely one orbital for each electron in the system). The kinetic energy of the non-interacting system can then be calculated as:

$$T_0 = \sum_i \int \psi_i(\mathbf{r}) \left(-\frac{1}{2}\nabla^2 \right) \psi_i(\mathbf{r}) d\mathbf{r}, \tag{2.3}$$

Using (2.3) it is possible to rewrite (2.1) as:

$$E[\rho] = T_0 + E_{\mathrm{Ne}}[\rho] + E_{\mathrm{ee}}[\rho]. \tag{2.4}$$

The first term on the right hand side of (2.4) is the kinetic energy of the fictitious electronic system. It should be noted that (2.4) is formally and conceptually different than (2.1). In (2.4), the kinetic energy term, T_0 refers to the kinetic energy of a non-interacting reference system, whereas in (2.1) the kinetic energy term refers to the true interacting system. Using the Born–Oppenheimer approximation, through which all nuclei are treated as fixed point charges, the nuclear–electron term is given by:

$$E_{\mathrm{Ne}}[\rho] = -\sum_{\mathrm{I}} \int \frac{Z_{\mathrm{I}}\rho(\mathbf{r})}{|\mathbf{R}_{\mathrm{I}} - \mathbf{r}|} d\mathbf{r}, \tag{2.5}$$

where the index I runs over all nuclei, Z_{I} is the nuclear charge of atom I, and $\mathbf{R}_{\mathrm{I}}$ are the nuclear coordinates of atom I. This relationship describes the Coulombic attraction between the electrons and the nuclei.

The last term in (2.4), $E_{\mathrm{ee}}[\rho]$, accounts for the interactions between electrons. In Kohn–Sham DFT (KS-DFT), this term is separated into two components:

$$E_{\mathrm{ee}}[\rho] = \frac{1}{2} \int\int \frac{\rho(\mathbf{r})\rho(\mathbf{r}')}{|\mathbf{r} - \mathbf{r}'|} d\mathbf{r}\, d\mathbf{r}' + E_{\mathrm{xc}}[\rho], \tag{2.6}$$

The first term on the right hand side of (2.6) represents the classical Coulombic electronic repulsion of an electron moving in the *average* electric field of the other electrons in the system. The second term, E_{xc}, is the so-called exchange correlation energy, which will be discussed further below.

When (2.5) and (2.6) are inserted into (2.4), one arrives at the following expressing for the electronic energy of the system:

$$E[\rho] = \sum_i \int \psi_i(\mathbf{r}) \left(-\frac{1}{2}\nabla^2\right) \psi_i(\mathbf{r}) d\mathbf{r} - \sum_I \int \frac{Z_I \rho(\mathbf{r})}{|\mathbf{R}_I - \mathbf{r}|} d\mathbf{r} + \frac{1}{2} \iint \frac{\rho(\mathbf{r})\rho(\mathbf{r}')}{|\mathbf{r} - \mathbf{r}'|} d\mathbf{r}\, d\mathbf{r}' + E_{XC}[\rho]. \quad (2.7)$$

This equation forms the basis of Kohn–Sham DFT. In order to solve this equation, one needs to find the Kohn–Sham wave functions Ψ_s that minimize the energy of the system, subject to the constraints that the total number of electrons in the system is conserved and the orbitals used to construct the fictitious electronic system are orthogonal. This involves self-consistently solving a set of one-electron equations. The details of this procedure, although straightforward, are beyond the scope of this discussion and will not be described here; however, the interested reader is directed to Chap. 7 of [11].

The first three terms on the right hand side of (2.7) are well-defined for a given set of orbitals. The accuracy of KS–DFT is thus dependent upon the expression for the last term, E_{xc}, which serves as a depository for contributions to the electronic energy that are not adequately described by the other terms in (2.7). These energy contributions include those arising from the correlated motion of electrons due to instantaneous electron–electron interactions (correlation), the correlated motion of electrons due to the Pauli exclusion principle (exchange), a correction for the difference between the kinetic energy of the fictitious electronic system of non-interacting orbitals and that of the true system, and a correction for the interaction of the electron with itself. The specific details of these energy contributions will not be considered in greater detail here.

The exchange-correlation energy, E_{xc}, is evaluated with the use of a functional, which is called the exchange-correlation (XC) functional. The form of this functional is unknown; however, if it was determined exactly, (2.7) would provide the exact ground state energy for the system. Although the true XC functional remains unknown, numerous attempts have been made to develop approximate XC functionals (e.g., see Chap. 6 of reference [11]). In general, approximate XC functionals are developed through either a consideration of the fundamental physics pertaining to electron–electron interactions or fitting parameters in various functional forms to experimental data. Although the former approach is more elegant, the latter often produces functionals that are very accurate for chemical applications.

The simplest XC functionals are based solely upon the value of the electron density [20,21]. This approach is termed the local density approximation (LDA) and is generally too inaccurate for use in most chemical applications. Significant improvements in accuracy are achieved if the XC functional depends on the value and gradient of the electron density [22]. This is termed the generalized gradient approximation (GGA). GGA XC functionals are used commonly in many chemical applications. More sophisticated functionals can be obtained by incorporating higher-order derivatives of the electron density and other quantities. However, for simulations of liquids and solids, GGA XC functionals are used in the vast majority of DFT calculations.

As noted above, DFT is currently the most popular method for routine quantum mechanical calculations. This popularity, which surged in the early 1990s, is due to the development of accurate XC functionals. These functionals offer accuracy which rivals that of high-level wave function-based methods. In terms of cost, DFT calculations can be performed on systems containing up to a few thousand atoms with modern computers. In terms of accuracy, studies [23, 24] show that LDA, BP86 [25, 26] (GGA), PBE [27] (GGA) and B3LYP [28, 29] functionals yield mean absolute errors of 36.4, 10.3, 8.6 and 2.2 kcal mol^{-1}, respectively, for the atomization energies of the molecules in the G-2 set, [30] which is a standard benchmark database for the assessment of the accuracy of theoretical methods. Calculations of reaction barriers for a set of 76 reactions showed that LDA, BP86 [25]; (GGA), PBE [27] (GGA) and B3LYP [26, 29] (hybrid) DFT functionals yielded errors of 14.9, 8.8, 8.7 and 4.3 kcal mol^{-1} with respect to the results of high-level calculations [31].

2.1.3 Plane Wave Basis Sets

In most practical DFT calculations the Kohn–Sham single electron orbitals, ψ_i., are expressed as a linear combination of basis functions, χ:

$$\psi_i(\mathbf{r}) = \sum_{\nu=1}^{K} c_{\nu i}\chi_\nu(\mathbf{r}), \tag{2.8}$$

where ν is the index of the basis function, K is the total number of basis functions and $c_{\nu i}$ is the mixing coefficient of basis function χ_ν, which determines the contribution of basis function χ_ν to orbital ψ_i. When optimizing the wave function, the coefficients $c_{\nu i}$ are variables that are altered to yield the 'best' wave function for the system, while the basis functions themselves are unaltered. The details of the procedure used to optimize these coefficients will not be discussed here; however, the interested reader is directed to Chap. 3 of [32] or Chap. 7 of [11]. Although any type of mathematical function can be used as a basis function, two types of basis functions have achieved great popularity: localized, atom-centered functions and delocalized plane waves. Localized, atom-centered basis functions are based on the notion that a molecular system is built up from atoms, and hence the total electronic wave function can be represented as a combination of atomic orbital-like functions, which are centered on specific nuclei. Atom-centered basis functions have achieved widespread use in the quantum chemistry community, where so-called Gaussian functions [33] are the most common type of basis function. For the work presented in this chapter, localized basis functions are used sparingly, and will not be discuss in further detail.

An alternative to atom-centered, localized basis functions involves the use of delocalized functions, which are not associated with specific nuclei. The use of such functions is inspired by the notion that electrons in periodic systems are only slightly perturbed by the presence of nuclei, and hence the electronic wave functions should most closely resemble those of free electrons. This description is particularly suited to periodic, conducting systems such as metals. With such systems in mind, one defines a periodic simulation cell (vide infra), e.g., a unit cell for a crystalline material, and calculates the electronic wave function within this cell. The natural basis set for such a calculation is a set of plane waves [34]:

$$\chi_{\mathrm{PW}}(\mathbf{r}) = \frac{1}{\sqrt{\Omega}} e^{i\mathbf{G}\cdot\mathbf{r}}, \tag{2.9}$$

where Ω is the volume of the simulation cell and $\mathbf{G}$ is given by the reciprocal lattice vectors of the cell. A plane wave basis set includes all plane waves up to some maximum value of $\mathbf{G}$.

Plane wave basis sets have several advantages over atom-centered basis functions. First of all, a plane wave basis set allows one to move between real and reciprocal spaces through the use of Fast Fourier Transforms. This allows one to calculate quantities in the most convenient space. For example, derivatives in real space are simply multiplications in reciprocal space, and hence it is convenient to perform such calculations in reciprocal space. The fact that plane waves are not associated with specific nuclei also simplifies certain types of calculations. Along the same lines, plane waves treat all regions of space equally, and hence a plane wave basis set is complete, which eliminates basis set superposition errors that occur when using atom-centered basis functions. Finally, plane waves are suited to calculations of periodic systems, which will be discussed ahead.

A significant drawback of plane wave basis sets is that a huge number of plane waves must be used to accurately describe the localized wavefunctions in regions around the nuclei. Specifically, the wavefunction and electron density vary rapidly as one approaches a nucleus. Accurately describing these rapid changes requires using a plane wave expansion up to very high values of $\mathbf{G}$, and hence a large number of plane wave basis functions must be considered in the calculation. In practice, tens or hundreds of thousands of plane wave basis functions must be used to achieve the same level of accuracy that can be attained with a few hundred Gaussian basis functions.

The accuracy and size of the plane wave basis set is determined by the highest value of $\mathbf{G}$ considered in the plane wave expansion. This is determined by taking advantage of the relationship between $\mathbf{G}$ and the kinetic energy of the plane wave:

$$E_{kin}(\mathbf{G}) = \frac{1}{2}|\mathbf{G}|^2, \tag{2.10}$$

which allows one to define a parameter E_{cut} that corresponds to the kinetic energy of the plane wave with the highest value of $\mathbf{G}$ in the basis set. The parameter E_{cut} is termed the kinetic energy cutoff. Assigning a value for

E_{cut} indicates that all plane waves where $\frac{1}{2}|\mathbf{G}|^2 \leq E_{cut}$ are included in the plane wave basis set. Without going into details, this leads to the following expression for the number of plane waves in the basis set:

$$N_{\mathrm{PW}} \approx \frac{1}{2\pi^2}\Omega E_{\mathrm{cut}}^{3/2}. \tag{2.11}$$

Equation (2.11) indicates that the number of plane waves considered in the calculation increases with both the volume of the simulation cell, Ω, and E_{cut}.

Thus in summary, plane waves are an alternative to atom-centered, localized basis sets. Plane waves are particularly suited to periodic systems and their use simplifies certain aspects of quantum chemical calculations through the flexibility offered by a dual space representation and the completeness of the basis set. The major drawback of plane waves is that a huge number of basis functions must be used to attain a level of accuracy that is comparable to that achieved with even a small number of localized basis functions.

A final point that is closely related to basis functions involves the use of pseudopotentials (also known as effective core potentials) to replace the core orbitals and to smooth the valence orbitals near the nuclei. For example, in carbon, which has a $1s^2 2s^2 2p^2$ electronic configuration, the 2s and 2p electrons would be treated with a basis set, while the 1s electrons would be represented by a pseudopotential. The motivation for eliminating the explicit treatment of core electrons stems from the notion that these electrons do not contribute significantly to chemical bonding, and hence the core states are altered minimally during chemical reactions. As such, the core states do not need to be treated explicitly in most quantum mechanical calculations. This dramatically decreases the number of basis functions needed for a particular calculation, which decreases computational cost. An additional advantage of pseudopotentials is that relativistic effects associated with core electrons in heavy atoms can easily be incorporated into the calculation without additional computational effort. In practice, all plane wave calculations use pseudopotentials to represent the core states of all atoms.

2.1.4 Periodic Boundary Conditions

It is often of interest to perform simulations of bulk liquids or bulk solids. By definition, such systems extend across length scales that are inaccessible through quantum mechanical (or even classical) simulation. One means of simulating these systems is to perform a calculation on a small system that is representative of the bulk material and repeat this small system infinitely in all three spatial directions. A two-dimensional example is shown in Fig. 2.3, where the central cell is repeated in the plane of the page. Representing a system in this manner is termed using periodic boundary conditions (PBCs). In what follows, a few key points regarding the use of PBCs in quantum mechanical calculations are qualitatively described.

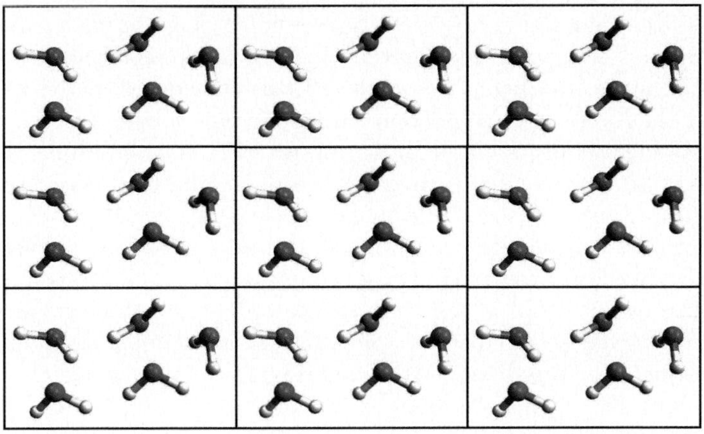

Fig. 2.3. Periodic boundary conditions in two dimensions. The central cell contains five water molecules and is repeated exactly in both spatial directions that define the plane of the page. Nine cells are shown explicitly; however, the system should be repeated infinitely in all directions

As mentioned above, the use of PBCs involves defining a simulation cell and repeating this cell periodically to form an infinite crystal. In this scheme, quantities such as the wave function are periodic and must be continuous across the walls of neighboring simulation cells. This introduces constraints into the calculations which ensure that the central simulation cell interacts with its periodic images, although the calculation is only performed on the central simulation cell. Long range electrostatic interactions between the periodic images, can be accounted for by using techniques such as Ewald summations [35]. This property of PBCs proves useful when simulating bulk materials, where any given atom should interact with those throughout the material. For example, in a crystal, which is composed of periodically repeated unit cells, the atoms in a given unit cell should interact with those in the other unit cells. In the case of non-crystalline systems, e.g., liquids or amorphous solids, it is desirable to study bulk systems; however, such systems should not be periodic. PBCs can still be used to approximate these systems by defining a large "central" simulation cell, which minimizes the effect of externally imposed periodicity on the behavior of the system. It should be noted that imposing PBC implies some constraints on the long-range structure and correlations in bulk systems. These must be carefully considered when simulating long-range properties such as low-frequency phonon modes.

2.1.5 Molecular Dynamics

In the preceding section, we have outlined how first principles density functional theory calculations can be used to calculate the potential energy surface. Exploring the PES is important because it provides insight into chemical processes that may occur. Conventional studies of materials and molecules that apply first principles quantum mechanical methods often do so in a static sense. That is, the behavior on the PES is considered in the zero Kelvin limit and then extrapolated to finite temperatures. Furthermore, static calculations focus on particular regions of the PES corresponding to relevant species such as the presumed reactants, products and transition state associated with a given processes. 'Static' calculations are extremely powerful and widely used. In this chapter, more novel ab initio molecular dynamics (MD) simulations are also presented. These ab initio MD simulations allow one to observe the evolution of a system in space and time at the molecular level subject to a given set of macroscopic external conditions, e.g., temperature and pressure. It is noted that MD simulations provide a great deal of qualitative insight into chemical processes, yet can also be applied quantitatively. In what follows, we discuss the MD methodology and its use in the study of chemical reactions.

Most MD calculations treat the nuclei comprising a chemical system as classical particles whose trajectories can be determined with Newtonian equations of motion:

$$\mathbf{F}_i = m_i \mathbf{a}_i = m_i \frac{d^2\mathbf{R}_i}{dt^2} \tag{2.12}$$

where the force on atom i, $\mathbf{F}_i$, is obtained by differentiating the potential energy with respect to the nuclear coordinates:

$$\mathbf{F}_i = -\frac{\partial E(\mathbf{R})}{\partial \mathbf{R}_i}. \tag{2.13}$$

Unfortunately, it is not possible to analytically integrate (2.12) for a many-body system, and one must numerically integrate the system of coupled differential equations representing the trajectory over a series of small time steps. The general procedure is as follows. At a certain time, t, the nuclei are described by a set of positions and momenta, $\mathbf{R}_i(0)$ and $m_i\mathbf{V}_i(0)$. The potential energy for the system is calculated and used to determine the forces on the nuclei, which are then used to update the nuclear positions and momenta to their values at a later time, $t + \Delta t$. The process is then repeated until a sufficiently long total period of time is simulated. The time intervals, Δt, used in this process must be short enough to capture the highest frequency motions in the system, which typically correspond to bond vibrations. As such, these intervals are on the order of 10^{-15} s, and hence $\sim 10^6$ force calculations must be performed to simulate one nanosecond of time. Despite the considerable computational expense associated with this procedure, MD simulation has been an exceedingly powerful and popular tool for over 50 years [36–38].

An important feature of MD simulations is the ability to study reactions under a given set of conditions. For instance, by properly scaling the nuclear kinetic energies it is possible perform simulations at a given temperature. Means of applying pressure to the system also exist. In a constant volume simulation with PBC, a simulation with high molecular densities will effectively lead to high simulation pressures. Alternatively, pressure and temperature can be added as independent dynamic variables in the equations of motion that are coupled to the positions and momenta of the molecules in the simulation [35]. These capabilities are quite important when studying the influence of external conditions on a reaction. By comparison, temperature and pressure can only be incorporated into static calculations in an *ad hoc* manner.

2.1.6 Ab initio Molecular Dynamics

In the preceding subsection, the classical MD methodology was briefly introduced in a generic sense, with the PES considered merely as a means of obtaining the forces on the nuclei. For the vast majority of MD simulations, the PES is evaluated with classical potentials, which are parameterized functions that subdivide the potential energy into distinct contributions associated with bond lengths, angles, dihedrals, electrostatics, etc. With the exception of a few specific cases, classical potentials cannot accurately describe changes in bonding, and hence are not suitable for use in the simulation of chemical reactions, i.e., bond formation and dissociation. In order to study such processes, it is necessary to explicitly consider the electronic structure of the system. A popular way to do this is ab initio molecular dynamics (AIMD) [39].

In the AIMD approach, the potential energy in (2.13) is derived from a QM calculation; typically at the density functional level of theory. The explicit consideration of the electronic structure of the system in this manner leads to a powerful simulation method that is capable of describing chemical reactions. In fact, because the system moves on the electronic PES according to well-defined equations of motion, i.e., Newton's equations, one can, in principle, simply initiate an AIMD simulation, follow the motion of the system and observe chemical reactions without having any preconceived notions of what the potential energy surface describing those reactions may be. The ability of AIMD simulations to investigate known reactions and identify new reactions has led to a dramatic increase in the popularity of this method in recent years.

AIMD simulation is not without its drawbacks, as these calculations suffer from many limitations due to the computational expense of the first principles QM calculations. If the potential energy is derived from a DFT calculation, it is typically only possible to perform simulations on systems composed of a few hundred atoms, and only sub-nanosecond time scales may be considered. Despite these limitations, AIMD simulation has recently experienced a tremendous increase in popularity and is commonly applied to a wide variety of systems [4, 5, 40–42].

Two popular types of AIMD simulations exist. The first approach is called Born–Oppenheimer AIMD (BO-AIMD) and involves optimizing the electronic structure of the system at each time step of the simulation. That is, a full wave function optimization is performed at each time step, and the resulting potential energy is used to determine the forces on the nuclei. BO-AIMD is an intuitive approach to AIMD simulation, yet is computationally expensive because of the tremendous number of wave function optimizations that must be performed.

The second main type of AIMD simulation is known as Car–Parrinello AIMD (CP-AIMD) [3] and involves treating the electronic states as effective dynamic variables which are propagated according to the classical Newtonian equations of motion. In a CP-AIMD calculation, the first-principles method used is almost exclusively DFT, where the Kohn–Sham orbitals are optimized at the very first step of the simulation only. Then, they are assigned a fictitious mass, μ, and propagated in time according to classical equations of motion through the following extended Lagrangian:

$$\mathcal{L}_{CP} = \underbrace{\frac{1}{2}\sum_{I=1}^{N_{atoms}} M_I \dot{\mathbf{R}}_I^2}_{\text{nuclear kinetic energy}} + \underbrace{\frac{1}{2}\sum_{i=1}^{n_{orbitals}} \mu_i \left\langle \dot{\psi}_i \,\middle|\, \dot{\psi}_i \right\rangle}_{\text{orbital kinetic energy}} - \underbrace{E^{DFT}[\rho]}_{\text{electronic potential energy}} + \underbrace{\sum_{i,j} \Lambda_{ij} \left(\langle \psi_i \mid \psi_j \rangle - \delta_{ij}\right)}_{\text{orthonormality constraints}} . \tag{2.14}$$

The first term of (2.14) is the classical kinetic energy of nuclei with masses $\mathrm{M_I}$, the second term is the fictitious 'classical' kinetic energy of orbitals due to the electronic degrees of freedom with inertial parameters (fictitious masses) μ_i, the third term is the electronic energy as described in (2.7), and the last term is the constraints to maintain orthogonality of orbitals. Upon substitution of (2.14) into Lagrange's equation of motion one obtains the Car–Parrinello equation of motion for both the nuclei and the KS orbitals:

$$M_I \ddot{\boldsymbol{R}}_I = -\frac{\partial E(\psi_i, \boldsymbol{R}_I)}{\partial \boldsymbol{R}_I} \tag{2.15}$$

$$\mu_i \ddot{\psi}_i = -\frac{\partial E(\psi_i, \boldsymbol{R}_I)}{\partial \psi_i} + \textit{orthogonolity constraints}. \tag{2.16}$$

In this approach, the Kohn–Sham orbitals are propagated as classical degrees of freedom at the same time the nuclear motion is treated. Therefore, there is no need to optimize the electronic structure at each time step as in the BO-AIMD approach. It is important to note that although the orbitals are treated as classical degrees of freedom, the electronic structure is still determined quantum mechanically. It is also noted that the mass, μ, is a fictitious quantity assigned to the orbitals that should not be confused with the mass of the electron. Moreover, in the limit that μ approaches zero, the Car–Parrinello equation converges to the KS equation so that E is minimized and the Born–Oppenheimer limit is recovered.

The key to successfully performing a CP-AIMD simulation lies in preventing the transfer of energy between the nuclear and electronic subsystems. This is achieved by altering the value of μ such that the power spectra of these two subsystems do not overlap, i.e., μ should be kept small to ensure that the orbitals oscillate at frequencies much higher than those of the nuclei. Furthermore, using a small value for μ allows the orbitals to adjust quickly to changes in the nuclear configuration, which keeps the system close to the ground state PES. In a properly performed CP-AIMD simulation the electronic structure will remain close to the true PES, and hence the dynamics observed in a CP-AIMD simulation closely follow those which would be determined in a BO-AIMD calculation.

As indicated above, the main advantage of the CP-AIMD approach is that the electronic structure is only optimized once during the course of a simulation. This can significantly decrease computational cost compared with a BO-AIMD simulation, where the electronic structure is optimized at each time step. However, properly integrating the orbital dynamics, which oscillate at high frequencies, requires the use of times steps, Δt, that are typically an order of magnitude smaller than those used in a BO-AIMD simulation. This decreases the computational advantage of CP-AIMD simulations over BO-AIMD simulations. Nonetheless, the former is faster in most cases and has become the most popular AIMD simulation technique for studying chemical systems.

2.2 Collision Dissociation of Nitromethane

Although numerous theoretical reports focusing on nitromethane (NM) have appeared in the literature [43–50] the decomposition mechanism of liquid NM under shock conditions remains elusive. The C—N bond is the weakest bond in NM and therefore unimolecular C—N bond scission has often been implicated as the dominant decomposition pathway. However, a number of high pressure studies do not support this mechanism, [51,52] vide infra. To gain some insight into these decomposition pathways bimolecular collision MD simulations have been performed by a number of researchers [53,54]. A summary [54] of bimolecular collisions of NM in a variety of orientations and a large range of incident collision velocities is shown in Table 2.1. For the offset anti-parallel molecular orientation, the critical velocity for successful dissociation was found to be 7.0 km s^{-1}, higher than the average atomic velocities expected at the shock front of a detonation [55]. Although these bimolecular collision simulations provide a qualitative glimpse into the shock-induced dissociation process, they cannot account for the effects that neighboring molecules may have on the dissociation mechanism. This issue is addressed in first principles Car–Parrinello molecular dynamics (CPMD) simulations of multimolecular collisions of NM.

2.2.1 Impact of a Single Molecule on Multiple Molecules

The previous bimolecular collision simulations are limited to collision-induced reactions without neighboring confinement and collision sequences beyond

Table 2.1. Threshold collision velocities (km s^{-1}) determined from multimolecular and bimolecular collision simulations

Orientation		Threshold collision velocity	
		Multimolecular	Bimolecular
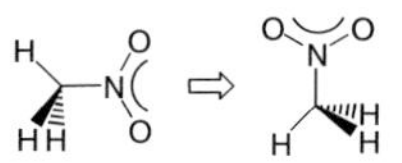	Perpendicular	12.0	8.0
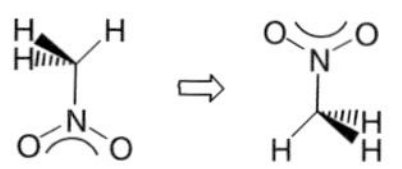	Anti-parallel	11.0	10.5
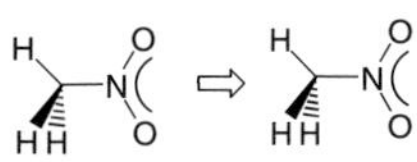	Head-to-tail	11.0	8.5
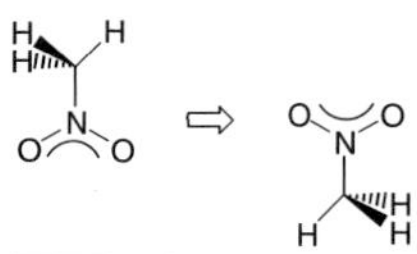	Offset Anti-parallel	12.0	7.0

the initial impact are meaningless due to only the two isolated molecules involved. Multimolecular collision simulations offer a substantial improvement since they allow for possible collision-induced reactions involving neighboring molecules beyond the initial impact. The effects of multiple collisions on the decomposition mechanism are important in that they determine whether the molecular decomposition occurs at the initial shock wave front or after further thermalization of the shock wave energy. This issue can have important implications on the behavior of the material as a high energy density material. Due to the very short time scales involved in passing of the initial shock wave and thermalization stages, information regarding the decomposition mechanism at these time scales can be experimentally difficult to obtain. This makes the use of simulation methods very useful for the study of this phenomenon.

The first study of multimolecular collisions involves collisions by the impact of a single molecule on multiple molecules. The simulations are conducted under the same four collision orientations of the initial colliding NM pair of molecules as the bimolecular collisions reported in Table 2.1. A total of 13 NM molecules are placed in an orthorhombic simulation cell of $19.0 \times 14.2 \times 14.2$ Å^3 such that one molecule is placed at one end of the simulation cell and the remaining 12 molecules are placed at the opposite end. The simulations are initiated by propelling the separated molecule into a stationary molecule, embedded in the cluster of NM molecules as illustrated in Fig. 2.4. The cluster of 12 molecules is randomly oriented and possesses a density of 1.14 g cm^{-3}, corresponding to the density of liquid NM at ambient pressure and temperature. The incident collision velocities range from 6 to 12 km s^{-1} for various orientations of the two colliding molecules. Each multimolecular collision simulation is run for 3,500 time steps or 0.507 ps.

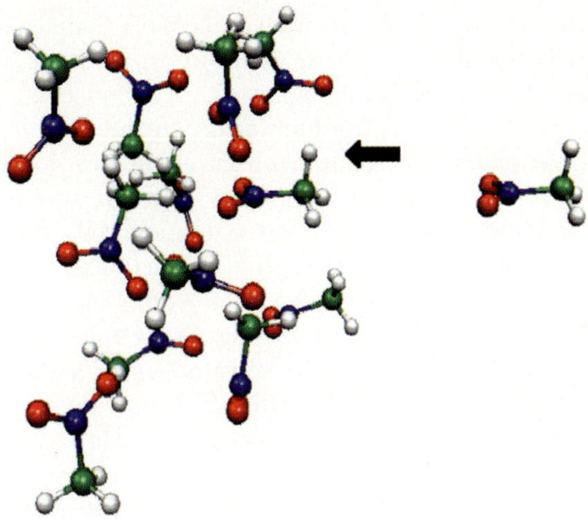

Fig. 2.4. Schematic of the single-molecule on multi-molecule nitromethane collision simulation with the head-to-tail orientation for the initially colliding pair

As evident from Table 2.1, the threshold collision velocities determined from the multimolecular collision simulations are significantly higher than those found previously from bimolecular collision simulations for each molecular orientation investigated. In all cases of multimolecular collisions, the sub-critical velocity collisions induce initial C—N bond scission of the incoming and stationary molecules. However, the neighboring molecules act as a trap to confine the CH_3 and NO_2 fragments produced, thereby enabling them to recombine to form intact NM molecules. The C—N bond then alternately breaks and reforms as the simulation continues until finally it remains intact when there is insufficient energy to break it. At the threshold collision velocities, the recoiling fragments produced during the initial collision possess enough translational energy to overcome the confinement forces of the neighboring molecules leading to permanent cleavage of the C—N bond to yield CH_3 and NO_2. Collisions of more than two molecules are observed, for instance, for the perpendicular molecular orientation at 12 km s^{-1}, where collisions occur between four NM molecules including the incoming, the stationary and two neighboring molecules. This results in permanent C—N bond cleavage of the stationary molecule, while the other molecules involved recombine to form intact NM molecules. Permanent C—N bond scission is observed in both the incoming and stationary NM molecules in the offset anti-parallel orientation at its threshold velocity of 12 km s^{-1}, while the anti-parallel and head-to-tail orientations display permanent C—N bond cleavage of the incoming NM molecule at the threshold velocity of 11 km s^{-1}. These velocities are not available under conventional detonation conditions of nitromethane with a shock velocity of 6–7 km s^{-1} where the molecular collision velocities are even less than this value.

While the threshold velocity of multimolecular collisions is higher than that of bimolecular collisions in all the collision orientations, the predominant mechanism of decomposition remains the C—N bond cleavage as in the bimolecular collisions. A different fragmentation mechanism, initial C—H bond scission of the stationary molecule is observed in one of the high velocity multimolecular collision simulations (12 km s^{-1} in a head-to-tail orientation). The initial C—H bond cleavage is followed immediately by the migration of this free H atom to one of the O atoms of the incoming NM molecule which induces N—O bond cleavage of this NM molecule yielding OH and CH_3NO. The N—O bond of the stationary molecule is then cleaved and the free O atom migrated to the CH_2 group of the stationary molecule, thereby inducing C—N bond cleavage to yield CH_2O and NO fragments. The final products produced from this cascade of reactions are: OH, CH_2O, NO, and CH_3NO.

2.2.2 Impact of Multiple Molecules on Multiple Molecules

The multimolecular collision simulations are further extended to the collisions between two groups of molecules as depicted in Fig. 2.5. The chemical system of 32 NM molecules is contained in a cubic simulation cell 14.2 Å on each side. The density of the 32 molecules in the cube corresponds to 1.14 g cm^{-3} for liquid NM. Periodic boundary conditions are employed to better represent the bulk material properties and effects of neighboring molecules. In an identical setup without using periodic boundary conditions, molecules leaving the cell will effectively be in a vacuum and eventually disperse as a result of Brownian motion. Initial positions for the simulation were first obtained from a classical MD simulation. The initial orientations and velocities of each of the NM molecules in the simulation cell are derived from an AIMD of a sample that was equilibrated for approximately 4.36 ps at 298.15 K with a Nose–Hoover thermostat. Eight NM molecules, occupying the most right-hand quarter of the simulation cell, are translated linearly forward to impact into the other 24 NM molecules occupying the rest of the cell as shown in Fig. 2.5. In each simulation, all eight NM molecules are assigned a uniform initial velocity ranging from 2 to 10 km s^{-1}.

The simulations are further characterized as *static* or *dynamic* simulations. The static simulations are similar to the previous simulations of bimolecular collisions and single-to-multiple molecular collisions, in which the kinetic energy and temperature of the molecules are initially set to zero. The eight molecules occupying the right-hand quarter of the cube are then translated forward with their assigned velocity. In the dynamic simulations, all 32 molecules initially possess equilibrium velocity and the total kinetic energy corresponds to the kinetic energy of the equilibrated sample of liquid NM at 298.15 K and 1 atm. The eight molecules occupying the right-hand quarter of the simulation cube are translated with an assigned velocity in addition to their equilibrium velocity. Thus, the dynamic simulations are considered more realistic than the static simulations. All simulations were performed for 5,000 time steps or 0.725 ps.

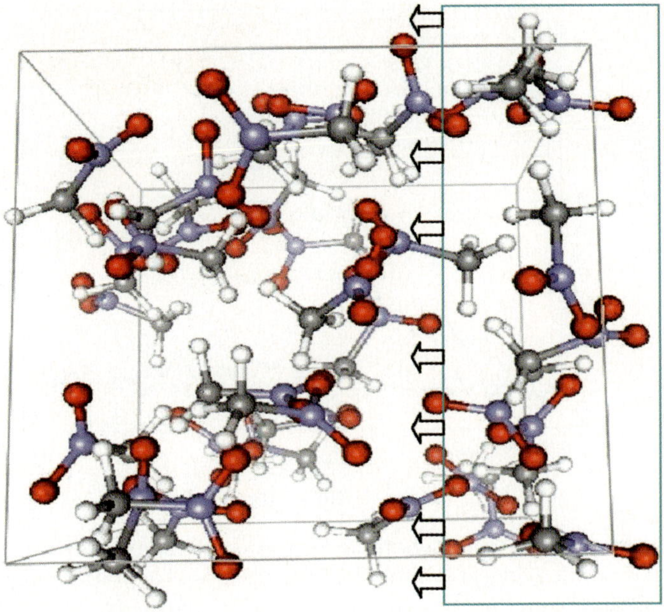

Fig. 2.5. Schematic of the multiple-molecule on multiple-molecule collision simulations

Figure 2.6 shows the dynamic simulation results for average C—N bond lengths of all 32 NM molecules as a function of time at collision velocities ranging from 2 to 10 km s^{-1}. At lower velocities, the bond lengths oscillate around the initial bond lengths. The initial bond length at 25°C is slightly higher than its ideal bond length, which is defined for 0 K in the gas phase and imparts the lowest total energy to the molecule. The ideal bond lengths for the C—N, N—O and C—H are 1.46, 1.22 and 1.10 Å, derived from geometry optimizations of the isolated nitromethane molecule at the PBE/6-31G(d,p) level. A bond is considered dissociated if the distance between the atoms of the bond is greater than 1.5 times the ideal bond length which corresponds to 2.19 Å for the C—N bond. No bond breakage was observed at velocities of 2–6 km s^{-1}. However, for a collision velocity of 8 km s^{-1}, there is a slight increase in the average C—N bond distance in the simulations approximately 500 time steps into the simulation, which further increases at around 1,250 time steps. Qualitative examination of the simulation reveals that the C—N bond of a single nitromethane molecule dissociated at approximately 1,250 time steps. However, the neighboring molecules confine the CH_3 and NO_2 fragments produced, thereby enabling them to recombine to form an intact NM molecule at nearly 1,600 time steps. The C—N bond then alternately breaks and reforms as the simulation continues until finally it remains intact when there is insufficient kinetic energy to overcome the bond strength and the confinement forces of neighboring molecules. At 8 km s^{-1}, while the average

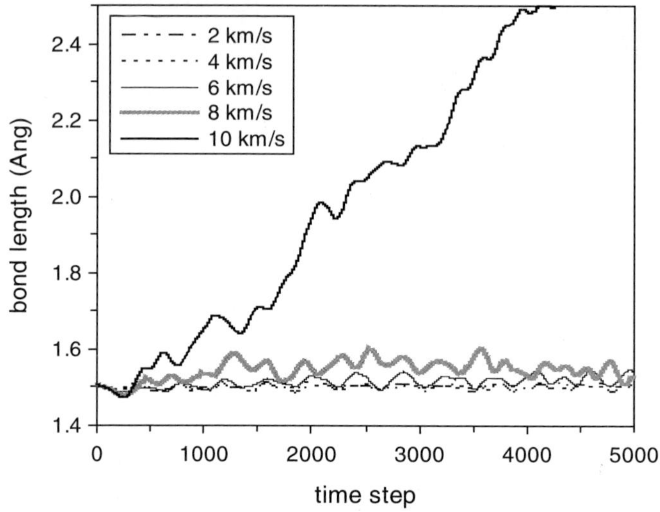

Fig. 2.6. Average C—N bond length as a function of time at 2–10 km s^{-1} velocity of impact of multiple molecules on multiple molecules. The initial temperature and pressure of the molecules in the simulation are 25°C and 1 atm

N—O and C—H bond lengths oscillate around their initial bond lengths at enlarged fluctuation amplitude, no N—O and C—H bonds are broken.

As the collision velocity is increased to 10 km s^{-1} the first C—N bond breakage is observed at 610 time steps, resulting in an increase in average bond length to 1.6 Å (Fig. 2.6). The time delay from the initial impact is ~0.05 ps, which is shorter than the time delay at the collision velocity of 8 km s^{-1} (~0.13 ps). The time delay between the initial impact and the C—N bond breakage and its dependence on impact velocity are likely associated with the conversion of translational energy to vibrational energy required for cleavage of the bond. As time proceeds, more and more C—N bonds are permanently broken as shown in Fig. 2.7. Breakage of the N—O and C—H bonds occurs only at 10 km s^{-1} and in a much delayed time. Nearly 2,600 time steps elapse till the first N—O bond dissociation and 3,000 time steps elapse till the first C—H bond dissociation. In all cases of N—O bond dissociation, bond cleavage occurs on an intact NM molecule and results in migration of the oxygen atom to a methyl fragment forming a formaldehyde intermediate. Also, the interaction of two methyl fragments repeatedly forms and breaks short-lived ethane. The C—H bond cleavage results in only a small number of hydrogen atoms that migrate back and forth to the nitro group of neighboring nitromethane.

The fact of the earlier C—N bond cleavage suggests that the predominant mechanism of decomposition remains the C—N bond breakage into the nitro and methyl groups at the threshold collision velocity between 8 and 10 km s^{-1}. As indicated in Table 2.1, this threshold velocity is less than the

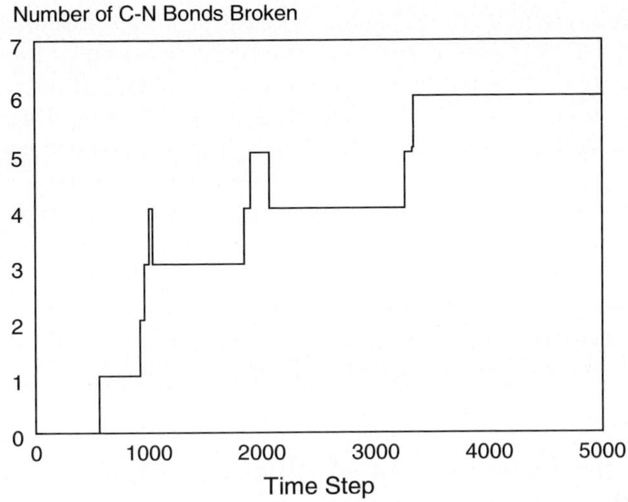

Fig. 2.7. Number of C—N bonds broken as a function of time at 10 km s^{-1} velocity of impact of multiple molecules on multiple molecules. The initial temperature and pressure of the molecules in the simulations are 25°C and 1 atm

values obtained from the collisions of a single molecule on multiple molecules. This is so because in the collisions of multiple molecules on multiple molecules, the total impact kinetic energy received by the receptor molecules is larger and the latent neighboring confinement becomes weaker due to the distributed impact kinetic energy.

The static simulations provide an identical decomposition mechanism and threshold collision velocity based on the C—N bond cleavage. The difference between the static and dynamic simulations lies in the time delay of the decomposition and the rate of bond length increase that becomes apparent at the impact velocity of 10 km s^{-1} as shown in Fig. 2.6. The difference observed can be explained by the additional kinetic energy at room temperature and pressure available in the dynamic simulations. This additional energy manifests itself vibrationally and promotes decomposition once initial bond breakage begins.

It is clear from these simulations that a strong shock can result in rapid decomposition of nitromethane, in times far short for thermal equilibrium to have been established. The simulations of this section indicate that the initial velocities required for the direct decomposition of nitromethane molecules is in the range of 8–10 km s^{-1}. Experimentally, it was found that the average speed of nitromethane molecules at the detonation shock front is about 2.5 km s^{-1} [56]. This shows that the nitromethane decomposition reaction mainly occurs in the thermalization stage behind the shock front of a conventional detonation. This conclusion also appears to be confirmed by recent experiments of the incident or multiple reflected shock compression initiation of detonation in liquid nitromethane, where the temperatures for incident and multiple reflected initiation upon shock compression are equal to about 700 K, while the multiple reflected shock pressure reaches 8 GPa, nearly two times that of the incident shock [57, 58]. The 8–10 km s^{-1} decomposition threshold collision velocity further suggests that the collision-induced detonation initiation in the shock front would likely be dominant for a detonation velocity beyond 15–20 km s^{-1} in a molecular condensed matter. This would result in a megabar detonation pressure since the detonation pressure is approximately proportional to the square of the detonation velocity. In order to examine the threshold pressure for mechanical dissociation under minimal thermal influence, first principles molecular dynamics simulations of static compression at low temperatures will be discussed in the next section.

2.3 Pressure Dissociation of Nitromethane

The reactions of high energy density materials under low temperature – high pressure conditions are of interest. For this purpose, the high pressure decomposition of nitromethane has been studied. To study the effects of isotropic pressure on the decomposition of NM, Car–Parrinello ab initio MD simulations were performed. Periodic boundary conditions were utilized with a simple cubic simulation cell and 32 nitromethane molecules such that the experimentally measured bulk density of 1.139 g cm^{-3} at 298 K and 1 atm pressure was reproduced. This simulation cell volume was then shrunk by a factor of 1.5–3.0. Well equilibrated classical MD simulations were used as the starting point.

Initial geometric configurations for the ab initio MD simulations were extracted from well equilibrated classical MD simulations. The cell volume was compressed isotropically by a factor of up to 3. Slow heat up was performed (50 K $\rightarrow$ 150 K $\rightarrow$ 300 K). At 300 K, equilibration was performed for 25,000 time steps (ca. 3.5 ps) and production runs were performed for 25,000 time steps. The R_{CN}, R_{NO}, and R_{CH} bond lengths at different compressions are shown in Fig. 2.8.

The ab initio MD simulations of compressed liquid NM show no molecular decomposition at compression factors of 2.0 and 2.5 at 300 K. However, when the liquid is compressed by a factor of 3.0 an interesting pressure-induced molecular decomposition is observed at 150 K after equilibration for about 3,300 time steps. An intermolecular H atom transfer event occurs between two very closely spaced molecules aligned in an anti-parallel alignment, as illustrated in the MD trajectory snapshots shown in Fig. 2.9. As illustrated in Fig. 2.9, the CH_3 group of one of the NM molecule re-orients itself such that two of its H atoms lie in the same plane as the NO_2 fragment and the other H atom is oriented nearly perpendicular to this plane directed at the O atom of the adjacent NM molecule. This H atom then transfers from the C atom of the first NM molecule to the O atom of the second NM molecule to yield the

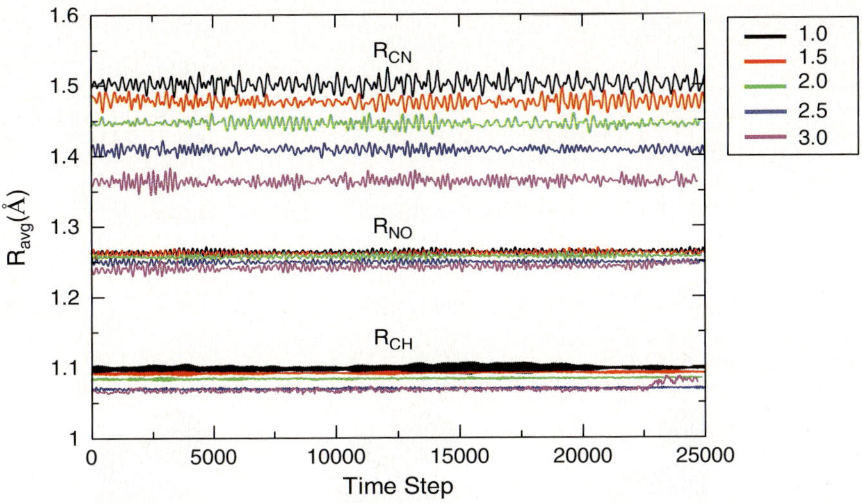

Fig. 2.8. Changes in intramolecular bond distances with pressure at 300 K. The colors correspond to different compression factors (effective pressures) of the simulation

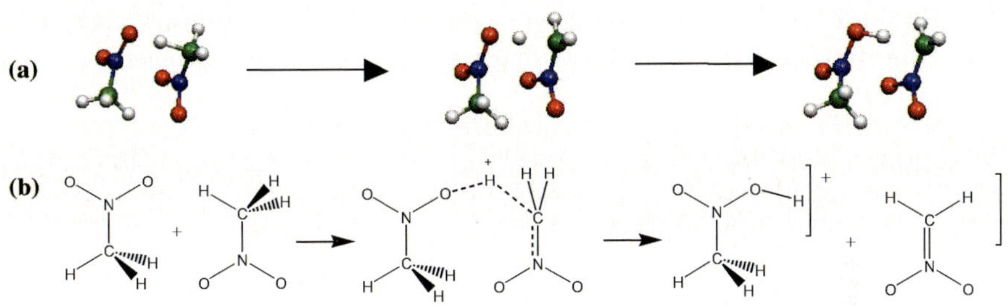

Fig. 2.9. (**a**) snapshots from ab initio MD simulation of highly compressed liquid nitromethane showing a proton transfer event. (**b**) Schematic of the proton transfer reaction observed

final products $CH_2NO_2^-$ and $CH_3NO(OH)^+$. The H transfer event occurs on a time scale of about 15 fs.

Since the C—N bond is the weakest covalent bond in nitromethane, one might think that the C—N bond should break as is commonly observed in the collision simulations of the previous sections. Indeed, Fig. 2.8 reveals that as the compression of the liquid is increased, there is a significant compression of the C—N bond, as compared to the CH and CO bonds. So one question to ask is why then does the C—H bond first break? To answer this question, we used molecular orbital theory and examined the electronic structure of nitromethane. As previously mentioned, it is notable that there is a significant compression of the CN bond as the pressure is increased. Examination of the frontier molecular orbitals of the nitromethane molecule as the C—N bond is shortened reveals that the methyl group proton of nitromethane becomes a

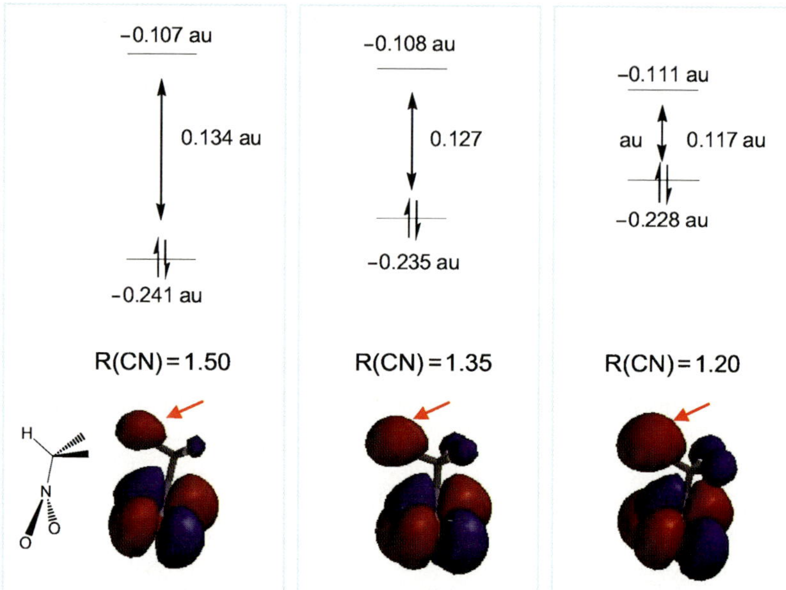

Fig. 2.10. Frontier molecular orbitals of nitromethane as a function of the CN bond length. The top portion of the figure show the calculated orbital energies of the highest occupied molecular orbital (HOMO) and the lowest unoccupied molecular orbital (LUMO). The bottom portion of the figure shows isosurface plots of the LUMO orbital of nitromethane. As the CN bond is compressed the LUMO orbital energy is stabilized, and the orbital lobe on one of the hydrogen atoms becomes larger, as highlighted by the red arrow

better Lewis acid. At the same time, the carbonyl oxygen becomes a better Lewis base. This is illustrated in Fig. 2.10 which shows the highest occupied molecular orbital (HOMO) and lowest unoccupied molecular orbital (LUMO) energies as a function of the C—N bond distance. The HOMO orbital of nitromethane (not shown) can be described as a pi-bonding orbital of the aromatic ONO fragment of nitromethane. This orbital is also polarized toward the oxygen, and as the C—N bond is compressed, it increases in energy, thereby making the oxygen atoms better Lewis base sites. The LUMO of nitromethane, shown on the bottom of Fig. 2.10, can be described as an anti-bonding orbital of the aromatic ONO fragment. However, there is a significant component of this molecular orbital localized on the hydrogen atoms as highlighted by the red arrow in Fig. 2.10. Moreover, this contribution becomes larger and the LUMO energy becomes stabilized as the CN bond is compressed. Both of these show that the nitromethane protons become more Lewis acidic as the CN bond is compressed.

From first principles molecular dynamics simulations, Manaa et al. [59] have also observed proton transfer as the first stage of the nitromethane decomposition reaction at high temperature (3,000 K) and high density

($\rho = 1.97\,\mathrm{g\ cm^{-3}}$) conditions. The samc observation was made under shock compression conditions with the shock speed of $7\,\mathrm{km\ s^{-1}}$ [60]. Some early, static high pressure experiments [61] show that the thermal detonation time for deuterated nitromethane is ten times longer than that of ordinary nitromethane. While more recent experiments comparing the detonation of nitromethane and deuterated nitromethane show less dramatic differences [56, 62], the measured particle velocities in the reaction zone of detonation reveal that changes upon deuteration are most distinct in the early part of the reaction zone. Consistent with these experimental results, our calculations do suggest that proton transfer might be important in the initial stages of the reaction. Moreover, the quantum chemical analysis of the proton transfer step as a function of the CN bond distance provides fundamental insight into why this process occurs more readily as the pressure is increased.

2.4 High Pressure Non-molecular Solid Phases of Polynitrogen

A covalent, single-bonded nitrogen phase, analogous to carbon in diamond, has long been thought to exist at high pressures. If such a material were metastable at ambient pressures, it would make an extremely powerful high-energy-density material (HEDM). Due to the uniquely large difference in energy between the N—N single and triple bonds, the energy density of single bonded polymeric nitrogen has been calculated [63] to be at least $0.4\,\mathrm{eV\ cm^{-3}}$, which is about three times that of the most powerful energetic materials known today.

The structure of single-bonded polymeric nitrogen was at first suggested to be that of the known single-bonded forms of nitrogen's group 15 congeners, namely, the black phosphorus (*BP*) structure and the *A7* structure of arsenic. In 1992, based on possessing gauche interactions among the lone pairs of adjacent nitrogen atoms, Mailhiot, Yang and McMahan [64] predicted a unique single-bonded structure that has no analogue in other natural structures that they called the *cubic gauche* (*CG*) structure. Based on first principles calculations, within a large range of atomic volumes, the *CG* structure was predicted to be more stable than the *BP* and *A7* phases. Further computational studies [65] showed that *CG* might be metastable at ambient pressures, thereby furthering the prospect of using polymeric nitrogen as a practical HEDM.

Experimentally, extensive efforts were made to produce polymeric nitrogen and evidence for non-molecular amorphous phases of nitrogen at 100 GPa and 1,000 K were reported [66–68]. In one such study [66], the amorphous material was even recovered at ambient pressures and low temperature. It is likely that the amorphous materials observed were mixtures of small clusters of non-molecular phases. In 2004, Erements and coworkers reported [69] the synthesis of a crystalline form of single-bonded polymeric nitrogen at pressures greater than 110 GPa and 2,000 K temperatures. Under these conditions, the molecular nitrogen is transformed into the *cubic gauche* crystal phase. The apparent high energy barrier between molecular and non-molecular phases was overcome by the high temperature synthesis. A strong pressure-dependent hysteresis was observed during the synthesis [66, 69]. The high sensitivity of the final phase on its pre-history was also noted in first-principles simulations [64, 70–72], and suggests the existence of many competing metastable high pressure phases separated by large energy barriers.

Recently, a number of new phases of polymeric nitrogen have been found by first principles molecular dynamics calculations at high pressures and temperatures [70–72]. Alemany and Martins [71] found a metastable metallic phase composed of zig-zag chains of nitrogen atoms packed in a body-centered orthorhombic structure. Through similar MD simulations using temperatures (up to 10,000 K), Martin and coworkers [70] also discovered a metastable non-molecular phase composed of zig-zag chains of alternating N—N single and double bonds. Although not single-bonded, this new zig-zag chain phase is lower in energy than *CG* at low pressures and competitive with *CG* at higher pressures.

2.4.1 Polynitrogen Phases from Simple Cubic Motifs

Although successful, the search for new phases by first principles MD simulations is *ad hoc* in nature and can be very computationally demanding. A systematic method to finding new polymeric nitrogen phases is obviously desirable. The fact that all single-bonded metastable phases of group 15 elements predicted or found in nature can be considered Peierls-like distortions [73] of the simple cubic (*SC*) structure hints at such a method. We have recently developed a systematic method of looking for new structures of solid, single-bonded polymeric nitrogen [74]. The method is a synthesis of modern first principles calculations with the geometrical model of crystal structure developed by Wells [75–77] and Burdett and McLarnan [78]. The procedure effectively exploits the connection of the other structures to the *SC* reference structure, allowing focus on the most physically meaningful structure candidates. The procedure is able to recover all the threefold connected nitrogen allotropes found before, (e.g., *A7, BP, CG*) and eight new metastable single-bonded allotropes.

The approach presented here is based on Peierls-like distortions of the *SC* reference structure in order to systematically generate new single-bonded polynitrogen structures consistent with a threefold connectivity pattern of nitrogen. In the undistorted *SC* structure, each nitrogen atom has six equidistant nearest neighbors. The distortion must be such that each nitrogen atom forms only three covalent bonds, thereby breaking the remaining three of the 'bonds' of the original *SC* structure. To obtain a complete set of possible *SC* distortions consistent with threefold connectivity we use the combinatorial analysis procedure of Burdett and McLarnan [78], which is based on a

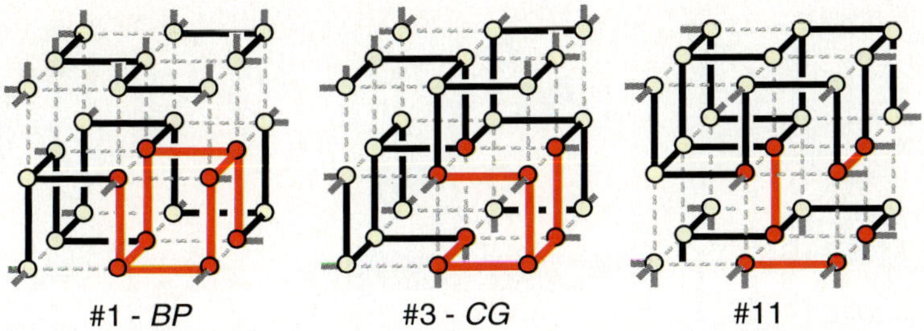

Fig. 2.11. Connectivity diagrams (or structure types) of a simple cubic reference structure consistent with a threefold coordination of nitrogen. Shown are the connectivity diagrams of structure #1, #3 and #11 of a possible 36. The structure numbering that is adopted is that of [78]. A single eight atom SC unit cell is highlighted with red atoms and bonds

generalization of Polya's enumeration theorem [79,80]. In particular, Burdett and McLarnan show that out of 4,096 possible combinations there are only 36 unique ways of labeling the edges of an eight atom *SC* unit cell as "bonds" and "non-bonds" consistent with the threefold connectivity. Those bond-breaking patterns can be described by connectivity diagrams or structure types, as depicted in Fig. 2.11. The complete specification of all the 36 structure types is provided in [78] and we adhere to the same numbering. Note that all the previously reported structures of single-bonded nitrogen, such as *CG, BP, A7*, and *LB* [72] are covered in the set of 36 structure types. Figure 2.11 shows the connectivity diagrams for *BP* (structure #1), *CG* (#3) and a new structure #11 that will be discussed in further detail later.

Once a structure type is chosen, the next step is to use the connectivity pattern to distort the *SC* structure to generate an initial geometry that can be optimized by first principles calculations. The most straightforward distortion is to displace each atom of the unit cell by a distance, d, along the body diagonal toward the three connected neighbors as shown in Fig. 2.12. Once all atoms are distorted in this way, a rough initial structure is generated as depicted on the r.h.s. of Fig. 2.12. This can be described as a Peierls-like distortion of the *SC* structure. This initial structure acts as a starting point for a series of first principles calculations. In our calculations, we utilized the Kohn–Sham density functional theory calculations with the Perdew–Burke–Ernzerhof (PBE) [27] exchange-correlation functional. The Vienna ab initio simulation package (VASP) [81] was used with the projector augmented wave (PAW) method of Blöchl [82] to treat the core states. A plane-wave cutoff of 39 Ry was used and Brillouin-zone integration was performed using a $8 \times 8 \times 8$ Monkhorst–Pack gamma point centered grid.

Although the generation of the initial structures may seem straightforward, the structures may not be stable at all pressures, if at all. Thus, a careful structure optimization procedure is needed in order to find the appropriate

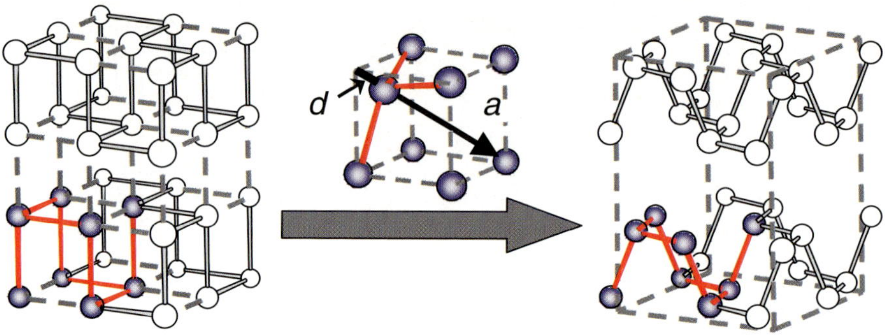

Fig. 2.12. Schematic illustration of the procedure. Left: a structure type of an eight atom cube is first chosen (blue atoms with red bonds). A Peierls-like distortion that is consistent with the connectivity is applied. Right: the structure after distortion of all atoms in an extended representation

ranges of pressure/volume conditions, as well as to prevent premature decomposition of structures before they get sufficiently close to the final optimized geometries.

Here we adopted the following computational procedure. Beginning with the initial structure generated by the aforementioned Peierls-like distortion, the lattice constant a is scanned from 4.1 Å downwards, with a step of 0.05 Å. For each value of a, the following optimization steps are performed.

- First the distortion parameter d is adjusted to minimize the ab initio calculated total energy.
- Then the unit cell shape is allowed to relax, fixing the fractional coordinates of the atoms and the unit cell volume.
- Finally, full relaxation of both the unit cell shape and the atomic positions at fixed unit cell volume a^3 are performed.

At each such relaxation stage the structure must still preserve the intended connectivity pattern. The scan of the lattice constant continues until either the first time a fully optimized structure with a required connectivity pattern is encountered or $a = 3.0$ Å is reached. In cases where a fully optimized structure that preserves the intended structure type is obtained, which happened for 26 out of 36 types, mechanical stability of the structure is verified by calculating the Phonon spectrum, and also the enthalpy-pressure curve. The mechanical stability is inferred from phonon densities of states for each of the 26 structures at various pressures. The stabilities are deduced from a force constant matrix calculated in a $2 \times 2 \times 2$ (64 atom) supercell by means of consecutive displacements of each atom of the original unit cell by 0.02 Å along each axis. In some cases this supercell size may be inadequate, giving rise to false negative frequencies. In this respect, only the affirmative outcomes of the metastability tests should be treated as conclusive.

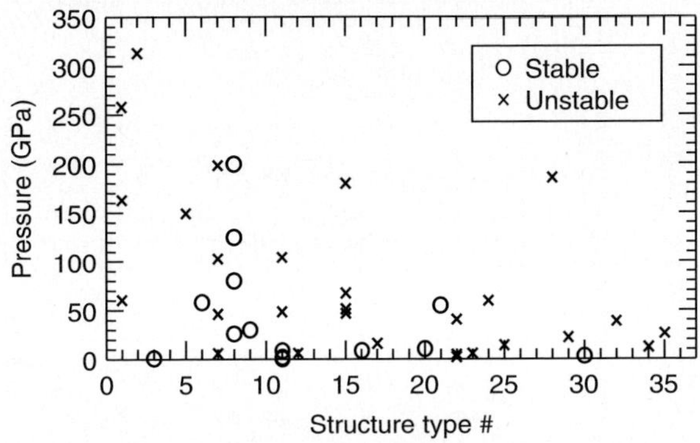

Fig. 2.13. Results of the phonon-spectra based mechanical stability tests for the 36 structure types and different pressures. Mechanically stable and unstable structures at different pressures are shown with circles and crosses, respectively [74]

Analysis of the mechanical stability, shows that at least 8 of the 26 new structures are mechanically stable at pressures less than 100 GPa (#6, 8, 9, 11, 16, 20, 21 and 30). Structures #11 and #30 are even found to be mechanically stable at zero-pressure. The tests of mechanical stability at various pressures are summarized in Fig. 2.13. Figure 2.14 reveals that the *cubic gauche* phase is still the lowest enthalpy structure over the whole pressure range studied. Moreover, it also shows pronounced mechanical stability in our phonon calculations, which is consistent with the same type of analysis in [65]. Leaving the *cubic gauche* case aside, the data in Fig. 2.13 suggest that new structures generated by Peierls distortions, such as structure types #8 and #11, have energies that are comparable or lower than those of the previously studied structures #1 (*BP*), #7 (*A7*), and #15 (*LB*). The 3-dimensional structures of structures #8 and #11 are depicted in Fig. 2.15.

To get an insight into the relative energy rankings of different structure types we analyze a correlation between the first-principles total energies and the number of *cis*-, *gauche*-, *trans*- dihedral angles, and four-member rings of those structure types. Based on such analysis, the lowest energy structures of the low pressure region, structures #3 (*CG*) and #11 (new) are recognized as structures that minimize the number of *trans*-dihedrals and four-member rings while maximizing the *gauche*-dihedrals. It is interesting to note that the stronger weight of *trans*- than the *cis*-dihedrals is in contrast with the simplified "lone-pair" orbital picture [64,78] gained from ab-initio calculations on small nitrogen molecules.

After mechanical stability screening, newly found structures #11 and #8 are the second lowest enthalpy structures after *cubic gauche* at low (<40 GPa) and moderate/high (∼>40 GPa) pressures, respectively. Selected views of structures #11 and #8 are shown in Fig. 2.16. Structure #11 is distinguished by the high R_{32} symmetry and its structure can be described by sets of

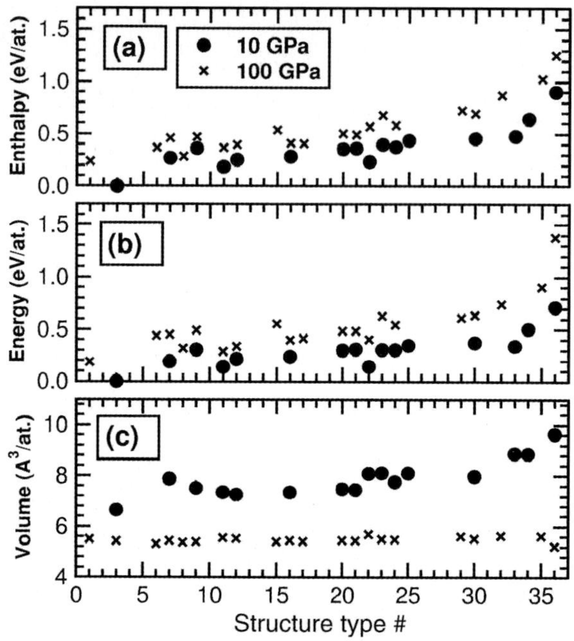

Fig. 2.14. Calculated (**a**) enthalpy, (**b**) total energy, and (**c**) volume per atom versus structure type # at pressures of 10 GPa (circles) and 100 GPa (crosses). The enthalpies and energies are with respect to those of the *CG* phase at the corresponding pressures. Note that structures #1, 3, 7, and 15 are the previously known phases – *BP, CG, A7*, and *LB*, respectively [74]

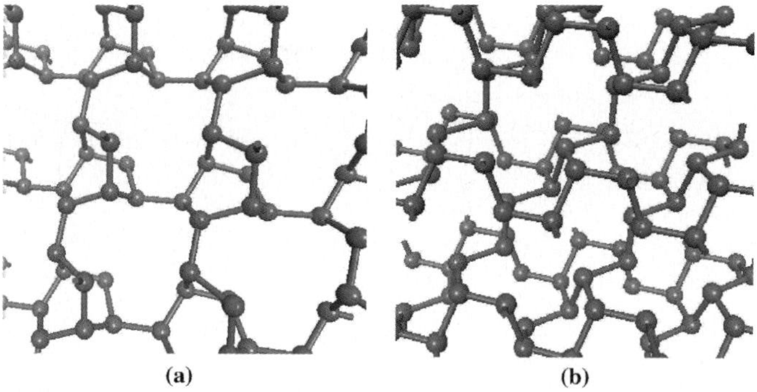

Fig. 2.15. Extended structures of two new low energy and metastable phases, (**a**) structure #11 and (**b**) structure #8

cis–trans chains directed along the three coordinate axes. In fact there are two mutually orthogonal sets of *cis–trans* chains along each axis. Structure #8 has P_{-1} symmetry and a very strong structural resemblance to structure #11, being similarly shaped by sets of *cis–trans* chains but along two out of

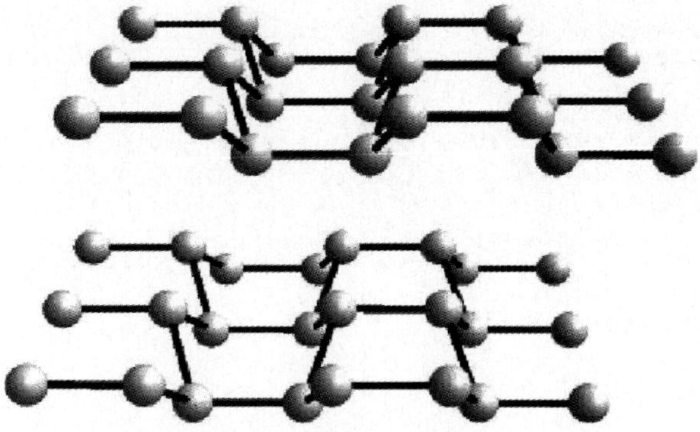

Fig. 2.16. A perspective view of the lattice structure of the Layered-Boat (*LB*) structure

the three axes, hence its lower symmetry. Structure #8 contains large voids in between those layers which may be responsible for the less favorable enthalpy at lower pressure. Structure #11 is an insulator with an indirect band gap decreasing from 2.8 eV at near zero pressure to 2.2 eV at 43 GPa. Structure #8 is also an insulator with an indirect gap at low pressures (e.g., 1.0 eV at 26 GPa) replaced by a higher energy direct gap at higher pressures (1.5 eV at 96 GPa).

2.4.2 Polynitrogen Phases from Chain Motifs

Using a series of high temperature and high pressure ab initio molecular dynamics simulations we have previously found [72] a new single bonded non-molecular phase that we have named the 'layered boat' or *LB* phase as shown in Fig. 2.16. The new phase consists of layers of fused six-membered rings in a boat conformation, and thus the layered boat name. *LB* has a $P2_1/m$ symmetry and consists of two-dimensional layers of fused N_6-rings where the rings are all in the boat conformation. A calculated enthalpy versus pressure phase diagram reveals that *LB* lies between *BP* and *A7* in enthalpy at pressures less than 200 GPa. At pressures higher than this, the enthalpy of *LB* edges above that of *A7*.

The structure of *LB*, initially obtained at high pressures, remains qualitatively unchanged when re-optimized at different pressures in the range of 0–300 GPa, indicating that, along with other polymeric phases of nitrogen, *LB* is also metastable for a wide range of pressures. An enthalpy versus pressure phase diagram for *LB* and various other phases is shown on Fig. 2.17 for pressures less than 200 GPa. At low pressures, from 0 GPA to 50 GPa, *zzCH* is predicted to be the lowest in enthalpy, in agreement with the work of Mattson et al. [70]. At pressures greater than 50 GPa and less than 200 GPa, *CG* becomes the lowest in enthalpy, and *BP* becomes lowest at even higher pressures. The new phase, *LB*, is higher in enthalpy than *CG* and *zzCH* for all observed pressures and is between *A7* and *BP* for pressures up to 210 GPa.

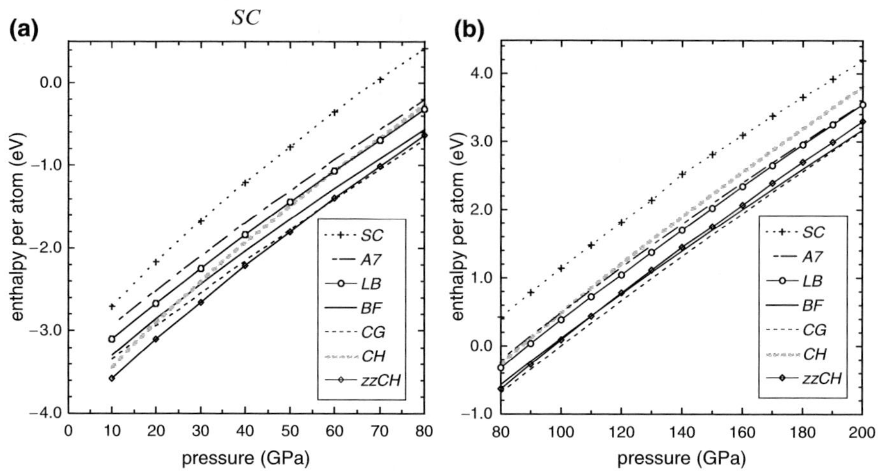

Fig. 2.17. Phase diagram of various polymeric nitrogen phases. The enthalpy per atom is relative to the enthalpy per atom of *CG* at 100 GPa. For clarity, the phase diagram is split into two pressure regions: (**a**) 0–80 GPa and (**b**) 80–200 GPa [72]

At pressures above 210 GPa, *LB* becomes slightly higher in enthalpy than *A7*. *SC* is the least favorable phase among all the phases discussed here. *CH* has the fastest growth of enthalpy with respect to the increasing pressure, and quickly becomes less favorable than all except *SC* as the pressure is increased.

The *LB* phase can be constructed using the *SC* starting structures as described in the previous section. Viewed from an alternative perspective, the structure of the *LB* phase is related to the *BP* and *A7* phases in that they are also composed of layers of fused six-member rings, however, in the latter two the rings are in the chair conformation. The *LB, BP* and *A7* phases can be thought of as different arrangements of the zig-zag chains in the *zzCH* phase approaching along given directions and forming new inter-chain bonds. This is shown in Fig. 2.18 where the *zzCH* phase is depicted with possible inter-chain bonds that would result in the formation of the *LB, BP* and *A7* phases. The two-dimensional layers emerge with a periodic six-ring pattern as a result of the chains in *zzCH* coming close together and forming bonds which change the coordination of the nitrogen atoms from two to three.

In this representation, the 'chains' of *A7* are aligned parallel to each other, and all are simultaneously tilted to one side. In Fig. 2.18 the chains are depicted tilted to the left. An enantiomeric 'right tilted' form of *A7* could be constructed from *zzCH* by choosing a different direction along which the zig-zag chains come together. In the same spirit, *LB* could be characterized as an arrangement of zig-zag chains tilted to the left and right in an alternating fashion. The relationship between the layered phases and the *zzCH* phase outlined here is not meant to show how the layered phases are most favorably formed from the *zzCH* phase. Nonetheless, it provides possible pathways

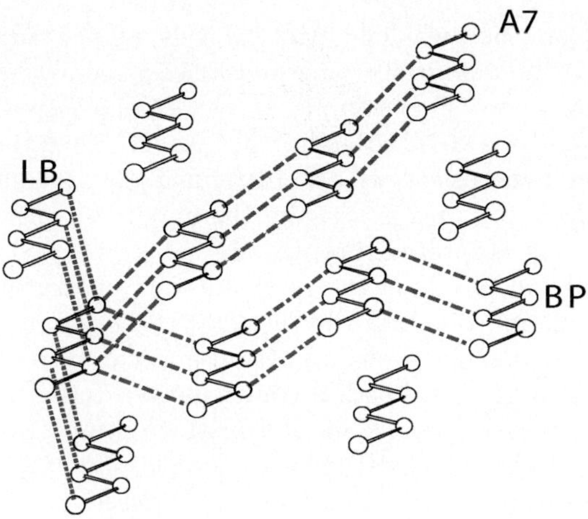

Fig. 2.18. The *zzCH* phase with the possible inter-chain bonding drawn: *BP* (*dash-dot lines*), *A7* (*dash lines*) and *LB* (*dot lines*)

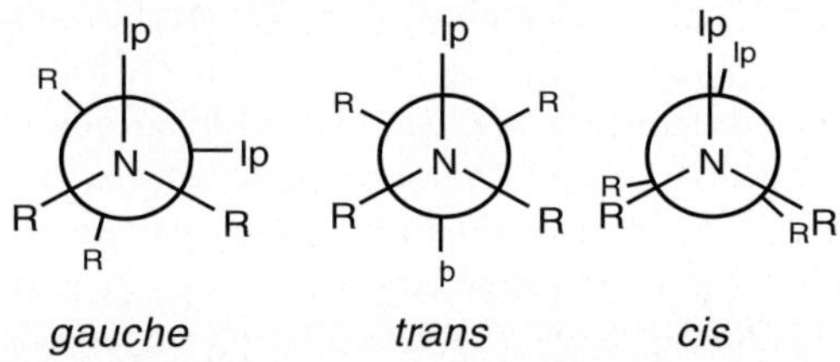

Fig. 2.19. Newman projections of the R_2N—NR_2 fragment in various conformations. 'lp' represents the lone-pair

for the formation of the layered phases from the *zzCH* phase that do not require significant structural rearrangement. We note that both *BP* and *LB* can also be considered in terms of *cis–trans* chains of the *CH* phase coming close together and forming inter-chain bonds.

The three layered phases, *A7, BP*, and *LB* all consist of fused six-member rings composed of nitrogen atoms that are sp^3 hybridized. The local structure of the three phases is distinct, for example, the rings in *A7* and *BP* are in the chair conformation whereas they are in the higher energy boat conformation in *LB.* Here we examine the relationship between the local structure and the relative internal energies of the phases by considering the dihedral angles between N—N single bonds.

Quantum chemical calculations of small molecules of the form R_2N—NR_2 ($R = H$, CH_3, etc) show that they have two minimum energy conformers – *gauche* and *trans* as depicted in Fig. 2.19. The *gauche* structure of R_2N—NR_2 is not ideal in that the lp—N—N—lp dihedral angle, θ, is not 60° but 90°. The 90° dihedral angle of the *gauche* conformer can be accounted for by the fact that this θ angle minimizes the two-orbital/four-electron destabilizing interaction between the two lone pair orbitals of the adjacent N atoms. (We note that gauche definition sometimes used is somewhat different than the convention used here. For example, the lp—N—N—lp dihedral angle, in the *cubic gauche* phase is 107° with a lp—N—N—R dihedral angle being close to zero.) For a later discussion we also introduce the *cis* conformation where the lp—N—N—lp = 0°. This is actually the maximum energy conformation as the lp—N—N—lp is rotated through 360°.

The energy of the N—N moiety depends on the mutual orientation of two adjacent lone-pairs as defined by the lp—N—N—lp dihedral angle. The structure of *CG* was postulated [64] as a solid phase of nitrogen that has all of its single bonds in a gauche dihedral conformation. Indeed, our calculations as well as many others [64, 70, 71, 83] demonstrate that above 50 GP, *CG* has the lowest enthalpy among all the known nitrogen phases.

The internal energy of the *A7, BP*, and *LB* phases can be related to the ratio of *gauche, trans* and *cis* N—N bonds there are in each phase. There are six such dihedral angles in each six-member ring of these phases. *BP* has a ratio of 4-*gauche*/2-*trans*/0-*cis*, while *A7* has a ratio of 0-*gauche*/6-*trans*/0-*cis* and *LB* has a ratio 0-*gauche*/4-*trans*/2-*cis*. With the *gauche* conformation being the most stable, followed by the *trans*-conformation and the *cis* conformation being the least stable, it is clear that this simple structural analysis predicts the following ordering of the three layered six-member ring phases: $BP < A7 < LB$. First-principles computations of the three phases are in agreement with this. The calculated internal energy per atom of the phases is $-268.6710\,\text{eV} < -268.6103\,\text{eV} < -268.5912\,\text{eV}$ at 20 GPa. Of course, the overall enthalpy of the phases contains a volume factor, which has substantial contribution at elevated pressures.

2.4.3 Polynitrogen Phases from Helical Motifs

Helical chains can also be considered as a natural structural theme in polymeric nitrogen phases. Recognition of helical structural motifs in the experimentally observed *CG* crystal lattice has lead to the discovery of a new single-bonded non-layered nitrogen structure named *chaired web* (*CW*) [84]. First-principles density functional theory calculations reveal that *CW*, which was originally identified at high pressures is metastable at ambient conditions as well. The metastability is demonstrated by both high-quality phonon dispersion calculations and finite temperature first-principles molecular dynamics simulations. In addition the new *CW* phase is thermodynamically more stable than the *CG* phase in the ambient pressure regime.

The results were obtained by first principles density functional theory calculations with the Perdew–Burke–Ernzerhof (PBE) [27] exchange-correlation functional. The SIESTA [85] package was used for most of the exploratory simulations, preliminary optimization of structures, and molecular dynamics simulations, while VASP [81] was used to calculate the final phase diagram, phonon dispersion and band structure. For the SIESTA calculations, the Troullier–Martins [86] norm-conserving pseudopotential with a nitrogen reference configuration of [He]$2s^22p^3$ and a cut-off radius of 0.98 Å was utilized. A custom numerical 'double-zeta with polarization' SIESTA-type basis set [87] was developed for the calculations. A 10 Å cut-off was employed for the k-point sampling. The calculations performed with VASP were based on the projector augmented wave (PAW) method of Blöchl [82] where a 39 Ry plane wave cutoff and Brillouin-zone integration with $12 \times 12 \times 12$ Monkhorst–Pack grid were used. In both programs a variable-cell-shape conjugate gradient method under constant pressure was used for the minimization of the geometries, and molecular dynamics was performed within the Nosé–Parinello–Raman scheme.

Ab-initio calculations on small Molecules of the type R_2—N—N—R_2 demonstrated that the most favorable dihedral angle is the so-called *gauche* dihedral, i.e., $\sim$120°, due to the effect of hyperconjugation and mutual repulsion of nitrogen lone electron pairs [64]. A chain of atoms constructed with all *gauche* dihedral angles results in the formation of right- or left-handed helices, depending on whether a dihedral of +120° or −120° is applied. Indeed, within the *CG* structure, helices that have four and eight atoms per turn can be identified along the [100] direction and threefold helices can be located along the [110] direction. Figure 2.20d highlights the eightfold helices in *CG*.

Pure helices are also observed in some high pressure phases of group 16 elements, such as sulfur, selenium and tellurium [88–90] as well as in scandium [91]. Unlike the group 16 elements, nitrogen cannot form stable helices on its own. In order to acquire stability a single helical chain of nitrogen atoms has to have its valence saturated (form three single bonds). If the valence is not saturated, then partial double bonds will form and the chain will flatten out to either a *cis–trans* or zig-zag chain. In the latter case the

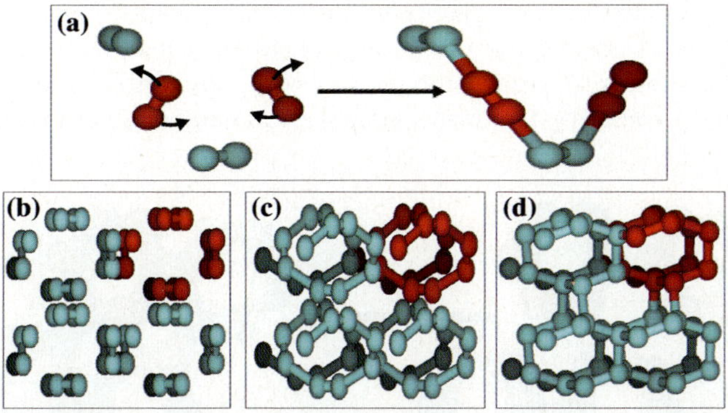

Fig. 2.20. (**a**) Flipping or 'inverting' two (highlighted in red) out of four N_2 molecules in the primitive unit cell of the molecular ζ-phase (l.h.s.) results in an eightfold helical chain structure (r.h.s.) (**b**) ζ-phase, (**c**) 'inverted' ζ-phase and (**d**) *CG*. For B, C and D a $2 \times 2 \times 2$ supercell is shown with one helical chain highlighted in red [84]

energy lowers by 0.5 eV per atom. The reverse tendency also holds, namely *cis–trans* or zig-zag chains acquire helicity once capped due to the above mentioned 'gauche-effect' and the energy is lowered by 0.6 eV per capped atom. It is interesting to note that in both cases the lowering of energy is achieved by changes in the local hybridization (from sp^3 to sp^2 and vice versa), symmetry breaking and splitting of degeneracy at the Fermi level. These are characteristic of a Peierls deformation. If a pure nitrogen structure is sought, the helices have to be brought close enough to each in order to create inter-helical bonds. The geometry of *CG* could be thought of as such a structure where interlinked nitrogen helices form a single-bonded network as illustrated in Fig. 2.20d.

The base helical structure of *CG* can be formed from what has been proposed to be the molecular ζ-phase, which itself is considered to be the immediate molecular precursor [92] of *CG* during its synthesis. Helices can be formed by changing the alignment of two out of the four N_2 molecules within the primitive unit cell of the ζ-structure. The necessary rotation about the two molecule's center of mass as to 'invert' their orientation is shown in Fig. 2.20a. Using this 'inverted' ζ-structure (Fig. 2.20b) as a starting geometry for a first-principles calculation, eightfold helical chains form after only a few optimization steps as depicted in Fig. 2.20c. Further optimization at high pressure leads to the formation of the single-bonded *CG* phase (Fig. 2.20d). In a similar manner the geometry optimization of properly arranged four- and threefold helical chains also result in the formation of *CG*.

It is natural to propose that certain arrangements of the helical nitrogen chains could form the basis of other polynitrogen structures. To achieve a structure where all nitrogen atoms are three-coordinate, bonds between adjacent chains must be made because the nitrogen atoms are only two-coordinate in the pure helical chains. Different arrangements of 3-, 4-, *etc.* fold helices or of their combinations could lead to alternative structures. For example, a slight rearrangement of the threefold helices within *CG* results in a structure that has recently been proposed as a new polynitrogen structure [93]. Arrangement and connection of sixfold helical chains as shown in Fig. 2.21, followed by geometry optimization, produced a new stable structure we have called *chaired-web* (*CW*). Connecting the helical chains together, results in the

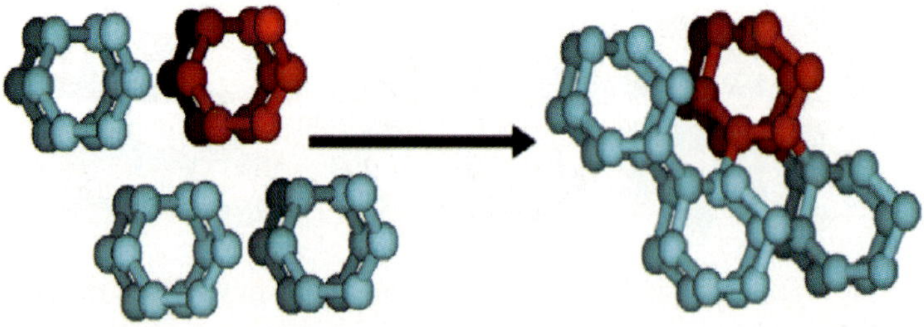

Fig. 2.21. A properly arranged set of sixfold helices (l.h.s.) forming the single-bonded *CW* phase (r.h.s) [84]

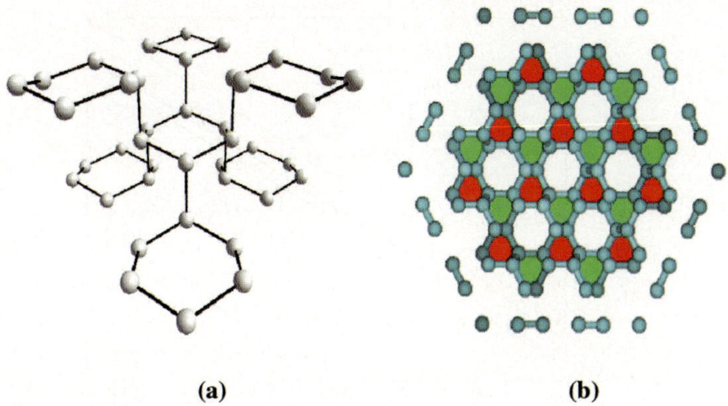

Fig. 2.22. Lattice structure of *CW*: (a) Perspective view, (b) A view along [111]. Left- (red) and right- (green) oriented sixfold helices with six-rings (white) lying between them within *CW* phase (3 × 3 × 3 supercell) [84]

formation of a network of six-member rings that are in the so-called 'chair' conformation, as the name given to the structure suggests. Figure 2.21a highlights the six-member rings of the *CW*.

The structure and symmetry of the new phase is better seen in Fig. 2.22. If the original helices turn clockwise (depicted in Fig. 2.22b in red) then new helices with counter-clockwise turns (depicted in Fig. 2.22b in green) are naturally formed in between the original helices due to the inter-helical bonds. The aforementioned six-member rings are formed in the voids between the helices as shown in white in Fig. 2.22b. These voids accommodate the nitrogen lone pairs. All of the six-member rings are in the so-called chair conformation. The interplay of helical chain and six-ring structural motifs in *CW* is reminiscent of *cis–trans* chain and six-member ring motifs interplay in the layered phases *BP*, *A7* and *LB* discussed above [72]. Additionally, we note that the structure of *CW* can be obtained by applying the Peierls distortion method for generating purely single-bonded polymeric nitrogen [74]. However, the 8-atom simple cubic reference cell first demonstrated [74] would not be large enough to generate *CW*. Rather a significantly expanded 48 atom (or larger) simple cubic reference structure would have to be employed.

The primitive unit cell of *CW* is rhombohedral. At an intermediate pressure of 28 GPa the unit cell vector length is 3.5 Å with an angle measuring 99.172° and the corresponding symmetry group is R $\bar{3}$ m. The six equivalent atomic positions obtained by the use of symmetry in fraction coordinates are (0.8756, 0.8756, 0.3206). There are two types of interatomic distances in the crystal lattice, a shorter N—N distance within a given six-member ring and slightly longer N—N distance between two adjacent six-member rings. At 28 GPa these distances are 1.36 Å and 1.45 Å respectively. There is a 2:1:0 ratio of *gauche-*, *trans-* and *cis-* dihedral angles in the crystal lattice. Band structure

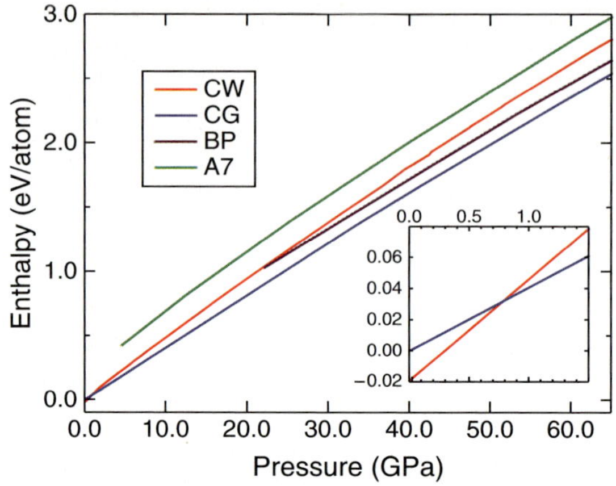

Fig. 2.23. Relative enthalpy versus pressure phase diagram of *CW, CG, BP* and *A7*. Enthalpy is relative to CG at 0° pressure. The inset in the lower right corner enlarges the low-pressure region [84]

calculations show that *CW* is an insulator at low pressure with a calculated band gap of ∼ 5 eV. The phase diagram shown on Fig. 2.23. demonstrates how close *CW* is to *CG* in terms of enthalpy. At zero pressure the enthalpy favors *CW* over *CG* by approximately 20 meV. The temperature-dependant free-energy and zero-point energy corrections within the harmonic approximation have been evaluated at zero pressure. The zero-point energy correction favors *CG* over *CW* by 54 meV thus reversing the energy order of the two phases at zero temperature. With the raise in temperature there is a cross-over of the two free-energy dependencies around 200 K. At ambient temperature the new *CW* phase is thermodynamically more stable than the *CG* phase with 83 meV in free energy.

The results of high-quality phonon dispersion calculations reveals that the *CW* phase is metastable at very low pressures. Although the lack of negative frequencies in the phonon dispersion calculation is theoretically rigorous verification of *CW*'s metastability (at least at low temperatures), three additional tests to probe the energy barrier of decomposition have also been performed: (i) a random displacement test in which the structure is optimized after the positions of atoms are randomly displaced by 5% of their inter-atomic distance; (ii) ab-initio molecular dynamics test at 0 GPa, 300 K for 10 ps. (iii) a hybrid test in which ten random snapshots of the initially equilibrated system are taken, the positions and velocities are randomly perturbed, and the system is further a subject to AIMD at 0 GPa, 300 K for an additional 4 ps. A rigorous study of *CW*'s metastability lifetime is substantially more challenging, but the tests indicate that the lifetime of *CW* may be of practical interest.

The structure of *CW* is, as are other purely single-bonded polymeric nitrogen phases, governed to a large extent by the repulsion of the lone pairs.

In *CW* the inherent stability of the initial *gauche* helices and the newly formed helices of reverse handedness, is offset by the repulsion of the lone pairs from neighboring helices. Importantly, in *CW* the lone pairs point into large void spaces, thereby reducing the repulsion. At high pressure the large voids are thermodynamically unfavorable because of the associated enthalpic penalty. On the other hand, at low pressure, the balance between these two contradictory influences of the lone pairs is such that *CW* is more stable than *CG* at ambient pressures.

In looking for a possible precursor of *CW* one has to note that the traditional understanding of N_2 phase diagram was recently questioned and preliminary results suggest the need for its extension and/or revision [94]. In particular, Gregoryanz et al. [95] found a molecular phase characterized by a configuration of three nitrogen molecules nearly forming a six-member ring. One could speculate that *CW* could be formed from such a phase under appropriately engineered physical conditions.

2.5 Final Remarks

The results presented in this chapter indicate that significant progress can be achieved in applying various first-principles computational methods for atomic-scale descriptions of energetic systems in gas, liquid or solid phases. Using well developed computational chemistry methods it has been possible to determine accurate information about the structural properties of various systems, the relative stabilities of different configurations in different phases, and the decomposition reaction mechanisms. These types of data allowed a direct correlation and interpretation of the corresponding experimental data. Moreover, in many instances the type of data obtained theoretically has substituted for the lack of information achievable through experimental means. This is particularly the case for example, for description of complicated reaction mechanisms where accurate identification of transition states for various pathways is experimentally very challenging.

Beside structural and energetic data, important steps have also been achieved computationally in description of the dynamic processes in condensed phases. Among the computational methods used, first principles calculation methods are the first choice to accurately evaluate the structural, energetic, electronic, and even spectral data.

Alternatively, the progress made in computational hardware and in computational algorithms should allow further development of combined quantum mechanical/molecular mechanical methods for description of energetic and dynamical problems, for systems of increased size and complexity. The use of ab initio molecular dynamics methods for description of the properties of ionic systems is another area which is expecting to grow in importance and relevance.

Acknowledgments

The authors would like to thank our coworkers who have contributed to the work presented here: Dr. Dongqing Wei, Dr. Steven Decker, Dr. Federico Zahariev, Dr. Nick Mosey, Dan Chau and James Hooper. The authors are grateful to Defence R&D Canada-Suffield, NSERC of Canada, the Canada Research Chair program, the Canada Foundation for Innovation, Ontario Innovation Trust and SHARCNet of Canada for a financial support.

References

1. Rahman, A.: Phys. Rev. A **136**, 405 (1964)
2. Verlet, L.: Phys. Rev. **1950**, 98 (1967)
3. Car, R., Parrinello, M.: Phys. Rev. Lett. **55**, 2471 (1985)
4. Parrinello, M.: Solid State Commun. **38**, 115 (1997)
5. Tse, J.S.: Annu. Rev. Phys. Chem. **53**, 249 (2002)
6. Dremin, A.N., Klimenko, V.-Y.: In: Bowen, J.R. (ed.) Progress in Astronautics and Aeronautics, pp. 253. AIAA, New York (1981)
7. Brenner, D.W., Robertson, D.H., Elert, M.L., White, C.T.: Phys. Rev. Lett. **70**, 2174 (1993)
8. Haskins, P.J., Cook, M.D., Fellows, J., Wood, A.: In: Proceedings of the 6th Symposium (International) on Detonation Snowmass, p. 897. Colorado (2000)
9. Barrett, J.J.C., Brenner, D.W., Robertson, D.H., White, C.T.: In: Schmidt, S.C., Tao W.C. (eds.) Shock Compression of Condensed Matter – AIP Conference Proceedings 370, pp. 191. American Institute of Physics, New York (1995)
10. Dremin, A.N.: Philos Trans R Soc Lond A **339**, 355 (1992)
11. Koch, W., Holthausen, M.C.: A Chemist's Guide to Density Functional Theory. Wiley-VCH, Weinheim (2001)
12. Abell, G.C.: Phys. Rev. B Condens. Matter Mater Phys. **31**, 6184 (1985)
13. Tersoff, J.: Phys. Rev. Lett. **56**, 632 (1986)
14. Brenner, D.W.: Phys Status Solidi B **217**, 23 (2000)
15. Hohenberg, P., Kohn, W.: Phys. Rev. **136**, B864 (1964)
16. Kohn, W., Sham, L.: J.Phys. Rev. **140**, A1133 (1965)
17. Fermi, E.: Rend. Accad. Lincei **6**, 602 (1927)
18. Thomas, L.H.: Proc. Camb. Phil. Soc. **23**, 542 (1927)
19. Teller, E.: Rev. Mod. Phys. **34**, 627 (1962)
20. Perdew, J.P., Wang, Y.: Phys. Rev. B **45**, 13244 (1992)
21. Vosko, S.J., Wilk, L., Nusair, M.: Can. J. Phys. **58**, 1200 (1980)
22. Adamo, C., di Matteo, A., Barone, V.: Adv. Quantum Chem. **36**, 45 (1999)
23. Ernzerhof, M., Scuseria, G.E.: J. Chem. Phys. **110**, 5029 (1999)
24. Bauschlicher, Jr. C.W.: Chem. Phys. Lett. **246**, 40 (1995)
25. Becke, A.D.: Phys. Rev. A **38**, 3098 (1988)
26. Perdew, J.P.: Phys. Rev. B **33**, 8822 (1986)
27. Perdew, J.P., Burke, K., Ernzerhof, M.: Phys. Rev. Lett. **77**, 3865 (1996)
28. Becke, A.D.: J. Chem. Phys. **98**, 5648 (1993)
29. Becke, A.D.: J. Chem. Phys. **98**, 5612 (1993)
30. Curtiss, L.A., Raghavachari, K., Trucks, G.W., Pople, J.A.: J. Chem. Phys. **94**, 17221 (1991)

31. Zhao, Y., Gonzalez-Garcia, N., Truhlar, D.G.: J. Phys. Chem. A **109**, 2012 (2005)
32. Szabo, A., Ostlund, N.S.: Modern Quantum Chemistry: Introduction to Advanced Electronic Structure Theory. McGraw-Hill, New York (1989)
33. Boys, S.F.: Proc. R. Soc. Lond. A **200**, 542 (1950)
34. Ashcroft, N.W., Mermin, N.D.: Solid State Physics. Holt Saunders, Philadelphia, PA (1976)
35. Allen, M.P., Tildesley, D.J.: Computer Simulation of Liquids. Oxford University Press, Oxford (1989)
36. Allen, M.P.: In: Attig, N. et al. (eds.) Computational Soft Matter: From Synthetic Polymers to Proteins, Lecture Notes, pp. 1. John von Neumann Institute for Computing, Jülich (2004)
37. Rapaport, D.C.: The Art of Molecular Dynamics Simulation. Cambridge University Press, Cambridge (2004)
38. Frenkel, D., Smit, B.: Understanding Molecular Simulation: From Algorithms to Applications. Academic Press, San Diego, CA (2002)
39. Marx, D., Hutter, J.: In: Grotendorst J (ed) Modern Methods and Algorithms of Quantum Chemistry, pp. 301. John von Neumann Institute for Computing, Jülich (2000)
40. Mosey, N.J., Muser, M.H., Woo, T.K.: Science **307**, 1612 (2005)
41. Alfè, D., Gillan, M.J., Price, G.D.: Nature **401**, 462 (1999)
42. Cavazzoni, C., Chiarotti, G.L., Scandolo, S., Tosatti, E., Bernasconi, M., Parrinello, M. Science **283**, 44 (1999)
43. Alavi, S., Thompson, D.L.: J. Chem. Phys. **120**, 10231 (2004)
44. Alper, H.E., Abu-Awwad, F., Politzer, P.: J. Phys. Chem. B **103**, 9738 (1999)
45. Liu, H., Zhao, J., Ji, G., Gong, Z., Wei, D.: Physica B **382**, 334 (2006)
46. Riad Manaa, M., Reed, E.J., Fried, L.E., Galli, G., Gygi, F.: J. Chem. Phys. **120**, 10146 (2004)
47. Siavosh-Haghighi, A., Thompson, D.L.: J. Chem. Phys **125**, 184711/1 (2006)
48. Sorescu, D.C., Rice, B.M., Thompson, D.L.: J. Phys. Chem. A **105**, 9336 (2001)
49. Tuckerman, M.E., Klein, M.L.: Chem. Phys. Lett. **283**, 147 (1998)
50. Zheng, L., Luo, S.-N., Thompson, D.L.: J. Chem. Phys. **124**, 154504/1 (2006)
51. Winey, J.M., Gupta, Y.M.: J. Phys. Chem. B, **101**, 10733 (1997)
52. Piermarini, G.J., Block, S., Miller, P.J.: J. Phys. Chem. **93**, 457 (1989)
53. Haskins, P.J., Cook, M.D.: In: Schmidt, Dandekar, Forbes (eds.) Shock Compression of Condensed Matter, p. 305 (1997)
54. Wei, D., Zhang, F., Woo, T.K.: Proceedings of the, Atlanta, GA. In: 12th Biennial International Conference of the APS Topical Group on Shock Compression of Condensed Matter, pp. 407. American Institute of Physics, Atlanta, GA, USA (2001)
55. Graham, R.A.: J. Phys. Chem. **83**, 3048 (1979)
56. Sheffield, S.A., Engelke, R., Alcon, R.R., Gustavsen, R.L., Robbins, D.L., Stal, D.B., Stacy, H.L., Whitehead, M.C.: In: Proceedings of 12th International Detonation Symposium, pp. 159. Office of Navy Research, Arlington, VA, San Diego, CA, USA (2002)
57. Higgins, A.J., Jetté, F.X., Yoshinaka, A., Zhang, F.: In: Proceedings of 12th International Detonation Symposium, pp. 993. Office of Navy Research, Arlington, VA, San Diego, CA, USA (2002)

58. Yoshinaka, A., Zhang, F., Petel, O.E., Higgins, A.J.: In: Proceedings of the Shock Compression of Condensed Matter Conference, pp. 1139. American Institute of Physics, Baltimore, MD (2005)
59. Manaa, M.R., Reed, E.J., Fried, L.E., Galli, G., Gygi, F.: J. Chem. Phys. **120**, 10146 (2004)
60. Manaa, M.R.: Chemistry at Extreme Conditions. Elsevier, Amsterdam (2005)
61. Shaw, R., Decarli, P.S., Ross, D.S., Lee, E.L., Stromberg, H.D.: Combust. Flame **35**, 237 (1979)
62. Sheffield, S.A., Davis, L.L., Engelke, R.: In: Furnish, M.D., Chhabildas, L.C., Hixson, R.S. (eds.) Shock Compression of Condensed Matter, pp. 789. American Institute of Physics, Snowbird, Utah (1999)
63. Uddin, J., Barone, V., Scuseria, G.E.: Mol. Phys. **104**, 745 (2006)
64. Mailhiot, C., Yang, L.H., McMahan, A.K.: Phys. Rev. B **46**, 14419 (1992)
65. Barbee, T.W., III, Phys. Rev. B **48**, 9327 (1993)
66. Eremets, M.I., Hemley, R.J., Mao, H.-k., Gregoryanz, E.: Nature **411**, 170 (2001)
67. Goncharov, A.F., Gregoryanz, E., Mao, H.-k., Liu, Z., Hemley, R.J., Phys. Rev. Lett. **85**, 1262 (2000)
68. Gregoryanz, E., Goncharov, A.F., Hemley, R.J., Mao, H.-k., Somayazulu, M., Shen, G.: Phys. Rev. B **66** (2002)
69. Eremets, M.I., Gavriliuk, A.G., Trojan, I.A., Dzivenko, D.A., Boehler, R.: Nat Mater **3**, 558 (2004)
70. Mattson, W.D., Sanchez-Portal, D., Chiesa, S., Martin, R.M.: Phys. Rev. Lett. **93**, 125501/1 (2004)
71. Alemany, M.M.G., Martins, J.L.: Phys. Rev. B **68**, 024110/1 (2003)
72. Zahariev, F., Hu, A., Hooper, J., Zhang, F., Woo, T.: Phys. Rev. B **72** (2005)
73. Peierls, R.: Quantum Theory of Solids. Pergamon, Oxford (1955)
74. Zahariev, F., Dudiy, S.V., Hooper, J., Zhang, F., Woo, T.K., Phys. Rev. Lett. **97** (2006)
75. Wells, A.F.: Models in Structural Inorganic Chemistry. Clarendon Press, Oxford (1970)
76. Wells, A.F.: Three-dimensional Nets and Polyhedra. Wiley, New York (1977)
77. Adams, D.M.: Inorganic Solids. Wiley, London (1974)
78. Burdett, J.K., McLarnan, T.J.: J. Chem. Phys. **75**, 5764 (1981)
79. Poya, G.: Acta Math. **68**, 145 (1937)
80. Hararay, F.: Graph Theory. Addison-Wesley, Reading, MA (1994)
81. Kresse, G., Furthmuller, J.: Phys. Rev. B **54**, 11169 (1996)
82. Bloechl, P.E.: Phys. Rev. B **47**, 17953 (1994)
83. Nordlund, K., Krasheninnikov, A., Juslin, N., Nord, J., Albe, K.: Europhys. Lett. **65**, 400 (2004)
84. Zahariev, F., Hooper, J., Alavi, S., Zhang, F., Woo, T.K.: Phys. Rev. B **75** (2007)
85. Soler, J.M., Artacho, E., Gale, J.D., Garcia, A., Junquera, J., Ordejon, P., Sanchez-Portal, D.: J. Phys. Cond. Matt. **14**, 2745 (2002)
86. Troullier, N., Martins, J.L.: Phys. Rev. B **45**, 1993 (1997)
87. Junquera, J., Paz, O., Sanchez-Portal, D., Artacho, E.: Phys. Rev. B **64**, 235111 (2001)
88. Hemley, R.J., Mao, H.-k., Struzhkin, V.: J. Synchrotron Rad. **12**, 135 (2005)
89. Crapanzano, L., Crichton, W.A., Monaco, G., Bellissent, R., Mezouar, M.: Nat Mater **4**, 550 (2005)
90. Decker, A., Landrum, G.A., Dronowski, R.: Z. Anorg. Allg. Chem. **628**, 295 (2002)

91. Akahama, Y., Fujihisa, H., Kawamura, H.: Phys. Rev. Lett. **94**, 195503 (2005)
92. Eremets, M.I., Gavriliuk, A.G., Serebryanaya, N.R., Trojan, I.A., Dzivenko, D.A., Boehler, R., Mao, H.-k., Hemley, R.J.: J. Chem. Phys. **121**, 11296 (2004)
93. Oganov, A.R., Glass, C.W.: J. Chem. Phys. **124**, 244704 (2006)
94. Gregoryanz, E., Sanloup, C., Bini, R., Kreutz, J., Jodl, H.J., Somayazulu, M., Mao, H., Hemley R.J.: J. Chem. Phys. **124**, 116102 (2006)
95. Sanloup, C., Gregoryanz, E., Somayazulu, M., Hemley, R.J., Mao, H.K., Goncharov, A.F.: In: Joint 20th AIRAPT – 43rd EHPRG, Karslruhe, Germany, 2005

3

Combined Compression and Shear Plane Waves

Z. Tang and J.B. Aidun

3.1 Introduction

The constitutive description of materials relates the six independent components of the stress tensor to the six independent components of strain, temperature, strain rate, and possibly other variables. This description, also called the constitutive equation, is very complicated in general and so it is quickly evident that one-dimensional longitudinal wave loading conditions, as achieved in traditional plane wave shock experiments, are not sufficient to construct the full constitutive equation [1]. Hence, one motivation for measuring planar shear waves, in addition to longitudinal waves, is the opportunity it provides for further characterizing the shocked state. Planar shear waves are also of interest because they can produce new loading states that are conducive to extending the study of the dynamic response of materials. As early as the 1930s, Bridgman [2–4] had explored the effects of high shearing stress combined with high hydrostatic pressure and shearing phenomena relevant to geology. To go further required developing an understanding of the propagation of combined compression and shear stress waves. Theoretical study of material properties under dynamic combined stress loading came first. Significant development of one-dimensional combined stress wave theory, and 2-D and 3-D stress wave theory dates to the 1950s. Representative achievements are the general characteristics theory of wave propagation [5,6] and investigations of combined stress wave behavior [7–11]. In particular, Ting [12] solved a series of combined stress wave propagation problems and built a unified theory for elastic and plastic combined stress wave propagation. Nowacki [13] systematically summarized plastic longitudinal–transverse waves, including simple waves in an elastic–plastic half-space, plane longitudinal–transverse waves in an isotropic elastic/viscoplastic medium, and longitudinal–transverse plane waves and radial cylindrical waves in an inhomogeneous elastic–viscoplastic medium.

In experimental investigations, Clifton and coworkers [10, 14, 15] generated one-dimensional, combined longitudinal and shear (torsion) waves by impacting a pre-torqued, long thin tube on which they made strain gauge measurements. Following the theoretical analysis of Johnson [16], Chhabildas and Swegle [17] employed an anisotropic crystal buffer in front of a sample to generate combined compression and shear loading. In the late 1970s, Clifton and coworkers [18], and Gupta and coworkers [19, 20] developed methods to produce combined planar compression and shear waves in samples by using an inclined parallel plate impact method. This has proved to be the most useful and adaptable of the several methods for generating planar compression and shear waves that have been demonstrated over the past four decades. The corresponding methods for measuring both longitudinal and shear waves are simultaneous use of either Normal Velocity Interferometry (NVI) combined with Transverse Displacement Interferometry (TVI), developed by Clifton and coworkers [21], or the Impulsive Measurement for Pressure and Shear Waves (IMPS) method developed by Gupta and coworkers [19, 20].

In the following, Sect. 3.2 describes the theory of combined stress waves, including combined compression–shear plane waves in linear and nonlinear elastic media and elastic–plastic media. Section 3.3 describes production and measurement methods for combined compression–shear plane waves. Finally, Sect. 3.4 introduces the applications of combined compression–shear plane waves. Since our main interest in this review is the use of the wave propagation characteristics to systematically reveal the response of materials under intense dynamic loading, this presentation focuses on the production and application of compression-shear plane waves using compressed light-gas guns (e.g., [20]).

3.2 Theory of Combined Stress Plane Waves

3.2.1 Basic Equations

Characteristics

Regardless of the material constitutive type (nonlinear elastic, elastic and plastic, viscoelastic or viscoplastic etc.), all 1-D stress wave propagation problems can be regarded as solving the following first order, quasi-linear partial differential equation

$$AW_t + BW_x = b, \tag{3.1}$$

where the subscripts t and x are, respectively, the time and the Lagrangian or Eulerian coordinate; the unknown variable $W = (W_1,\ W_2,\ \ldots,\ W_n)^T$ is an n-dimensional vector whose components are normally the components of stress, strain, and particle velocity; and the n-dimensional vector b and the nth order matrices A and B are functions of W, t, and x. Thus the eigenvalues c, corresponding to the characteristic wave speeds, are determined by the roots of the determinant

$$\|cA - B\| = 0. \tag{3.2}$$

While the characteristic condition is given by [22]

$$l^T B \frac{dW}{dt} = c l^T b, \qquad (c \neq 0) \tag{3.3a}$$

or

$$l^T A \frac{dW}{dt} = l^T b, \qquad \text{for all } c \tag{3.3b}$$

along $dx/dt = c$, where

$$\frac{dW}{dt} = cW_x + W_t \qquad \text{and} \tag{3.4}$$

$$l^T(cA - B) = 0, \tag{3.5a}$$

where dW/dt is the total derivative along the characteristic curve, l is the left eigenvector of the system. Similarly, the right eigenvector, r, can be defined by

$$(cA - B)r = 0. \tag{3.5b}$$

Equation (2.5b) is useful in finding simple wave solution. $\boldsymbol{r}$ is also related to the discontinuity in W_x and W_t across the characteristics.

Simple Wave Solution

If W is constant along a family of curves $\zeta(x,t)$, then the solution of the equation

$$AW_t + BW_x = 0 \tag{3.6}$$

is defined as a simple wave solution of (3.1). By using $\zeta(x,t)$ (3.6) can be reduced to an ordinary differential equation as

$$(cA - B)\frac{dW}{d\varsigma} = 0, \tag{3.7}$$

where

$$c = dx/\,dt = -\varsigma_t/\varsigma_x. \tag{3.8}$$

Comparing with (3.5b) shows that curves $\zeta(x,t)$ must be the characteristics, hence

$$dW/d\varsigma \propto r \tag{3.9}$$

or, in component form,

$$\frac{dW_1}{r_1} = \frac{dW_2}{r_2} = \cdots = \frac{dW_n}{r_n} = k d\varsigma, \tag{3.10}$$

where k is the proportionality constant. It is worth noting that the characteristics for a simple wave problem are straight lines.

The next two sub-sections present the applications of the general and simple wave theories.

3.2.2 Combined Compression Shear Waves in Nonlinear Elastic Solids

Abou-Sayed and Clifton [23] derived the properties of combined compression–shear plane wave propagation in a hyperelastic material. Their analysis is summarized here.

Governing Equations

Let the surface $x = 0$ of a half space be subjected to motion in both the x and y directions. Then the only non-zero components of the strain tensor, $E(x,t)$, are E_{11} and E_{12}, which will be denoted by ε and γ, respectively. Assume that the material in the half space is isotropic and hyper-elastic [24] with initial density ρ_o and an adiabatic strain energy function per unit mass $\Theta(E)$ that is described by

$$\Theta = a_0 + a_1 J_1 + a_2 J_1^2 + a_3 J_1 + a_4 J_1^3 + a_5 J_1 J_2 + a_6 J_3, \tag{3.11}$$

where J_1, J_2 and J_3 are the three invariants of the strain tensor E. Since ε and γ are the only nonzero strain components, (3.11) can be reduced to

$$\Theta = a_2\varepsilon^2 - a_3\gamma^2/4 + a_4\varepsilon^3 + (a_2/2 + a_3/8 - a_5/4)\varepsilon\gamma^2. \tag{3.12}$$

The corresponding non-zero components of the stress tensor can be derived as

$$\sigma = \rho_o \frac{\partial\Theta}{\partial\varepsilon} = \rho_0 \left\{2a_2\varepsilon + 3a_4\varepsilon^2 + (a_2/2 + a_3/8 - a_5/4)\gamma^2\right\} \tag{3.13a}$$

and

$$\tau = -\rho_o \frac{\partial\Theta}{\partial\gamma} = -\rho_o\{-a_3\gamma/2 + 2(a_2/2 + a_3/8 - a_5/4)\varepsilon\gamma\}, \tag{3.13b}$$

where σ and τ denote the stress components σ_{11} and σ_{12}, respectively. Let $u(x,t)$ and $v(x,t)$ be the particle velocity components in x and y directions, respectively. Then the governing equations for unknowns σ, τ, u, and v form a system of four quasi-linear first order partial differential equations and can be written in the form of (3.1)

$$AW_t + BW_x = b, \tag{3.14}$$

where

$$A = \begin{vmatrix} \rho_o & 0 & 0 & 0 \\ 0 & \rho_o & 0 & 0 \\ 0 & 0 & c_2^2/\rho_o d & b\gamma/\rho_o d \\ 0 & 0 & b\gamma/\rho_o d & c_1^2/\rho_o d \end{vmatrix}, \qquad W = \begin{vmatrix} u \\ v \\ \sigma \\ \tau \end{vmatrix}, \tag{3.15}$$

$$B = \begin{vmatrix} 0 & 0 & -1 & 0 \\ 0 & 0 & 0 & -1 \\ -1 & 0 & 0 & 0 \\ 0 & -1 & 0 & 0 \end{vmatrix}$$

and

$$c_1^2 = 2a_2 + 6a_4\varepsilon, \quad c_2^2 = -a_3/2 + b\varepsilon,$$
$$b = (4a_2 + a_3 - 2a_5)/4, \text{ and } \quad d = c_1^2 c_2^2 - b^2\gamma^2. \tag{3.16}$$

Since A and B are both symmetric and A is positive definite, (3.14) is a hyperbolic system. Thus, the characteristic wave speed, c, can be determined from (3.2) by substituting for A and B in (3.15)

$$c = \pm c_f, \qquad c = \pm c_s, \tag{3.17}$$

where

$$c_f = \left\{ \frac{1}{2} \left(c_1^2 + c_2^2 + \left[\left(c_1^2 - c_2^2 \right)^2 + 4b^2\gamma^2 \right]^{1/2} \right) \right\}^{1/2}$$

and

$$c_s = \left\{ \frac{1}{2} \left(c_1^2 + c_2^2 - \left[\left(c_1^2 - c_2^2 \right)^2 + 4b^2\gamma^2 \right]^{1/2} \right) \right\}^{1/2}$$

Substituting for c_1^2 and c_2^2 from (3.16) and neglecting terms higher than second order in ε and γ yield the approximate equations

$$c_f \approx \left(2a_2 + 6a_4\varepsilon + e^2\gamma^2\right)^{1/2} \quad \text{and} \tag{3.18a}$$

$$c_s \approx \left(-a_3/2 + b\varepsilon - e^2\gamma^2\right)^{1/2}, \tag{3.18b}$$

where $e^2 = 2b^2/(4a_2 + a_3)$. According to (3.5b), the right characteristic vector r can be determined as

$$r_f = \begin{bmatrix} 1 \\ -\phi \\ \mp\rho_o c_f \\ \pm\rho_o c_f \phi \end{bmatrix}, \qquad \text{for} \quad c = \pm c_f \tag{3.19a}$$

and

$$r_s = \begin{bmatrix} \phi \\ 1 \\ \mp\rho_o c_s \phi \\ \pm\rho_o c_s \end{bmatrix}, \qquad \text{for} \quad c = \pm c_s, \tag{3.19b}$$

where $\phi = (c_f^2 - c_1^2) b\gamma = (c_2^2 - c_s^2)/b\gamma$. The coefficient ϕ reflects the coupling of longitudinal and transverse motion. For small values of ε and γ, $\phi \approx 2b\gamma/(4a_2 + a_3) + O(\gamma^2, \varepsilon^2)$. Thus, ϕ vanishes as γ approaches zero. For linear elastic materials $b = 0$ and we have $c_f = c_1$, and $c_s = c_2$, based on (3.16) and (3.17). Correspondingly, $\phi = 0$, which means there is no coupling of longitudinal and transverse motion.

Simple Wave Solutions

The simple wave method can be used to determine the main features of the wave profiles analytically [23]. By using (3.10) with W from (3.15) and r from (3.19), we have

$$\frac{-d\gamma}{d\varepsilon} = \frac{dv}{du} = \frac{d\tau}{d\sigma} = -\phi \approx \frac{-2b\gamma}{4a_2 + a_3} \tag{3.20a}$$

and

$$\frac{du}{d\sigma} = \frac{dv}{d\tau} = \frac{-1}{\rho_o c_f} \tag{3.20b}$$

Equation (3.20a) shows that the normal and transverse components of motion are coupled in the Fast Simple Wave (FSW) solution. Furthermore, the components indicate that for a small value of γ, the coupling is weak and linear in γ. Similar analysis for the Slow Simple Wave (SSW) solution leads to the relations

$$\frac{-d\varepsilon}{d\gamma} = \frac{du}{dv} = \frac{d\sigma}{d\tau} = \phi \approx \frac{2b\gamma}{4a_2 + a_3} \tag{3.21a}$$

and

$$\frac{du}{d\sigma} = \frac{dv}{d\tau} = \frac{-1}{\rho_o c_s}. \tag{3.21b}$$

Hence, the normal and transverse motion is coupled in the SSW solution, also.

Equations (3.20) and (3.21) result in $(d\sigma/d\tau)_{c_f} \times (d\sigma/d\tau)_{c_s} = -1$, which indicates that the stress paths for the FSW and SSW constitute a network of orthogonal curves in the σ-τ plane. A similar result exists for the strain paths in the ε-γ plane. The trajectory of strain paths in the ε-γ plane for FSW can be obtained by integrating the corresponding terms in (3.20), which yields

$$\gamma = \gamma_o \exp\left\{2b(\varepsilon - \varepsilon_o)/(4a_2 + a_3)\right\}, \tag{3.22}$$

where $(\gamma_o, \varepsilon_o)$ is an initial state in front of the wave. Similarly, for a SSW, integrating (3.21) gives

$$\varepsilon = \varepsilon_1 - b\left(\gamma^2 - \gamma_1^2\right)/(4a_2 + a_3), \tag{3.23}$$

where ε_1 and γ_1 are the values of strain ahead of the SSW region. Alternately, the normal stress σ and particle velocity u across a SSW can be obtained as

$$\sigma \approx \sigma_1 + \left[-\rho_o c_2^2 b/(4a_2 + a_3)\right]\left(\gamma^2 - \gamma_1^2\right) \tag{3.24}$$

and

$$u \approx u_1 + \left[c_2^2 b/c_s(4a_2 + a_3)\right]\left(\gamma^2 - \gamma_1^2\right), \tag{3.25}$$

where σ_1 and u_1 are the normal stress and particle velocity ahead of the SSW region.

Solution for Step Loading

Step loading is experimentally achievable through inclined, parallel plate impact. Assuming a half-space is initially at rest, stress and strain free, and subjected to step loading of combined pressure and shear at the boundary $x = 0$ and time $t = 0$, the initial conditions are $u(0,t) = uH(t)$ and $v(0,t) = vH(t)$ in the x and y directions, respectively, where $H(t)$ is the Heaviside step function. This loading produces centered FSW and SSW separated by a region of constant state, as shown in Fig. 3.1 [23]. In the FSW region, γ, τ, v and ϕ are zero, so we have

$$\begin{aligned} c_f(\varepsilon) &= x/t, \\ \varepsilon(x,t) &= (x^2/t^2 - c_o^2)/6a_4, \\ u(x,t) &= (c_o^3 - x^3/t^3)/9a_4, \text{ and} \\ \sigma(x,t) &= \rho_o(x^4/t^4 - c_o^4)/12a_4, \end{aligned} \tag{3.26}$$

where $c_o^2 = 2a_2$. In the constant state region $c_2 < x/t < c_f^*$, we have

$$\begin{aligned} u &= u_1, \\ \varepsilon &= \varepsilon_1 = (c_f^{*2} - c_o^2)/6a_4 \qquad \text{and} \\ \sigma &= \sigma_1 = \rho_o(c_f^{*4} - c_o^4)/12a_4, \end{aligned} \tag{3.27}$$

where

$$\begin{aligned} c_f^* &= (c_o^3 - 9a_4u_1)^{1/3} \\ c_2 &= (-a_3/2 + b\varepsilon_1)^{1/2}. \end{aligned} \tag{3.28}$$

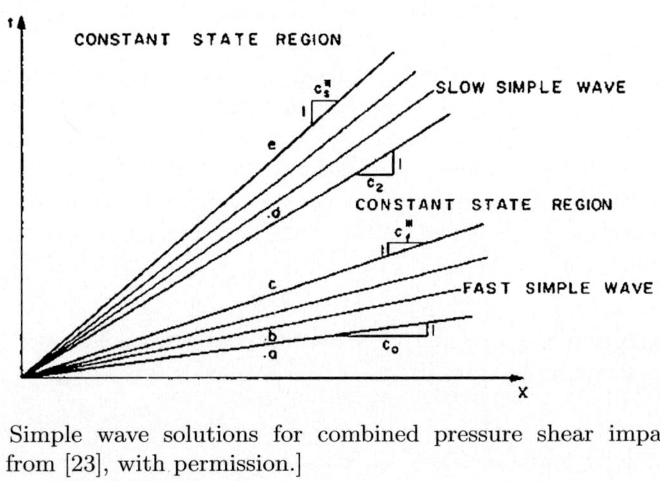

Fig. 3.1. Simple wave solutions for combined pressure shear impact loading. [Reprinted from [23], with permission.]

For the SSW region $c_s^* < x/t < c_2$, we have

$$\begin{aligned}
&\gamma = (c_2^2 - x^2/t^2)^{1/2}/e, \\
&\tau = -\rho_o c_2^2 \gamma + 0(\gamma^3), \\
&v = c_2 \gamma + 0(\gamma^2), \\
&\varepsilon = \varepsilon_1 - b\gamma^2/(4a_2 + a_3) = \varepsilon_1 - 0(\gamma^2), \\
&\sigma = \sigma_1 - \rho_o c_2^2 b \gamma^2/(4a_2 + a_3) = \sigma_1 - 0(\gamma^2), \text{ and} \\
&u = u_1 + b c_2^2 \gamma^2 / c_s(4a_2 + a_3) = u_1 + 0(\gamma^2).
\end{aligned} \tag{3.29}$$

The region $x/t < c_s^*$ has a constant solution

$$\begin{aligned}
&u = u_o \approx u_1, \\
&\sigma = \sigma_o \approx \sigma_1 = \rho_o(c_f^{*4} - c_o^4)/12a_4, \\
&\varepsilon = \varepsilon_o \approx \varepsilon_1 = (c_f^{*2} - c_o^2)/6a_4, \\
&v = v_o, \\
&\tau = \tau_o \approx -\rho_o c_2 \gamma_o, \text{ and} \\
&\gamma = \gamma_o \approx v_o/c_2,
\end{aligned} \tag{3.30}$$

where $c_s^* = (c_2^2 - e^2 v_o^2/c_2^2)^{1/2} \approx c_2(1 - e^2 v_o^2/2c_2^4)$.

Comparison of Theory with Experiment

Abou-Sayed and Clifton [23] measured wave propagation in fused silica under combined pressure–shear plate impact with a keyed gas gun (see Sect. 3.3). The material constants applied in corresponding model calculation are $c_o = 5.876\,\text{mm}\,\mu\text{s}^{-1}$ (longitudinal elastic wave speed), $c_s = 3.649\,\text{mm}\,\mu\text{s}^{-1}$ (shear elastic wave speed), $\rho_o = 2.203\,\text{g}\,\text{cm}^{-3}$; a_2, a_3, a_4, a_5, a_6, b and e^2 are 17.26, −26.63, 60.08, −115.13, −20.4, 68.17 and 219.1 $(\text{mm}\,\mu\text{s}^{-1})^2$, respectively. Figures 3.2 and 3.3 show very good agreement of the model predictions for the particle velocity profile at the rear free surface with the measurements [23]. The calculations used imposed velocity boundary conditions of $u_o = 0.0915\,\text{mm}\,\mu\text{s}^{-1}$ and $v_o = 0.031\,\text{mm}\,\mu\text{s}^{-1}$ with the rear free surface located at $x = 6.35\,\text{mm}$. Furthermore, the measured transverse velocity profile exhibits a very short rise time (<10 ns), which is consistent with the predicted value of about 2 ns.

3.2.3 Combined Compression Shear Stress Plane Waves in Elastic–Plastic Materials

The first work on combined plastic stress waves appears to be that of Rakhmatulin [7] and Cristescu [8] who suggested impacting inclined plates to produce the combined state of stress. Bleich and Nelson [9] obtained

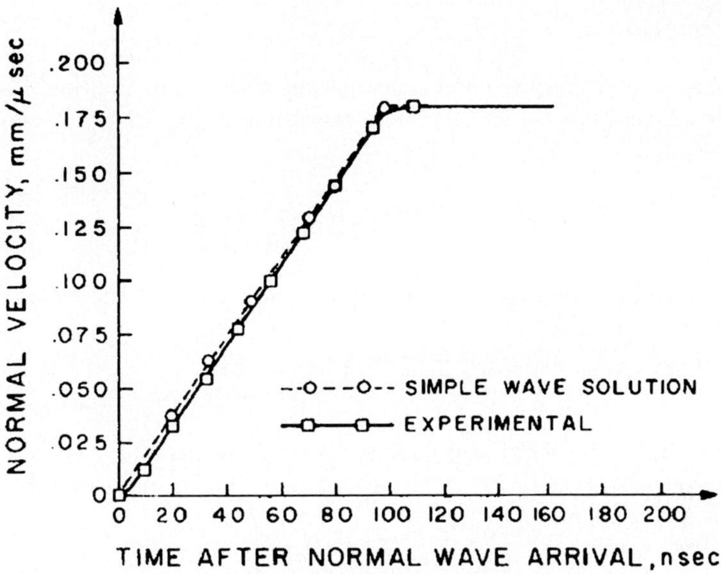

Fig. 3.2. Normal velocity-time profile. [Reprinted from [23], with permission.]

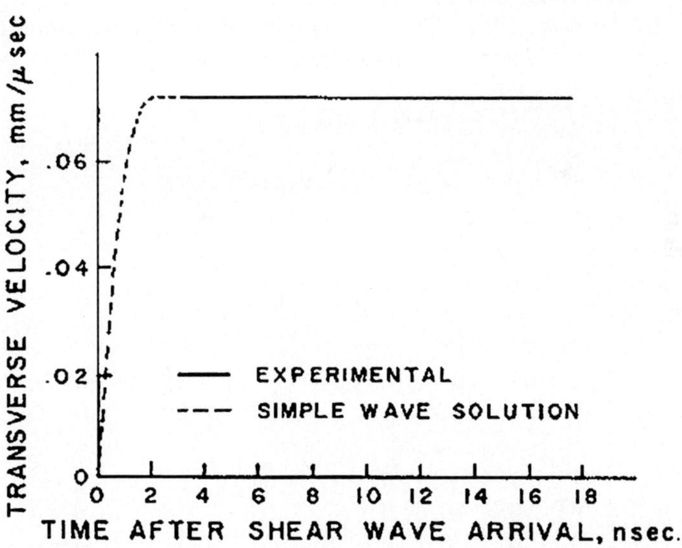

Fig. 3.3. Transverse velocity-time profile. [Reprinted from [23], with permission.]

the closed-form solutions for the case of an elastic, perfectly plastic half space subjected to combined surface pressure and shear. Ting and Nan [11] and Ting [25] extended the study to general elastic–plastic materials. Abou-Sayed and Clifton [26] analyzed combined pressure–shear waves in an elastic–viscoplastic material. The theory of Ting and Nan [11] is introduced here.

Basic Equations

The geometry and loading conditions are the same as in the Sect. 3.2.2. The stress–strain relation of an isotropic, work-hardening material can be expressed as [27]

$$\frac{\partial \varepsilon_{ij}}{\partial t} = \frac{1+\nu}{E}\frac{\partial \sigma_{ij}}{\partial t} - \frac{\nu}{E}\delta_{ij}\frac{\partial \sigma_{kk}}{\partial t} + \frac{\partial f}{\partial \sigma_{ij}}\frac{\partial \lambda}{\partial t}, \tag{3.31}$$

where E is Young's modulus, ν is Poisson's ratio, and f is the yield condition

$$f = \left(\frac{\sigma_1 - \sigma_2}{\theta}\right)^2 + \tau^2 = k^2. \tag{3.32}$$

Here $\sigma_1 = \sigma_{xx}$, $\sigma_2 = \sigma_{yy} = \sigma_{zz}$, k is the yield limit, and θ equals $3^{1/2}$ for von Mises yield criterion [24]. The parameter λ can be expressed as

$$\frac{\partial \lambda}{\partial t} = \frac{\theta^2 \alpha(k)}{4k^2 E}\frac{\partial f}{\partial \sigma_{kl}}\frac{\partial \sigma_{kl}}{\partial t}, \tag{3.33}$$

where $\alpha(k)$ characterizes work-hardening.

Similar to the Sect. 3.2.2, the governing equations for the combined stress wave problem can be written in the matrix form of (3.1)

$$AW_t + BW_x = b, \tag{3.34}$$

where

$$A = \begin{bmatrix} \rho_o & 0 & 0 & 0 & 0 & 0 \\ 0 & \rho_o & 0 & 0 & 0 & 0 \\ 0 & 0 & \frac{1}{E} & \frac{-2\upsilon}{E} & 0 & s \\ 0 & 0 & \frac{-2\upsilon}{E} & \frac{2(1-\upsilon)}{E} & 0 & -s \\ 0 & 0 & 0 & 0 & \frac{1}{\mu} & 2\tau \\ 0 & 0 & s & -s & 2\tau & \frac{4k^2 E}{\theta^2 \alpha} \end{bmatrix},$$

$$B = \begin{bmatrix} 0 & 0 & -1 & 0 & 0 & 0 \\ 0 & 0 & 0 & 0 & -1 & 0 \\ -1 & 0 & 0 & 0 & 0 & 0 \\ 0 & 0 & 0 & 0 & 0 & 0 \\ 0 & -1 & 0 & 0 & 0 & 0 \\ 0 & 0 & 0 & 0 & 0 & 0 \end{bmatrix}, \text{ and } W = \begin{bmatrix} u \\ v \\ \sigma_1 \\ \sigma_2 \\ \tau \\ \lambda \end{bmatrix},$$

where μ is the shear modulus and $s = 2(\sigma_1 - \sigma_2)/\theta^2$.

The characteristics c of (3.34) are the roots of $|cA - B| = 0$. Expanding the determinant leads to

$$c^2 D(c) = 0, \tag{3.35a}$$

where

$$D(c) = s^2\left(\frac{c^2}{c_2^2}-1\right)\left(\frac{3}{\beta}\frac{c^2}{c_2^2}-1\right)+\frac{4\tau^2}{\beta}\frac{c^2}{c_2^2}\left(\frac{c^2}{c_2^2}-\frac{c_1^2}{c_2^2}\right)$$
$$+\frac{12k^2}{\theta^2\alpha(\beta+1)}\left(\frac{c^2}{c_1^2}-1\right)\left(\frac{c^2}{c_2^2}-\frac{c_1^2}{c_2^2}\right) \tag{3.35b}$$

and $c_1 = [(1+\nu)E/\rho_o(1-\nu)(1-2\nu)]^{1/2}$ is the elastic longitudinal wave speed, $c_2 = (\mu/\rho_o)^{1/2}$ is the shear wave speed, and $\beta = 2(1-\nu)/(1-2\nu)$.

It can be seen that if $\pm c_f$ and $\pm c_s$ denote the roots of (3.35b), we have

$$0 \le c_s \le c_2 \le c_f \le c_1, \tag{3.36}$$

where c_f and c_s correspond to the fast wave speed and slow wave speed, respectively. There are two extreme cases: (1) In the elastic region $\alpha = 0$, then $D(c) = 0$ gives $c = c_1$ and $c = c_2$; (2) in the ideally plastic region $\alpha = \infty$, then $c = \pm c_2$.

Since A and B are symmetric the left and right eigenvectors are identical:

$$r = \begin{bmatrix} \Psi \\ 1 \\ -\rho_o c\Psi \\ \Phi \\ -\rho_o c \\ \Omega \end{bmatrix}, \tag{3.37}$$

where

$$\Psi = \frac{s}{\tau}\frac{c^2-c_2^2}{c^2-c_1^2},$$
$$\Phi = \frac{\rho s(c^2-c_2^2)(c^2-\beta c_2^2)}{2\tau c(c^2-c_1^2)}, \text{ and} \tag{3.38}$$
$$\Omega = \frac{1}{2\tau c}\left(\frac{c^2}{c_2^2}-1\right).$$

Generalized Simple Wave Solution

Based on (3.10), we have

$$\frac{du}{\Psi} = \frac{dv}{1} = \frac{d\sigma_1}{-\rho_o c\Psi} = \frac{d\sigma_2}{\Phi} = \frac{d\tau}{-\rho_o c} = \frac{d\lambda}{\Omega}. \tag{3.39}$$

From (3.39) one obtains

$$\frac{d\sigma_1}{d\tau} = \frac{s}{\tau}\frac{c^2-c_2^2}{c^2-c_1^2} = \frac{2(\sigma_1-\sigma_2)}{\tau\theta^2}\left[\frac{1-\frac{c_2^2}{c^2}}{1-\frac{c_1^2}{c^2}}\right]. \tag{3.40a}$$

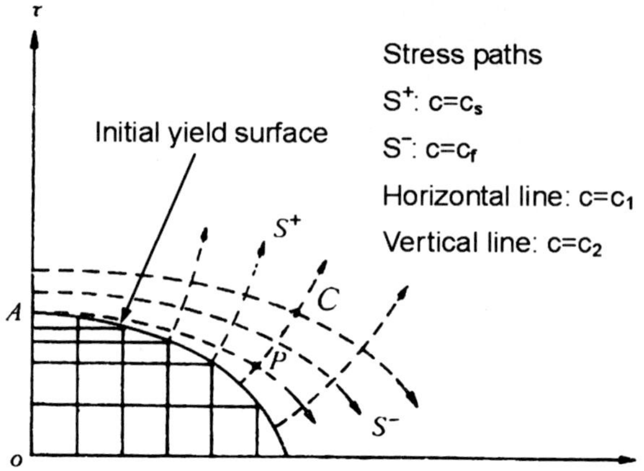

Fig. 3.4. Stress paths.

For a 1-D strain loading condition, σ_1 and σ_2 are not independent. Letting $\sigma_1 - \sigma_2 = f(\sigma_1)$, (3.40a) can be written as

$$\frac{d\sigma_1}{d\tau} = \frac{2f(\sigma_1)}{\tau\theta^2}\left[\frac{1-\frac{c_2^2}{c^2}}{1-\frac{c_1^2}{c^2}}\right]. \tag{3.40b}$$

In the σ_1-τ stress plane, (3.40b) can describe different stress paths, as shown in Fig. 3.4. In the elastic region disturbances can propagate only with the two wave speeds c_1 and c_2, which correspond to the horizontal and vertical path lines, respectively, in Fig. 3.4. For $c = c_f$ in the plastic region, we have $c_2 \leq c_f \leq c_1$ according to (3.36). Then (3.40b) gives $(d\sigma_1/d\tau)_{c_f} < 0$, which means that the slope of the stress path across the characteristic line c_f deceases. On the other hand, one has $c_s \leq c_2$ for the SSW, which yields $(d\sigma_1/d\tau)_{c_s} > 0$; that is, the slope of the stress path across the SSW in the σ_1-τ plane is increasing, as shown in Fig. 3.4 [28*]. Furthermore, the stress paths of the FSW and SSW are orthogonal to each other; i.e.,$(d\sigma_1/d\tau)_{c_f}(d\sigma_1/d\tau)_{c_s} = -1$.

The simple wave stress paths depend on the initial stress state (σ_1^i, τ^i) and the final state (σ_1^o, τ^o). We discuss this briefly here assuming the half space is at rest and stress free at time $t = 0$, that there is a step compression and shear stress loading $\sigma_1^o H(t)$ and $\tau^o H(t)$ on the whole surface, and that $\left(\frac{\sigma_1^o}{\theta}\right)^2 + \tau^{o^2} = k^2$. Three possible solutions exist, depending on the location of the final state (σ_1^o, τ^o) in the $\sigma_1 - \tau$ plane. For example, if the initial state is at the original point O and the final state is at the point A, shown in Fig. 3.5a, the stress path will first follow line OM with wave speed c_1, then follow lines MN with the FSW speed c_f, and NA with the SSW speed c_s (speeds shown in Fig. 3.5b). If the final state is at point B, then the stress path will be along

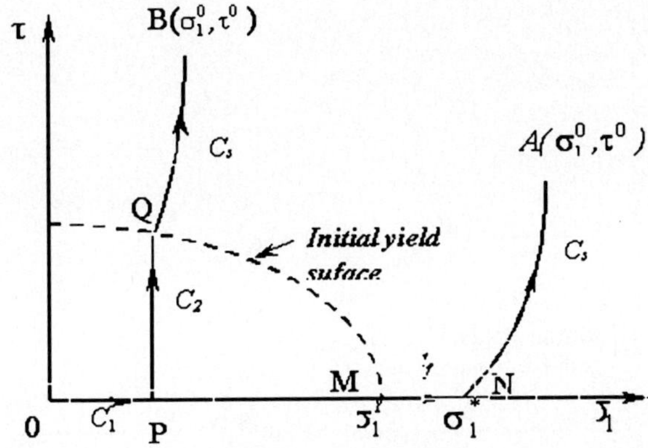

(a) $\sigma_1 - \tau$ plane

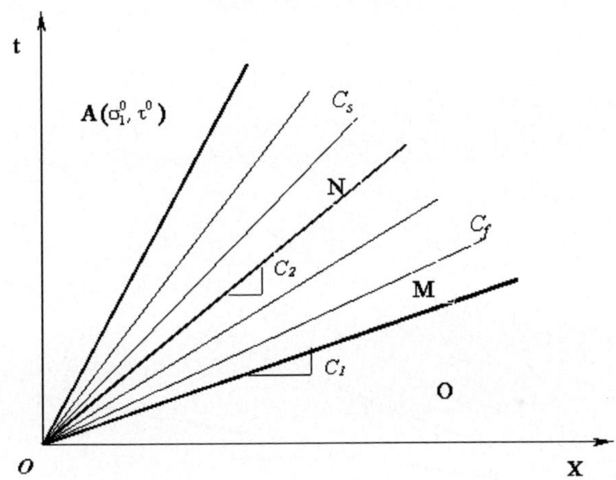

(b) Wave structure for final state A

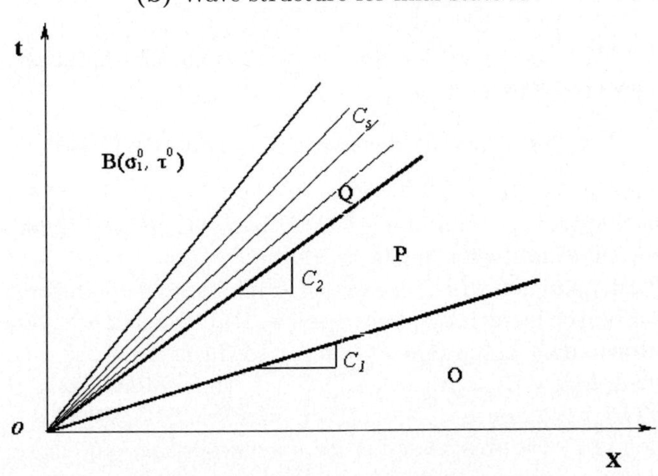

(c) Wave structure for final state B

Fig. 3.5. Stress paths for different final stress states.

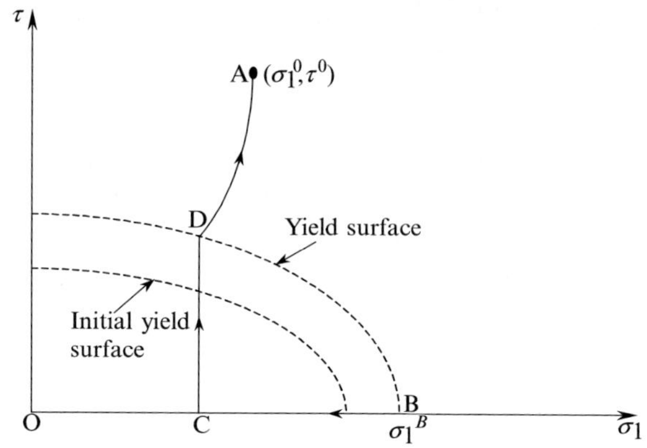

(a) $\sigma_1 - \tau$ plane

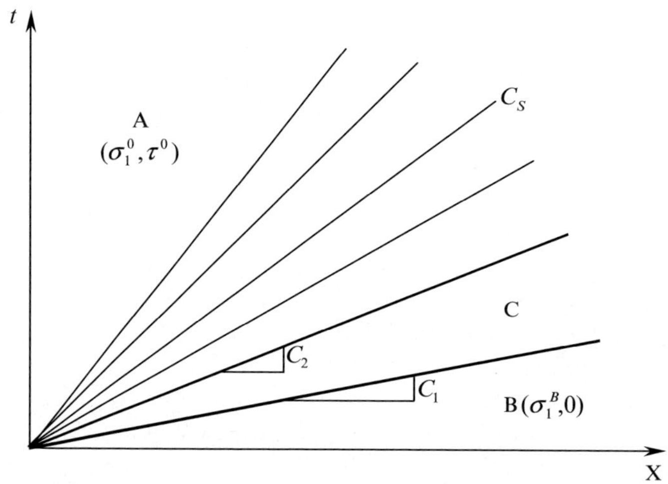

(b) Wave structure for final state A from the initial prestressed state B

Fig. 3.6. Unusual phenomenon for prestressed case.

OP with wave speed c_1, then along PQ with speed c_2, and finally along QB with speed c_s (speeds shown in Fig. 3.5c).

Unusual phenomenon will occur for some prestressed conditions. For example, if the half-space is initially prestressed to state B with $\sigma^B \geq \sigma_Y$ (Fig. 3.6) and then subjected to a step impact loading to the state A (σ_1^o, τ^o), the actual stress path will follow BC with wave speed c_1, then follow CD with speed c_2, and finally DA with speed c_s, as shown in Fig. 3.6 [28]. This means that an impact loading of a prestressed state may experience an elastic unloading process first for σ_1, then loading to the final state.

3.3 Experimental and Diagnostics Methods

The idea of producing combined compression and shear waves by impacting parallel inclined plates was first suggested in the late 1950s [7, 8]. Some theoretical work regarding one-dimensional pressure-shear waves in nonlinear solids started at the same time as discussed in Sect. 3.2. Experimental work in this area began about two decades later. Compression-shear wave experiments were first reported by Abou-Sayed and coworkers [18] using oblique-plate impact on a keyed gas gun and optical measurement of the rear free surface. Gupta and coworkers [19, 20] built a similar keyed gas gun facility to generate the compression and shear loading in the target plate, and developed the electromagnetic particle velocity gauge to measure the pressure and shear components inside of insulating sample materials. Mashimo and coworkers [29] constructed a keyed powder gun capable of accelerating a 30–40 g projectile to a velocity of 2 km s^{-1} for oblique-impact shock study of solids in the several tens of GPa region.

An alternative technique is to use wave propagation in anisotropic solids such as Y-cut quartz, which was first suggested by Johnson [16]. Chhabildas and Swegle [17] demonstrated dynamic pressure-shear loading using this technique.

There are other methods for producing compression-shear waves such as the "Inclined Impact Method" [30] and the "Reflected Shear Wave Technique" [31]. The former is a skewed impact method in which a flyer plate impacts the non-parallel, inclined target plate so that the flyer surface takes a finite time to sweep over the entire target surface. The sweep rate depends on the flyer velocity and the angle of inclination. The resulting longitudinal wave front is not parallel to the shear wave front. For the latter technique, the incident longitudinal shock wave is obliquely reflected from a free surface creating a reflected compression and shear wave. If the initial wave is elastic, the amplitude of the reflected shear wave is determined by the angle of the reflection of the incident wave and the Poisson's ratio of an isotropic material. The reflected shear wave front is not parallel to the incident longitudinal wave front. The lack of parallelism of the waves produced in these two innovative techniques complicates quantitative investigation sufficiently that they are not commonly used. For this reason, we do not discuss them further here.

3.3.1 Experimental Methods to Generate Combined Pressure–Shear Plane Waves

Inclined Parallel Impact

Figure 3.7 shows an oblique plate impact light gas gun facility for producing combined compression and shear waves [32]. This technique is the most popular for compression-shear plane wave investigations due in large part to its production of parallel longitudinal and shear wave fronts, which greatly

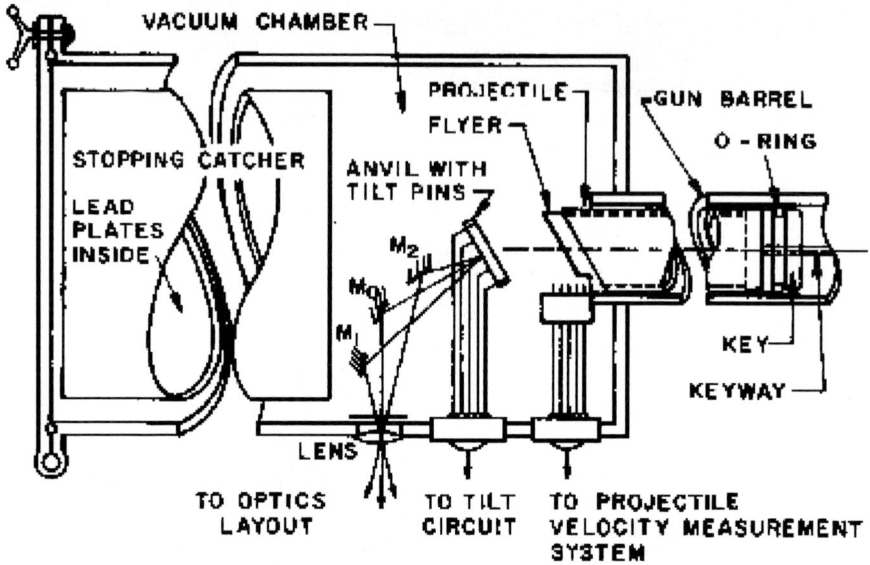

Fig. 3.7. Schematic diagram of pressure-shear experiment.

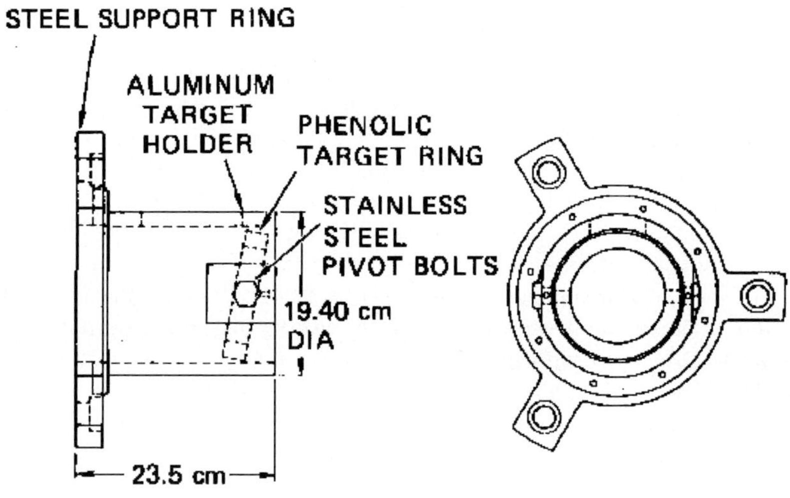

Fig. 3.8. Diagram of a target holder. [Reprinted from [20], with permission.]

simplifies quantitative analysis. The essential difference from the conventional normal parallel plate impact experiment is that the two impacting plates are equally inclined to the axis of the barrel. This is achieved by mounting and inclining the flyer and specimen plates to the desired angle (Fig. 3.8 [20]), and constraining the sabot from rotating about its axis. For this purpose, a square channel slot, or keyway, runs the length of the gun barrel and the sabot has

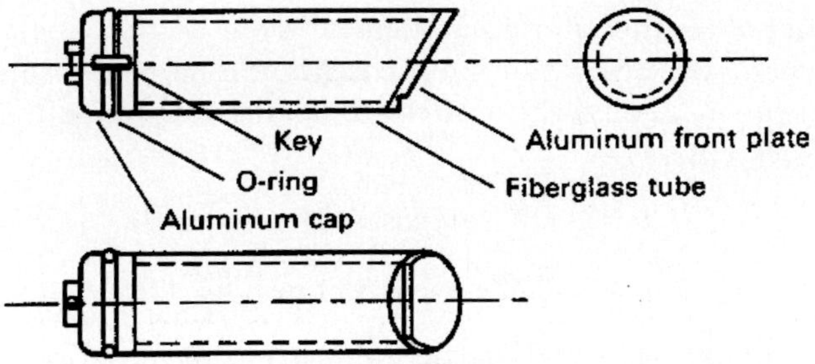

Fig. 3.9. Projectile used in the plate impact test.

a protruding tab or "key" that slides along the keyway as it travels down the gun barrel (Fig. 3.9 [32]). The target chamber is evacuated to a pressure of 0.12×10^{-3} Torr or less to prevent the formation of a gas cushion between the impacting plates, which would be detrimental due to both cushioning the impact and lubricating the impacting surfaces. The ratio of the longitudinal to shear particle velocities is controlled by varying the inclination angle, α, of the impacting plates within a range from 0° (normal impact) to perhaps 25° or 30°, depending on the materials.

The particle velocity transmitted into the target plate during inclined parallel impact depends on the inclination angle, the frictional character of the surfaces, and the material dynamic properties of the plates. If only a single shock wave is generated in the flyer and target plates, respectively, and there is no sliding on the impact interface, then the transmitted longitudinal and transverse components of the particle velocity, u_p and u_s, can be expressed as [33].

$$\begin{aligned} u_p &= u_0 \ \cos\ \alpha/(1+\mu_p)\,, \\ u_s &= u_0 \ \sin\ \alpha/(1+\mu_s), \end{aligned} \tag{3.41}$$

where u_0 is the speed of projectile, the subscripts p and s denote the longitudinal and shear components. The wave impedance ratios of the target (T) and flyer (F) are given in terms of the density, ρ, and the wave speeds, D, by

$$\mu_p = \rho_T D_T^p / \rho_F D_F^p, \quad \text{and} \quad \mu_s = \rho_T D_T^s / \rho_F D_F^s. \tag{3.42}$$

Thus, we obtain the transmitted particle velocity, u, on the target surface and its direction α'

$$u = u_0 \sqrt{\left(\frac{\cos\alpha}{1+\mu_p}\right)^2 + \left(\frac{\sin\alpha}{1+\mu_s}\right)^2}, \text{ and } \tan\alpha' = \frac{u_s}{u_p} = \frac{1+\mu_s}{1+\mu_p} \cdot \tan\alpha. \tag{3.43}$$

Notice that α' and α are not equal, in general. For symmetric impact $\mu_p = \mu_s = 1$ and so

$$u = u_0/2, \quad \alpha' = \alpha, \quad u_p = u_0 \cos\alpha/2, \quad \text{and} \quad u_s = u_0 \sin\alpha/2. \tag{3.44}$$

When sliding occurs on the impact interface, then (3.41) for u_s is not valid. Instead, the transmitted shear component depends on the maximum friction,

$F_{\max} = p \cdot \tan\phi$, on the interface, where ϕ is the friction angle and p is the normal pressure. Based on the jump condition, one can obtain the normal pressure and the maximum shear stress component, $\tau_{\max}$, on the interface for the single shock case:

$$\begin{aligned} p &= \rho_T D_T^p \cdot u_0 \cos\alpha / \left(1+\mu_p\right), \\ \tau_{\max} &= \rho_T D_T^p \cdot u_0 \cos\alpha \cdot \tan\phi / \left(1+\mu_p\right). \end{aligned} \tag{3.45}$$

Then the maximum transverse particle velocity transferred across the sliding interface will be

$$u_{s\max} = u_0 \cos\alpha \cdot \tan\phi / D_{sp}(1+\mu_p), \tag{3.46}$$

where $D_{sp} = D_T^s / D_T^p$. Hence the maximum wave inclination in the target will be

$$\tan\alpha'_{\max} = u_{s\max}/u_p = \tan\phi / D_{sp}. \tag{3.47}$$

The maximum impact inclination without sliding is provided by (3.43),

$$\tan\alpha_{\max} = \frac{1+\mu_p}{1+\mu_s} \cdot \frac{\tan\phi}{D_{sp}}. \tag{3.48}$$

For symmetric impact, $\alpha_{max} = \phi$.

For nonlinear and inelastic materials, there will be more complicated wave structures during inclined impact. Then the above equations do not apply, but one can use the stress–strain particle velocity relation of the target and flyer to calculate the transmitted components if their constitutive relations are known. Conversely, the measured longitudinal and transverse waves in the sample can be used to determine constitutive information for the sample material, which is precisely the intended purpose for the inclined impact investigation.

Two qualitatively different types of pressure-shear experiments have been developed as shown in Fig. 3.10 [34]. One is the conventional plate impact (Fig. 3.10a), which uses thicker sample plates and measures the wave profiles inside of the target. The other is a sandwich impact (Fig. 3.10b), where a thin specimen (e.g., 0.003–0.3 mm) is sandwiched between two hard elastic plates. The principle of this configuration is same as the split Hopkinson pressure bar introduced by Kolsky [35], i.e., the stress and strain state in the specimen becomes nominally uniform after a few reverberations of the waves in the thin specimen. Then, a dynamic stress–strain curve can be deduced by measuring the elastic waves in the bounding plates. From one-dimensional elastic wave analysis in the bounding anvil plates, the compression traction, σ, and shear traction, τ, on the specimen are given by [32],

$$\sigma(t) = \rho c_1 u_{fs}(t)/2, \text{ and} \tag{3.49a}$$

$$\tau(t) = \rho c_2 v_{fs}(t)/2, \tag{3.49b}$$

where u_{fs} and v_{fs} are, respectively, the measured normal and transverse components of the free surface velocity of the anvil plate, and ρc_1 and ρc_2 are

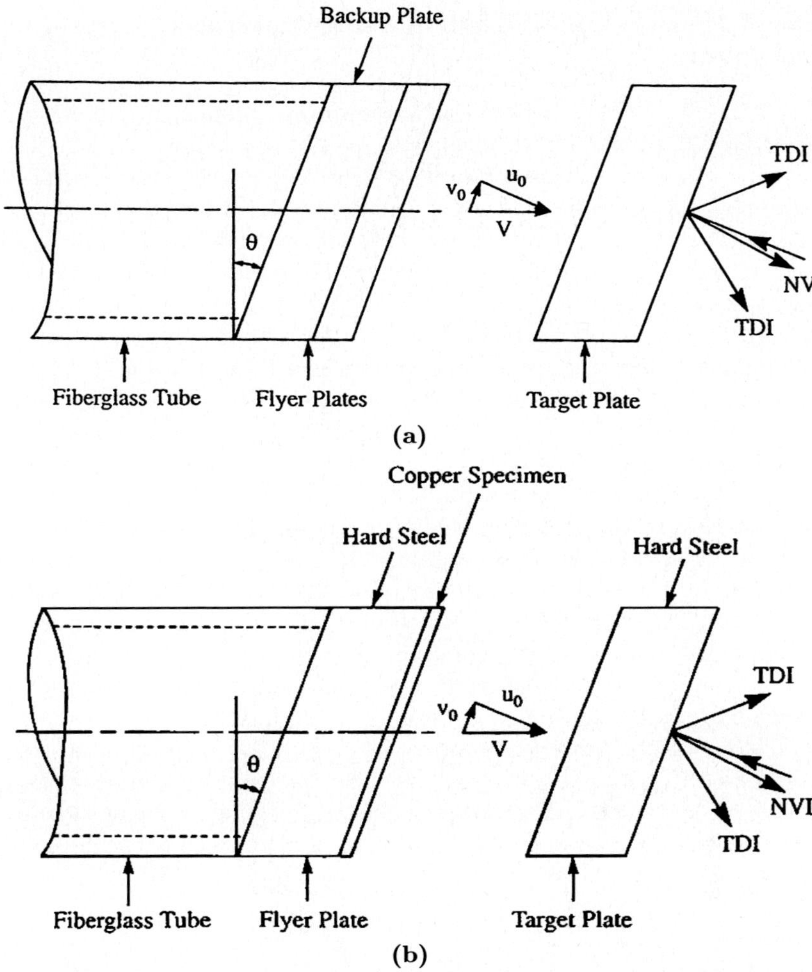

Fig. 3.10. Schematic configurations for the conventional plate impact (**a**) and the sandwich plate impact (**b**).

the elastic longitudinal and shear wave impedances of the anvil plates. The nominal compression and shear strain rates can be expressed as

$$\dot{\varepsilon}(t) = (u_0 - u_{fs}(t))/h \tag{3.50a}$$

and

$$\dot{\gamma}(t) = (v_0 - v_{fs}(t))/h, \tag{3.50b}$$

where u_0 and v_0 are the normal and transverse components of the projectile velocity. By this method, shear strain rates of $10^5\,\mathrm{s}^{-1}$ or higher can be attained to enable investigation of dynamic response of materials at very high strain rates.

Dynamic Pressure-Shear Loading Using Anisotropic Crystal Y-Cut Quartz

Figure 3.11 [17] is the experimental configuration proposed by Johnson [16] for using normal impact on an anisotropic crystal to produce compression and shear plane waves in a sample of interest [17,36]. A y-cut quartz crystal buffer is placed between the flyer and the target sample. The principle underlying this method is that normal impact onto an anisotropic single crystal will generally produce a quasi-longitudinal wave (QL) and a quasi-transverse wave (QT) propagating into the crystal, as shown in Fig. 3.12 [17], except for impact along certain "specific" directions [16]. The particle velocity vector of the QL wave is mainly along wave propagation direction (x-axis in Fig. 3.12), but it has transverse component, as well. Similarly, the particle velocity vector of the QT wave has a small longitudinal component, but it is orthogonal to that of the QL wave. Interaction of the QL wave with the interface between y-cut quartz and an isotropic sample simultaneously generates a plane longitudinal and a plane shear wave in the sample. If the longitudinal wave amplitude exceeds the yield strength of the sample material, the longitudinal motion will consist of an elastic precursor E, followed by a plastic wave P. The transverse motion in the QL wave will generate only a single shear wave S, as shown in Fig. 3.12, because the transverse amplitude is small.

The advantage of this technique is that a normal impact can produce a coupled longitudinal and shear motion in the target sample and the wave fronts of longitudinal and shear waves are parallel to each other to simplify the analysis of the experimental results. The disadvantages are (a) the ratio of the longitudinal and shear components is fixed for a given crystal and orientation, and (b) there might be sliding on the interface between the buffer

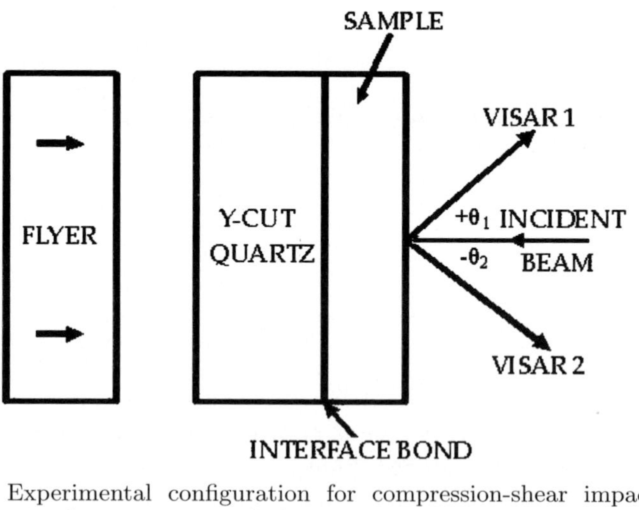

Fig. 3.11. Experimental configuration for compression-shear impact with an anisotropic buffer. [Reprinted from [17], with permission.]

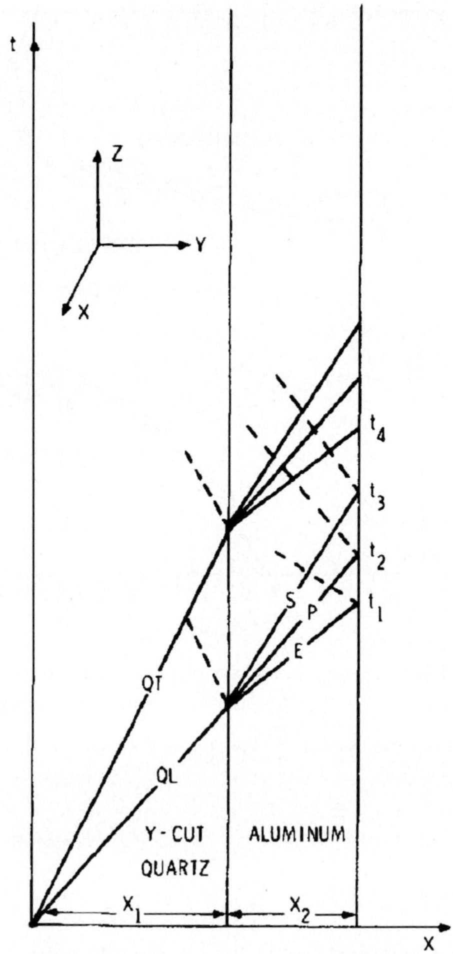

Fig. 3.12. Lagrangian x-t diagram for aluminum sample bonded to a y-cut quartz crystal. The QL wave produced an elastic precusor E, plastic longitudinal wave P, and shear wave S. [Reprinted from [17], with permission.]

and sample for the large amplitude shear component. The first disadvantage can be reduced if different crystals and orientations are selected. The second disadvantage might be partially mitigated by using an epoxy bond at the interface. In this case the maximum shear stress will be limited to the strength of the bond layer, which may not exceed the frictional stress between the dry surfaces of the crystal and sample.

Figure 3.13 compares the longitudinal and transverse free surface velocity profiles, as calculated using an elastic-perfectly plastic model, with those measured for a QL wave transmitted into an aluminum sample [17]. The close agreement validates this technique for combined compression–shear plane wave investigation.

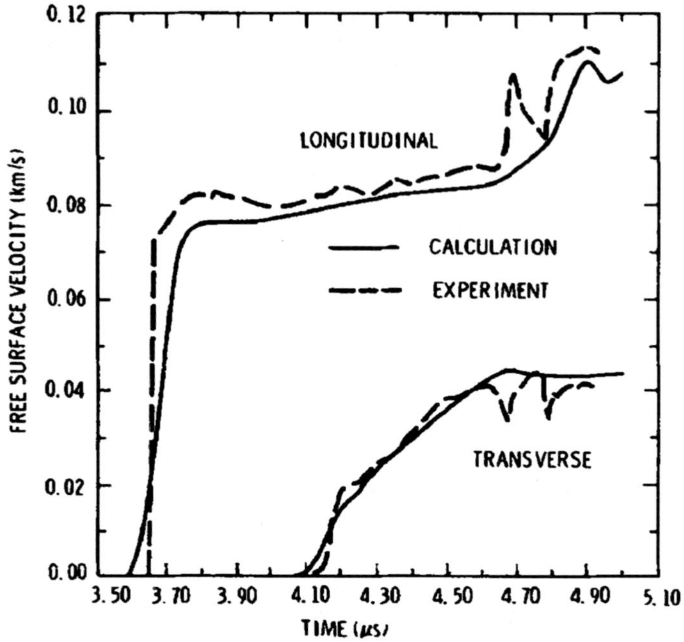

Fig. 3.13. Calculated (elastic-perfect plastic model) and experimental longitudinal and transverse free surface velocity profiles for aluminum sample due to QL.

3.3.2 Diagnostics

Electromagnetic Particle Velocity (EMV) Gauges

Measurement Principle

The first in-material measurements using electromagnetic particle velocity (EMV) gauges for a combined compression–shear plane wave study were reported by Gupta and coworkers [19,20], which they called the Internal Measurement for P and S waves (IMPS) method.

For an EMV gauge of length l moving with velocity $u(t)$ in a steady uniform magnetic field of strength H, as shown in Fig. 3.14 [33], the induced voltage of the gauge obeys Faraday's law:

$$E(t) = \overrightarrow{l} \cdot \left(\vec{u}(t) \times \vec{H} \right) . \tag{3.51}$$

Applying a homogenous magnetic field, H, at angle θ with respect to the projectile direction, X, as shown in Fig. 3.14, the voltages induced in an EMV gauge due to the P wave, E_p, and the S wave, E_s, will be

$$\begin{aligned} E_p &= H u_p l \sin\left(\theta - \alpha'\right), \quad \text{and} \\ E_s &= H u_s l \cos\left(\theta - \alpha'\right), \end{aligned} \tag{3.52}$$

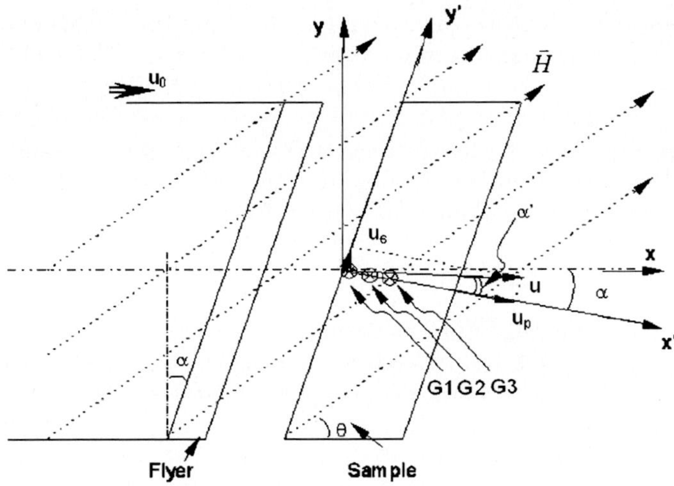

Fig. 3.14. Schematic diagram of IMPS method for an inclined plate impact. x-y: coordinates with x along impact axis, $x' - y'$: coordinates with x' perpendicular to the impact surface. G1-3: particle velocity gauges 1–3. H and θ: magnetic field strength and its angle respect to the x-axis. U_0 and α: impact speed and the inclined angle. U and α': the transmitted particle velocity in the target and its angle respect to x'. u_p and u_s: the x' and y' components of transmitted particle velocity.

Table 3.1. Induced voltage versus magnetic field angle for symmetry impact with no sliding (α is the inclined angle)

Magnetic field angle θ	0	α	2α	π/2 + α
Ep/E_0	$-\sin\alpha\cos\alpha$	0	$\sin\alpha\cos\alpha$	$\cos\alpha$
Es/E_0	$\sin\alpha\cos\alpha$	$\sin\alpha$	$\sin\alpha\cos\alpha$	0
Schematic Diagram	Ep Es	Es	Es Ep	Ep

where α' is the direction of the transmitted particle velocity at the target surface, defined in (3.43). For symmetric impact without sliding ($\alpha' = \alpha < \alpha_{max}$), the magnitudes for compression and shear signals will be

$$\begin{aligned} E_p &= E_0 \cos\alpha \sin(\theta - \alpha), \text{ and} \\ E_s &= E_0 \sin\alpha \cos(\theta - \alpha), \end{aligned} \tag{3.53}$$

where $E_0 = H u_0 l/2$. Equation (3.53) shows that rotating the magnetic field can change E_p and E_s, and their ratio. Some special cases of θ are listed in Table 3.1. In particular, one can measure only the compression or only the shear component when $\theta = \pi/2 + \alpha$ or $\theta = \alpha$, respectively. Applying just a horizontal magnetic field ($\theta = 0$ or $\theta = \pi$) offers an optimal resolution for both the compression and shear components at the same time. This configuration is additionally appealing for only requiring a solenoidal magnet. It was used with very good effect in Conner's study of fused silica [35].

Swegle [37] described the possibility to simultaneously measure both pure longitudinal motion and pure transverse motion in a sample containing EMV gauges oriented perpendicular to one another. This suggestion was subsequently demonstrated by Aidun and Gupta [38].

Magnet System

An important element for EMV measurement is the magnet system as shown in Fig. 3.15 [20], which has a two-axis electromagnet with magnetic fields along the X-axis and Y-axis. A horizontal magnetic field is produced by a solenoid with its axis parallel to the gun barrel axis. A vertical field normal to the solenoid field is produced by a dipole. The magnet is powered by two DC power supplies with maximum current outputs of 1,000 and 500 A, respectively. The desired field orientations are set by adjusting the current supplies.

Fig. 3.15. Dipole magnet positioned around the target holder. [Reprinted from [20], with permission.]

Gauge Description and Facility Modification

EMV gauges embedded in the sample can be made of fine copper wire or foil. In making shear wave measurements, the use of side-lead gauges is not desirable due to the edge release wave effects. The gauge leads must lie along the direction of wave propagation to avoid spurious signal coming from their motion in the magnetic field [39].

To avoid disturbance of the magnetic field for IMPS measurements, some modifications are made for the impact facility: one makes the target holder out of nonmagnetic materials (aluminum alloy, nonmagnetic stainless steel or polymers as shown in Fig. 3.8), and one uses a plastic sabot or adds a plastic (usually polymethyl methacrylate, PMMA) extension on an aluminum sabot so that the metal does not enter the magnetic field during the experiment.

Figure 3.16 is a typical experimental wave profile from a PMMA sample impacted at a projectile velocity of 450 m s^{-1} and an inclination of 12° [40]. The gauge is located 5.64 mm from the impact surface and the magnetic field was aligned to give signals from both waves. The compression, shear, and longitudinal release waves can be recognized in the record and their wave speeds can be obtained from a single measurement. The ability to measure shear and release waves in the shocked state have provided new information on shock behavior of materials. In particular, the propagation speed of the onset of the shear wave and the start of longitudinal release, respectively, correspond to the elastic shear and longitudinal wave speeds in the shocked state into which these waves propagate.

The main advantages of the EMV gauge method are (a) it is based on Faraday's law and so requires no assumptions or calibration; (b) using EMV gauges at multiple locations in one experiment can provide greater and more accurate constitutive information. The main shortcoming is it cannot be used

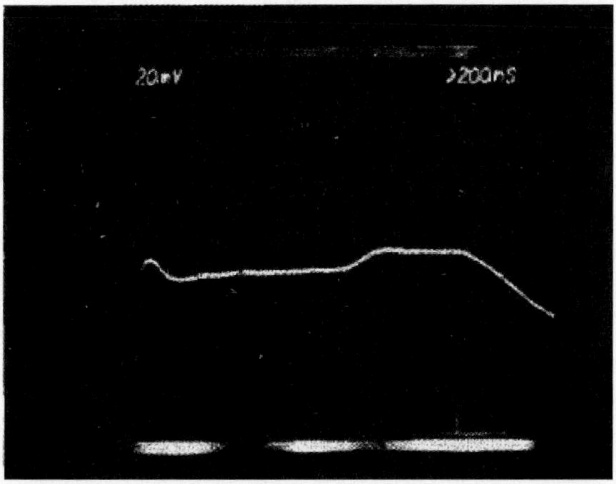

Fig. 3.16. Voltage-time profile from PMMA compression-shear experiment. Projectile velocity 450 m s^{-}, inclined angle 12°, gauge location 5.64 mm.

for magnetic materials and is also problematic in conductors because the mechanical effects of insulating coverings degrade the gauge response. In general, the epoxy bonds joining the layers of the sample may influence the compression wave profiles, especially for thick epoxy bonds, poorly fitting layers. The epoxy bonds may also influence the shear wave profiles for strong shear waves.

Transverse Displacement Interferometer (TDI) Method

Kim and coworkers [21] developed a new optical method of transverse displacement interferometry (TDI). With this method the motion of the stress-free rear surface of the target plate can be measured by a combined normal velocity and transverse displacement interferometer system, as shown in Fig. 3.17 [32]. In the figure, a normal incident laser beam impinges on a plane diffraction grating, which is affixed to the rear surface of the target. This produces a reflected beam normal to the plane of the grating, which is used for the normal velocity measurement, and diffracted beams at angles θ_N, which are used for the TDI measurement. For the normal velocity interferometer (NVI), each "fringe" (one peak–peak variation in intensity) corresponds to a change, Δu, in the normal component of the free surface velocity given by [41]

$$\Delta u = \lambda/2\,T, \tag{3.54}$$

where λ is the wavelength of the laser light source, and T is the time required for the light to traverse the delay leg of the NVI.

For the TDI, the Nth order diffraction angle θ_N satisfies the diffraction equation

$$d \sin\theta_N = N\lambda, \qquad N = \pm 1,\ \pm 2,\ \pm 3, \ldots, \tag{3.55}$$

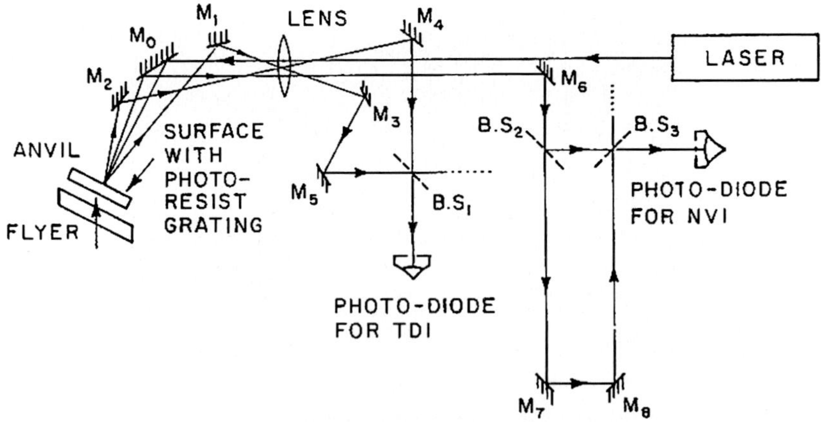

Fig. 3.17. Combined normal velocity interferometer and transverse displacement interferometer.

where d is the pitch of the grating and λ is the wavelength of the laser source. The two diffracted beams shown in Fig. 3.17 are recombined at the beam splitter BS_1, and the resulting superimposed collinear beams are monitored at the photodetector for the TDI.

The change in phase $\Delta\phi$ due to the transverse motion of grating is

$$\Delta\phi = 2\pi \frac{2N}{d} \Delta y(t), \tag{3.56}$$

where Δy is the transverse component of the displacement of the grating. Equation (3.56) demonstrates that each fringe corresponds to a change in the transverse displacement (i.e., its sensitivity) given by

$$\Delta d_y = d/2N. \tag{3.57}$$

For a 5-Watt argon ion laser with a single-mode, $\lambda = 0.5144$ µm, a typical delay time is $T = 7\,ns$. Then the fringe constant of (3.54) is 0.0367 mm µs^{-1} per fringe. The diffraction grating can be better than 1,000 lines per mm with d less than 1 µm. For the first order fringes (i.e., $N = 1$), which are normally used for measurement, (3.57) gives the sensitivity of the transverse displacement measurement better than 0.5 µm per fringe.

Figure 3.18 is the transverse velocity history at the rear surface of a 6061-T6 aluminum sample plate under combined pressure–shear impact loading [42]. The target thickness is 3.2 mm and the projectile velocity is 215 m s^{-1}. The zeroth order reflected beam is used for the NVI measurement, which recorded the normal particle velocity history shown in Fig. 3.19 [42].

Related to these developments, Espinosa [43] developed a so-called "in-material interferometric measurement", i.e., the grating is embedded

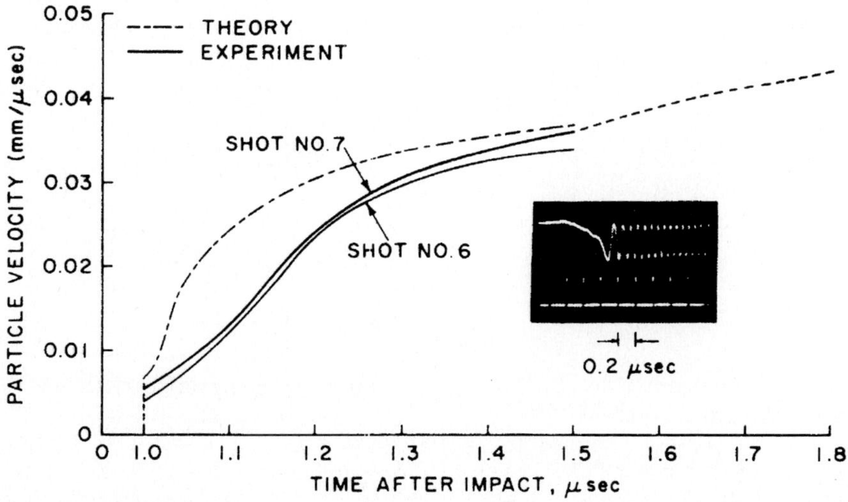

Fig. 3.18. Transverse velocity-time profiles at the rear surface.

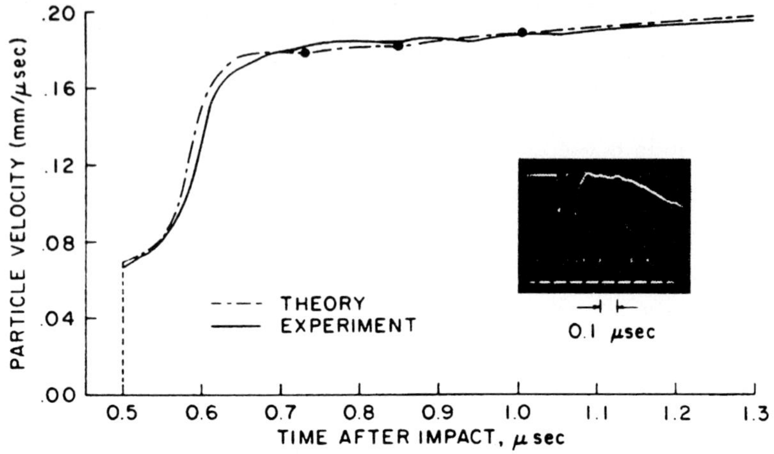

Fig. 3.19. Normal velocity profiles at the rear surface.

between the sample plate and the window plate. Also, Mello and coworkers [44] developed the multi-point interferometer for monitoring two-dimensional wave motions.

Other Methods

VISAR Transverse Velocity Interferometry (TVI)

Chhabildas and Swegle [17] applied a VISAR system developed by Barker and Hollenbach [41] to measure the transverse particle velocity at the free surface of the sample, as shown in Fig. 3.11. By using one incident laser beam and monitoring two beams at angles $+\theta_1$ and $-\theta_2$, the longitudinal velocity $U(t)$ and the transverse velocity $V(t)$ can be determined simultaneously. This is possible because the velocities, $u(\pm\theta, t)$, measured by the two VISARs are related to $U(t)$ and $V(t)$ by

$$u(+\theta_1, t) = \frac{1}{2}U(t)(1 + \cos\theta_1) + \frac{1}{2}V(t)\sin\theta_1 \tag{3.58}$$

and

$$u(-\theta_2, t) = \frac{1}{2}U(t)(1 + \cos\theta_2) - \frac{1}{2}V(t)\sin\theta_2. \tag{3.59}$$

The velocity measured by each VISAR is related to the number of fringes $F(t)$ by the relation

$$u(\theta, t) = \lambda F(t)/2\tau(1 + \delta), \tag{3.60}$$

where λ is the wave length of the incident laser, τ is the delay time and δ is the optical correction.

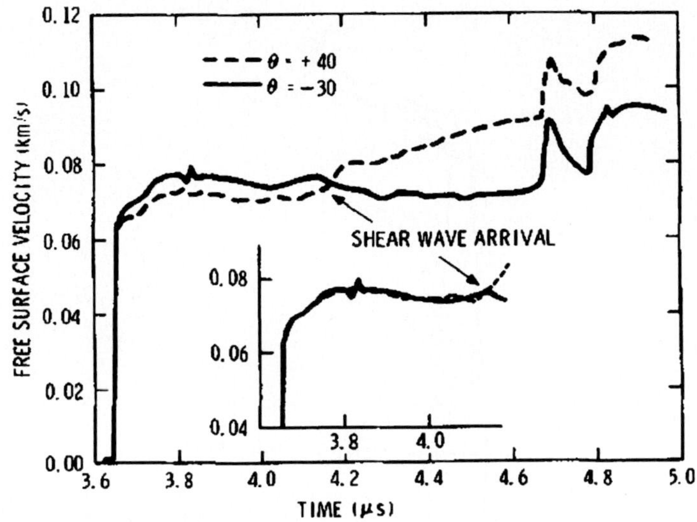

Fig. 3.20. Free-surface velocity profiles obtained for aluminum sample. [Reprinted from [17], with permission.]

Free surface velocity profiles obtained for beams collected at $+40°$ and $-30°$ in an measurement on a 6061-T6 aluminum sample, impacted at a velocity of $105.1\,\mathrm{m\,s^{-1}}$, are shown in Fig. 3.20 [17]. The difference in the two velocity profiles before shear wave arrival is due to the asymmetry in the angles of the two beams. If the two profiles are normalized to a $30°$ diffraction angle, they overlap until the shear wave arrives, as shown in the inset of Fig. 3.20. The free surface longitudinal velocity profile $U(t)$ and the transverse velocity profile $V(t)$ are then obtained by solving (3.58) and (3.59). Typical experimental longitudinal and transverse free surface velocity histories obtained by using this method are shown in Fig. 3.13.

"Shadow" Technique

In early studies of the combined pressure–shear method, Abou-Sayed and coworkers developed a "shadow" technique (the dynamic Moiré technique) to monitor the transverse motion at the target free surface (Fig. 3.21) [18]. It consists of a grating with 50 lines per mm printed on the rear surface of the target and a stationary replica grating printed on a high resolution photographic plate. A laser beam passes through the stationary grating, reflects from the target grating and goes through the stationary grating again. A photodiode is used to monitor the intensity of one of the returning diffracted beams. The intensity of the detected beam is related to the change in relative position of the two gratings, so this method can be used to monitor the transverse displacement at the rear surface of the target. However, this method is sensitive

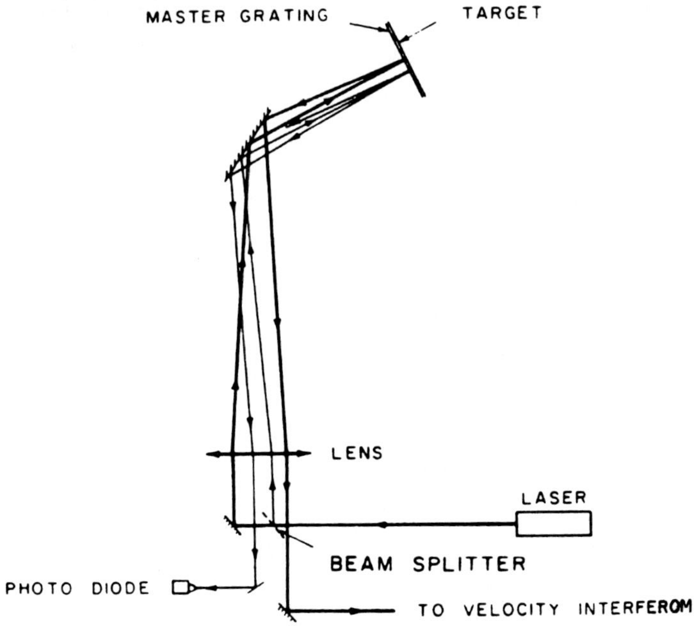

Fig. 3.21. Optical arrangement for simultaneous normal and transverse motion measurements.

to the shape of the grating and misalignment of the two gratings. Also, normal motion can produce an intensity change. In later studies, the NDI method mentioned above was used instead of this method.

Piezoelectric Shear Stress Gauge

Gupta [45] demonstrated the feasibility of using piezoelectric $LiNbO_3$ as a shear transducer analogous to quartz stress gauges commonly used for longitudinal stress measurement. His innovation was to look for an orientation that exhibited a strong piezoelectric response to shear stress and, simultaneously, negligible response to normal stress. For such an orientation, a crystal affixed to the rear surface of a sample subjected to compression-shear loading would register a signal solely in response to the shear stress transmitted into it from the sample. Based on crystal elastic and piezoelectric constants reported in the literature, Gupta identified 137° Y-cut $LiNbO_3$ as a suitable orientation for a shear stress gauge according to these criteria. He subsequently demonstrated the efficacy of this shear stress gauge by fabricating and using several 137° Y-cut $LiNbO_3$ gauges in compression-shear experiments. The resulting gauge signals clearly showed a strong preferential sensitivity to shear stress. However, the observed response to normal stress demonstrated both a pronounced sensitivity to uncertainties in the reported crystal properties and the excessive challenge of precisely orienting the crystals.

3.4 Applications

3.4.1 Plasticity Under Combined Compression and Shear Loading

Clifton and coworkers [42, 46, 47] used the pressure-shear sandwich plate impact technique (Fig. 3.10b) to examine the validity of different constitutive equations for high strain rate combined stress states in metals. They investigated the rate-dependent plastic behavior of 6061-T6 Aluminum and commercially pure α-titanium, conducting experiments with various angles of inclination and impact velocities. The experiments were conducted with a slotted gas gun; the normal and transverse components of the particle velocity of the rear surface of the specimen were measured with the NVI and TDI techniques mentioned in the previous section.

Figure 3.22 shows the normal and transverse velocity histories at the rear surface of two 6061-T6 aluminum specimens (shots #A-16 and #A-24) with the same 26.6° inclination, 3.2 mm thickness, and almost the same projectile velocity of 220–213 m s^{-1} [47]. Computed velocity histories are shown for a visco-plasticity model with two conventional hardening models; kinematic hardening, corresponding to a pure translation of the yield surface in the direction of the plastic strain rate, and isotropic hardening, corresponding to a pure expansion of the yield surface. For the normal velocity histories shown in Fig. 3.22a, the agreement between theory and experiment is quite good. Since the calculated histories for these two models are almost identical, it is evident that the normal velocity is relatively insensitive to the choice of the plastic flow potential. For the transverse components (Fig. 3.22b), the histories computed using the modified strain rate function show improved agreement with the experimental results.

Figure 3.23 shows the normal and transverse velocity histories at the rear surface of two α-titanium specimens (shots #A-18 and #A-22) with the same 26.6° inclination angle, 4.6 mm thickness and 208 m s^{-1} projectile velocity [47]. The agreement between computed and measured wave histories suggests that the flow stress increases strongly with plastic strain rates above $10^4\,s^{-1}$ or that the hydrostatic pressure up to 2 GPa has a significant effect on plastic flow in 6061-T6 aluminum, but only a minor effect on plastic flow in α- titanium.

Gilat [48] later obtained improved agreement between theory and experiment by introducing a new surface hardening model that embodies some of the features of both the kinematical and isotropic hardening models. More recently, Frutschy and Clifton [49, 50] modified the sandwich inclined plate impact facility, shown in Fig. 3.10b, to test the plastic response of materials at high-temperature up to 700°C. Together with the high strain rate characteristic of this experiment ($10^6\,s^{-1}$), the high- temperature capability allows the shearing resistance of materials to be inferred from the elastic response of

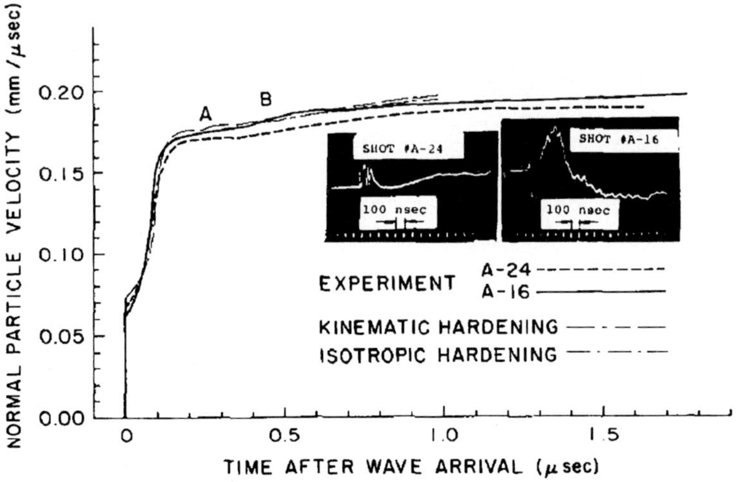

(a)

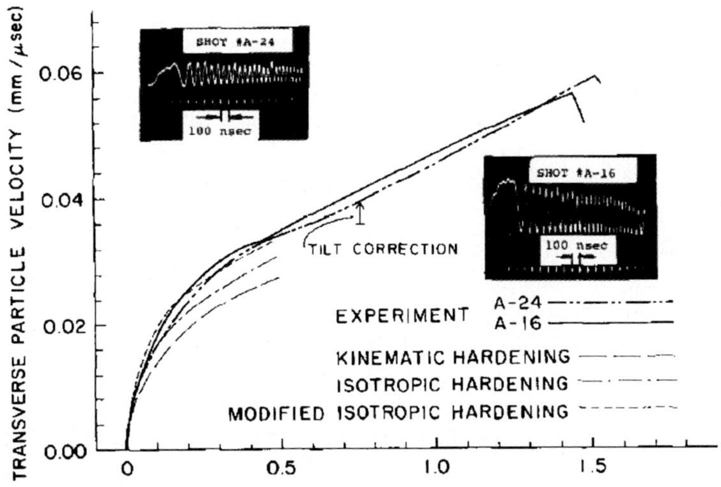

(b)

Fig. 3.22. Experimental and computed normal and transverse velocity-time profiles at the rear surface of two 6061-T6 aluminum specimens (shots #A-16 and #A-24): (**a**) Normal components, (**b**) Transverse components. [Reprinted from [47], with permission.]

adjacent plates under conditions unattainable with other testing equipment. The three main modifications made in this study are: (a) a high temperature grating up to 700°C; (b) pure tungsten carbide as the impactor, which is elastic up to 700°C; and (c) the use of an induction heater to heat the target sample, as shown in Fig. 3.24 [49].

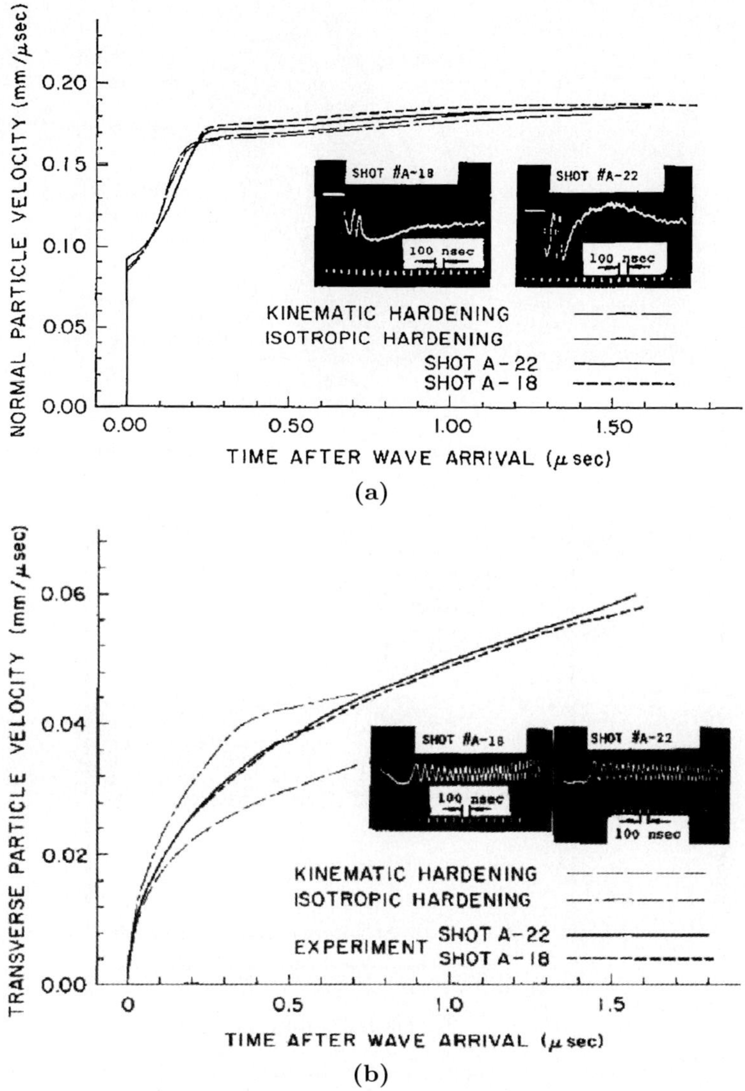

Fig. 3.23. Experimental and computed normal and transverse velocity-time profiles at the rear surface of two α- titanium specimens (shots #A-18 and #A-22): (**a**) Normal components, (**b**) Transverse components. [Reprinted from [47], with permission.]

Figure 3.25 shows the measured shear stress–strain plot for OFHC copper at 691°C (plot KF9708 in the figure) [49,50]. Comparing to the plot TW8902 at 22°C, from [51], the evolution of the flow stress at high temperature shows some initial hardening, then softening, and finally stabilization. (The jump at $\gamma \approx 2$ corresponds to the arrival of the normal waves at the free surface).

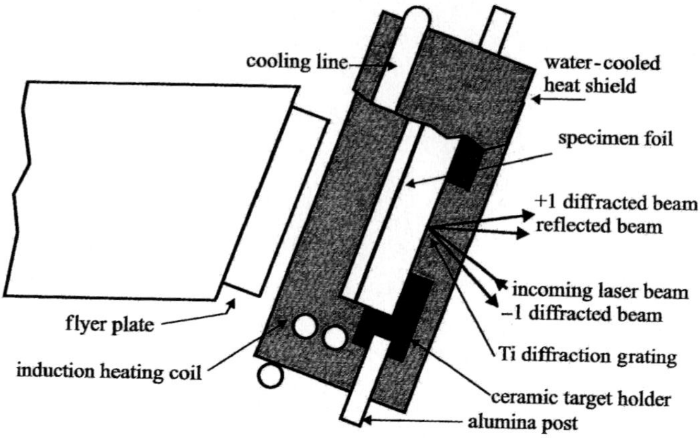

Fig. 3.24. Impact schematic for high temperature pressure-shear impact.

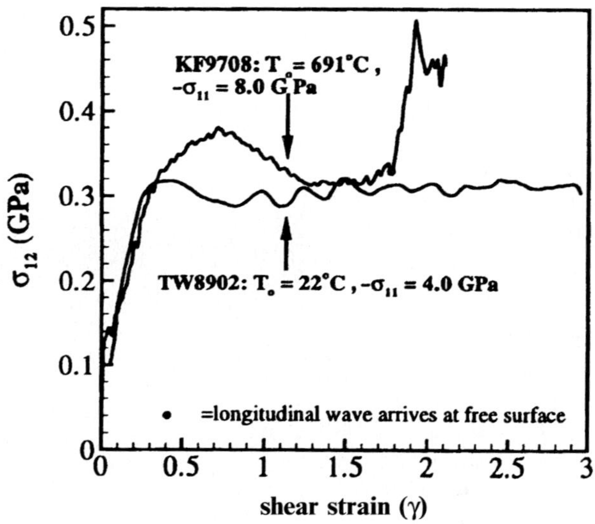

Fig. 3.25. Shear stress – shear strain for OFHC copper (KF9708). [Reprinted from [50], with permission.]

3.4.2 Investigations of Post-Yield Material Behavior

Phase Transformations in Calcium Carbonate

Structural phase transformations under shock loading are of considerable interest for understanding the response of solids under non-hydrostatic stresses and at high strain rates [52]. However, the conventional shock experiments cannot separate the deviator stresses from the longitudinal measurements. Aidun [53], and Aidun and Gupta [38, 54, 55] applied the combined compression–shear loading experiments to examine the phase transformation in Carrara marble, which served as a convenient source of high density polycrystalline calcite ($CaCO_3$). They used the in-material EMV measurements (see Electromagnetic Particle Velocity (EMV) Gauges), which permit determination of the elastic moduli of the shocked material. These values are determined from the measured speeds in the peak shocked states of the onset of the shear wave and of the longitudinal release wave. These wave speeds are combined with the density deduced from the multiple, simultaneous longitudinal particle velocity measurements. The measured particle velocity profiles of the longitudinal and shear components are shown in Fig. 3.26 [55]. The measured speeds of the shear and longitudinal waves are shown in Fig. 3.27 [55]. Both the shear and longitudinal wave speeds in the shocked marble show decreases at small compression that are associated with the calcite-$CaCO_3$ (II) transformation; the metastable $CaCO_3$(II) phase has lower modulus values than does calcite.

The new and unique information provided by the shear wave measurements in a shock-induced, high-pressure phase warrants special emphasis. These measurements in $CaCO_3$ show that the shear wave speed varies non-monotonically with compression. The increase in the shear modulus above approximately 4.5% compression occurs after the material has both yielded and undergone the transformation to $CaCO_3$ (II). The complexity of the elastic shear response deserves particular attention given the tendency in pure longitudinal experiments to interpret a reduction in compression wave speed as an indication of either loss of strength or melting. Indeed, the bulk modulus values deduced from the in-material measurements permit a more discriminating comparison between the shock and hydrostatic data. Figure 3.28 shows the bulk modulus versus peak shock density [55]. The density variation of the bulk modulus in the shocked state agrees reasonably well with the static data up to 8% compression in Fig. 3.28. The transformation to $CaCO_3$ (II) is marked by a large reduction in the bulk modulus. At 11% compression, the bulk modulus in the shocked marble lies far below the value from the hydrostat, which suggests the onset of a further transformation to a phase other than $CaCO_3$ (III). Aidun and Gupta [55] ultimately showed that the aragonite phase is generated by 15% compression by the longitudinal shock wave.

Pressure-Shear Behavior of Ceramic Materials Under Shock Loading

Gupta [58] used the compression-shear plate impact facility and EMV gauges to study elastic shear and compression wave propagation in polycrystalline Al_2O_3 shocked to 90 kbar. The measured shear wave velocity data agreed well with the extrapolation of the ultrasonic measurements. The bulk and shear moduli in the shocked state were determined directly from the observed shear and longitudinal release wave speeds. Gupta additionally showed

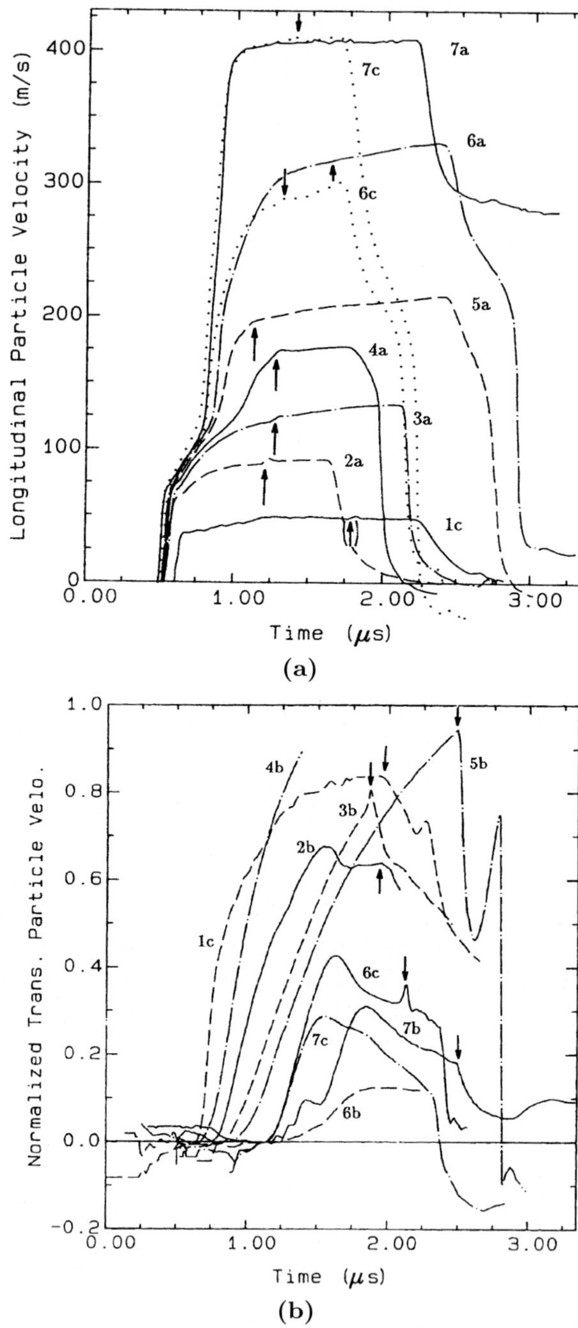

Fig. 3.26. EMV gauge records of the particle velocity profiles (**a**) Longitudinal component histories for different peak states at approximately 3.3 mm depth (**b**) Transverse particle velocity histories measured at a nominal depth of 3.9 mm for different peak states.

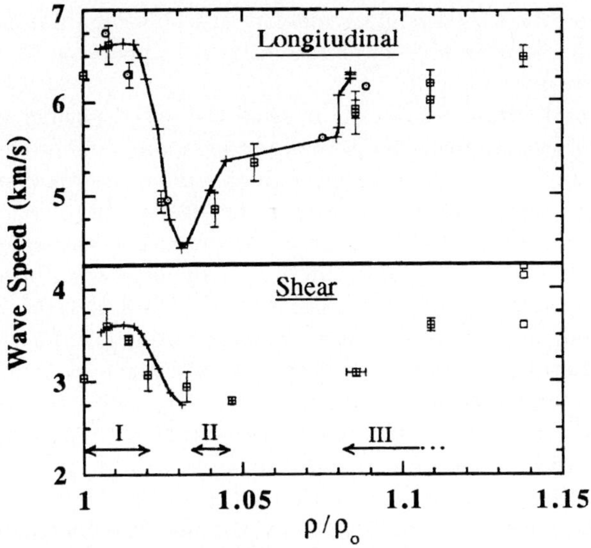

Fig. 3.27. Longitudinal release and shear wave speeds in shocked carrara marble versus normalized density (squares). Also shown are data for release speeds [56] (circles) and hydrostatic [57] (pluses) for Oak Hall limestone.

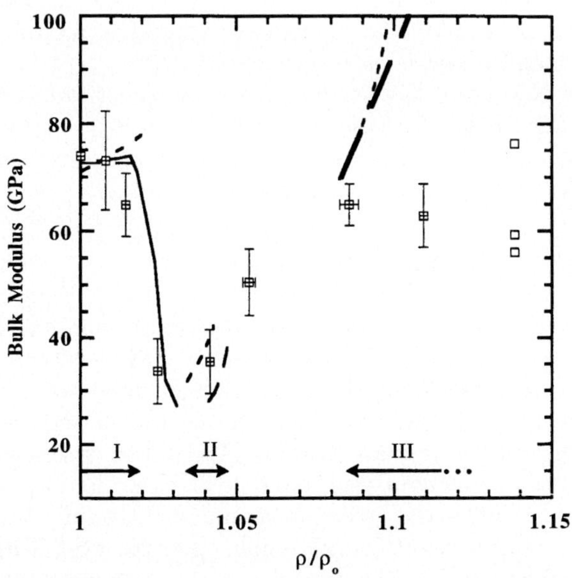

Fig. 3.28. Bulk modulus in shocked Carrara marble versus normalized density (squares). Other lines are hydrostatic compression results for comparison.

that integrating the experimentally determined bulk modulus as a function of density provides a mean stress-volume relation comparable to static data for non-rate-dependent materials.

Mashimo and coworkers [59, 60] applied the keyed powder gun and EMV gauges to generate and measure the compression and shear waves over 20 GPa in sapphire samples, and found that sapphire preserves a considerable shear strength under shock compression even in the plastic region up to over 30 GPa.

Yuan and coworkers [61] conducted compression-shear wave experiments on polycrystalline silicon carbide (SiC) instrumented with in-material EMV gauges. The peak compression stresses ranged from 3–18 GPa. The measured longitudinal and shear wave velocities were used to determine the elastic moduli of the material as functions of density compression in the shocked state. The data were further analyzed to obtain the mean stress response of the SiC under uniaxial-strain state. The results show that the Poisson's ratio of the material increases with elastic shock compression from an ambient value of 0.161–0.192 at the Hugoniot elastic limit of 11.5 GPa. Above the elastic limit, the maximum shear stress supported by the material increases from 4.5 to 6.4 GPa at a peak stress of 18 GPa.

Shear wave measurements to determine the nonlinear elastic response of fused silica under shock loading reported by Abou-Sayed and coworkers [24] were discussed above at the end of Sect. 3.2.2. Conner [62] used EMV gauge measurements of compression and shear waves in fused silica to determine its nonlinear elastic constants from the density-dependent moduli obtained from the shear and longitudinal release wave speeds.

Walley and coworkers [63] studied the response of thermites to dynamic high pressure and shear. Zavattierti and Espinosa [64] conducted an examination of the competition between bulk behavior and interfacial behavior of ceramics subjected to dynamic pressure-shear loading.

Pressure-Shear Behavior of Polymers

Gupta performed large amplitude, one-dimensional compression-shear wave measurements on polymethyl methacrylate (PMMA) with imbedded EMV gauges [40]. Measurements of shear and longitudinal wave velocities were used to determine the shear modulus, bulk modulus, and the mean stress-volume relations under impact loading. The mean stress-volume relation determined from impact data was found to be considerably stiffer than the static hydrostat. Unlike the quasi-static-uniaxial strain results, the shock data imply a strength reduction at higher stresses, even with allowing for the rate dependent mechanical response of the polymer. Measurements of shear wave amplitudes were consistent with the strength loss interpreted from the wave velocity analysis. However, using lateral stress gauge measurements Gupta later demonstrated that PMMA does not lose strength in shock loading [unpublished, *cf.* 55].

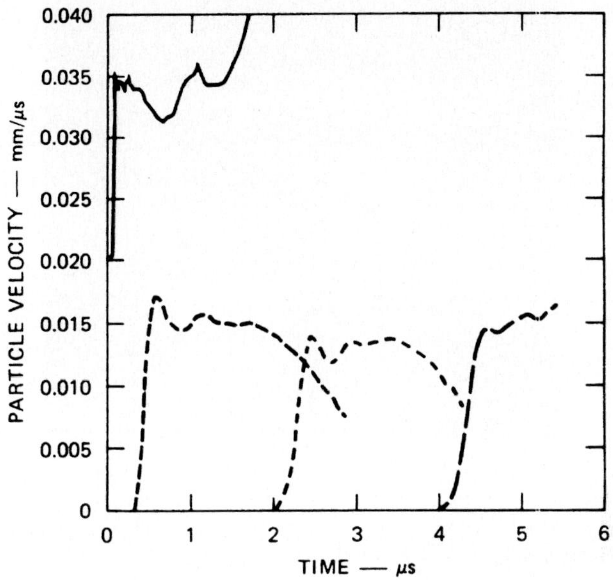

Fig. 3.29. Shear particle velocity profiles from all four gauges (projectile velocity $370\,\mathrm{m\ s^{-1}}$, $\theta = 10°$.)

Gupta's measurements of the shear particle velocity show a rapid attenuation between the impact surface and the interior gauges (Fig. 3.29), which suggests a time-dependent shear failure near the impact surface [40]. Further investigation of the surface response discloses that increasing the normal pressure and shear component by increasing the projectile velocity and the inclination angle leads to a dramatic decrease of the shear particle velocity measured on the impact surface. This demonstrates that the shear particle velocity at the impact surface is governed not only by the frictional response at the surface, but also by the shear strength in the vicinity of the impact surface.

To explore the physical mechanism regarding this "shear failure" of polymers, Li and coworkers selected nylon-66 for a recent experimental investigation [65]. Nylon-66 was chosen for its spherical grain structure, which can be observed under a polarized microscope. The shear wave amplitude attenuates rapidly between the impact surface and the interior gauges when the impact velocity and inclination angle reach critical values of about $130\,\mathrm{m\ s^{-1}}$ for $\theta = 20°$. Figure 3.30 shows the original configuration of spherical grains near the sample surface before the experiment [65]. Figure 3.31a and b are polarized micrographs of recovered sample of shot #0402 having a $134\,\mathrm{m\ s^{-1}}$ projectile velocity and a 20° inclination angle [65]. The impact surface is near the top of the micrographs. It can be seen in Fig. 3.31a that near the impact surface there is a melting layer of about 6–8 μm in thickness, which impedes the shear stress propagating into the interior of the sample. The interesting

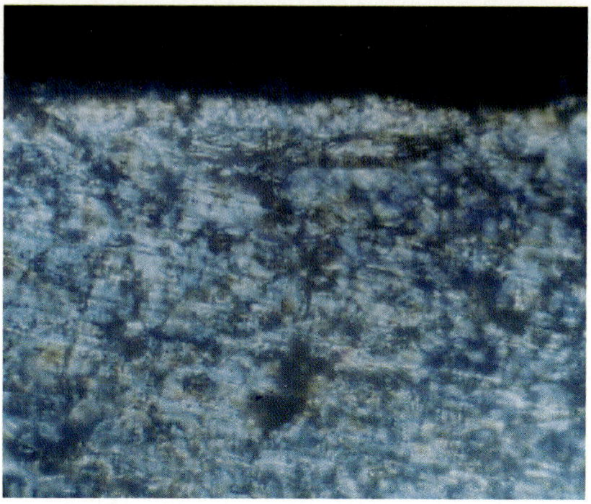

Fig. 3.30. Spherical grain configuration of nylon-66 sample before experiment.

thing is that there is a discontinuous crystalline layer about 2–3 μm thick above the melting layer, which means the melting may not be directly caused by the friction on the impact surface and the heat may be produced inside of the sample and near the surface. Figure 3.31b, taken from the same sample as Fig. 3.31a, but near the lateral side, shows an adiabatic shear band about 1–2 μm thick parallel and close to the surface, which reveals shear failure of the material. The heat produced by the localized thermal softening and plastic flow will cause the material to melt first in a shear band, since nylon-66 has a low melting point (250–260°C) and low heat conductivity.

Figure 3.31c is the polarized microscopic picture of the recovered sample of shot #0404 having a 156 m s^{-1} projectile velocity and a 20° inclination angle. It shows the same characteristics as in Fig. 3.31a, but the unmelted layer at the impact surface has less continuity. A one-dimensional heat transfer simulation shows that the melt layer produced by the shear flow energy is about 7 μm thick, which is consistent with the experimental results.

These microscopic level observations can support Gupta's experimental findings and conjectures [40], i.e., the shear particle velocity at the impact surface is also governed by the material response in the vicinity of the impact surface (e.g., shear strength), which is different from the mechanism of metals discussed in Sect. 3.4.1. Formation of the adiabatic shear band and melt layer underneath the impact surface causes a large attenuation of the shear wave propagating into the sample. With the increase of impact velocity the unmelted layer near the surface becomes thinner and less continuous, and the surface gauge signal becomes unstable and greatly decreases.

Gupta also used the EMV gauge technique and the keyed gas gun facility to produce and measure large-amplitude, one-dimensional compression and shear waves in two elastomers, the polyurethane elastomer Solithane 113 [66] and the energetic elastomer BAMO/THF (50/50), a copolymer of

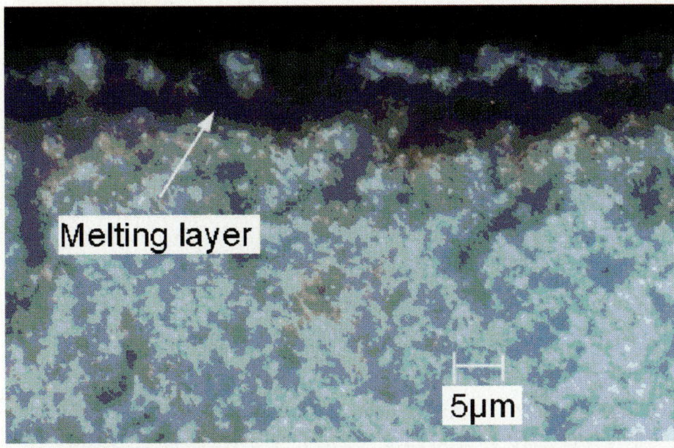

(a) Central part of shot 0402 (134 m s^{-1}, 20°)

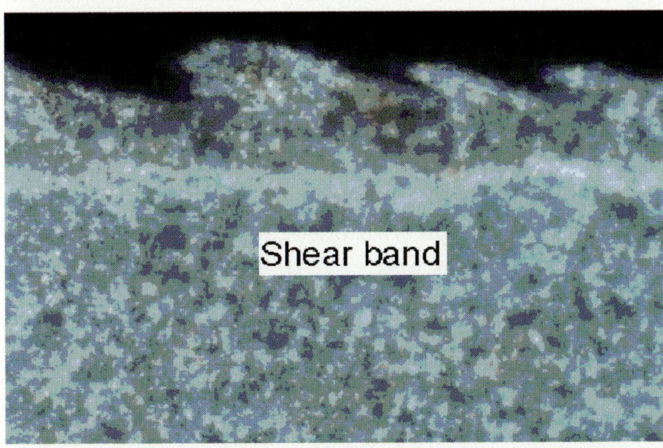

(b) Lateral part of shot 0402 (134 m s^{-1}, 20°)

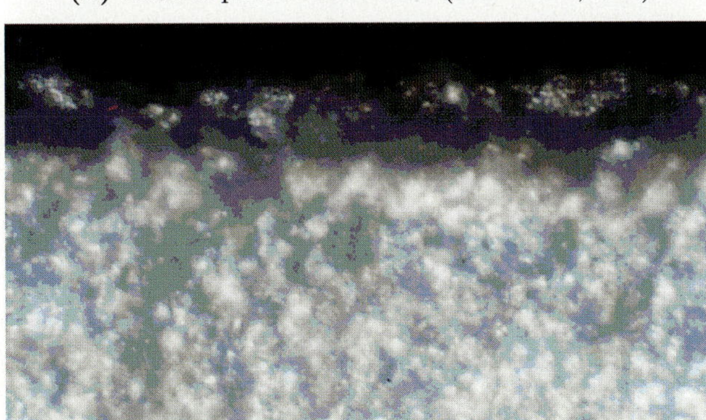

(c) Central part of shot 0404 (156 m s^{-1}, 20°)

Fig. 3.31. Polarized microscopic pictures of recovered samples of shot 0402 with projectile velocity 134 m s^{-1} (**a, b**), and shot 0404 with projectile velocity 156 m s^{-1}. The inclination angle was 20°.

3,3-bis (azidomethyl) oxetane (BAMO) and tetrahydrofuran (THF) [67]. The Solithane data were analyzed to provide longitudinal and shear wave velocities, and the corresponding stress–strain curves on the microsecond time scale. The shear modulus varies between 3 and 9 kbar for compressive stresses ranging between 2 and 14 kbar. In contrast, shear modulus values inferred from quasi-static measurements of bulk and Young's modulus under static high pressure are negligible. These comparisons suggest a high rate-dependence of shear modulus in the glassy state. The shear stress–strain curves suggest an elastic–plastic response with yield strength from 0.12 kbar to 0.25 kbar, with increasing compression, for the compressive range investigated.

In BAMO the shear to bulk modulus ratio was found to be very small, $G/K \ll 1$, over the entire compressive stress range; BAMO/THF behaves like a viscous fluid under shock loading. This result was applied to examine the shock-induced phase transition of CdS. By comparing the shock response of single-crystal CdS with that of CdS particles dispersed in the fluid-like matrix of BAMO/THF, Tang and Gupta [68] examined the role of deviatoric stress on the phase transition. Clifton [34] also suggested using shear waves to explore the effects of strength on phase transformations.

Boyle et al. [69] looked for effects of combined pressure and shear on reactivity in explosives. They performed parallel/inclined impact experiments in the pressure range from 0.33–31 GPa on thin (0.6–1.0 mm) explosive samples whose response was monitored with a high-speed framing camera through a transparent anvil to detect evidence of reaction. No explosive reaction was detected for the 15 μs duration of these tests.

3.4.3 Damage and Failure Investigations for Cementious Composites

To reveal the intrinsic rules of dynamic damage and failure for brittle materials subjected to shock loading, a so-called transversal Shear Wave Tracing technique (SWT) was proposed by Tang and coworkers [33] based on wave propagation theory and the combined compression–shear impact technique. The idea of the SWT method is to characterize the material state in real time by measuring the propagation features of the loading and unloading shear waves.

A Brief Introduction to SWT Method

In inclined, planar impact between an isotropic impact plate and an isotropic target plate with impact velocity u_0 and an inclination angle α to the x-axis (Fig. 3.14), there are normally four wave fronts moving successively in the target sample: loading longitudinal compression wave (P^+), loading transversal shear wave (S^+), unloading longitudinal wave (P^-), and unloading shear wave (S^-), as shown in Fig. 3.32 [33]. It should be noted that S^- is not generated

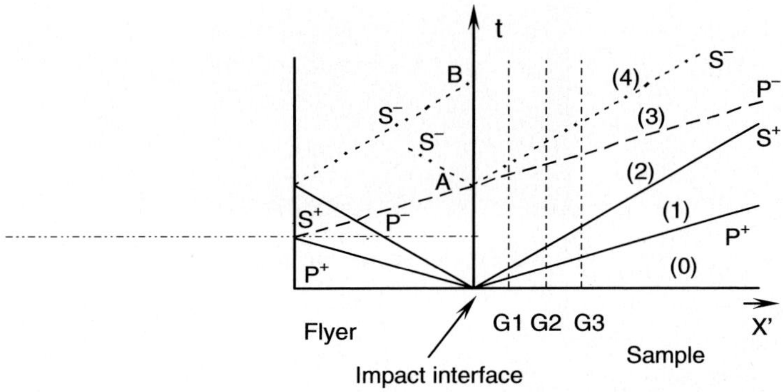

Fig. 3.32. Schematic diagram of wave fronts in inclined parallel plate impact experiment. G1, G2 and G3 denote the locations of in-material particle velocity gauges 1, 2 and 3, respectively (cf Fig. 3.14). P^+, S^+, P^- and S^- are loading compression longitudinal wave, loading shear transverse wave, unloading longitudinal wave and unloading shear wave, respectively. (0) $\sim$ (4) are regions defined by these four wave fronts.

from point B in Fig. 3.32, but is generated as soon as P^- reaches the impact surface and reduces the normal stress there to zero.

These four waves separate the sample into five regions, (0) to (4) in Fig. 3.32. According to stress wave propagation theory, the propagation feature of each wave depends on the wave amplitude and the real-time material characteristics of the region ahead of it, which result from the effect of each preceding wave. Hence, tracking the S wave fronts behind each P wave permits characterization of the state produced by each P wave. Since these wave profiles are recorded in one impact experiment, the measurements are *in situ* and time resolved. The SWT method can even detect the attenuation of the P^+ wave and the variation of damage or failure with distance, which makes it especially useful for studying impact damage.

Experimental Investigation

Cement mortar was made up with the weight proportioning: water/cement/standard sand = 220 g/500 g/1,250 g. Each cylindrical specimen of 50 mm diameter and 30 $\sim$ 40 mm in length was sawn in half down its axis. A multi-EMV-gauge was affixed to the axial surface, then the two halves were rejoined with epoxy resin (Fig. 3.33) [33].

Typical records from experiments with only the horizontal magnetic field applied are shown in Fig. 3.34 ($\theta = 0$, *cf*. Table 3.1), in which the impact velocity is 76 m s^{-1} in (a) and 243 m s^{-1} in (b) [33]. In Fig. 3.34a the complete S^- wave shows the elastic response. In contrast, in Fig. 3.34b the S^- wave totally vanishes and indicates the material has substantially failed.

Fig. 3.33. Specimen preparation.

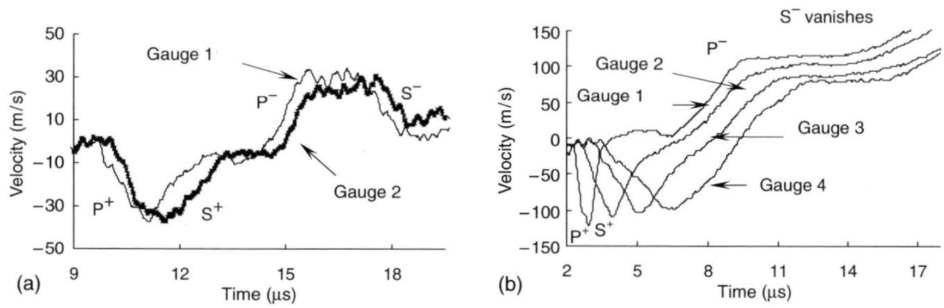

Fig. 3.34. Experimental records of particle-velocity-time profiles (**a**) Impact velocity $u_o = 76\,\mathrm{m\,s^{-1}}$ and (**b**) $u_o = 243\,\mathrm{m\,s^{-1}}$. S^- vanishes.

Figure 3.35 shows the P-V Hugoniot data for cement mortar based on both the normal impact (*) and inclined impact (♦) experiments [33]. Four critical points are shown in the figure, A, B, C and D. Point A corresponds to the elastic limit of the material; B corresponds to the vanishing of the S^- wave. Therefore we can take point B as the critical point of vanishing shear strength and the transition from a damage state to a failure state. It should be emphasized that there is no obvious cusp for the Hugoniot curve at B because the hydrostatic pressure during the loading process makes the critical point B very difficult to detect. Nonetheless, the SWT method can reveal clearly the occurrence of the failure state.

A remarkable phenomenon in the figure is the cusp at C; the Hugoniot shows a large volume reduction after C. The observed volume decrease from point C to point D corresponds well to collapse of the original 10.5% porosity of the material. Hence, we interpret point C as the critical point of void collapse. Point D is then on the compression curve of solid material.

3.4.4 Inclined Impact Surface Behavior Investigations

In recent times much attention has been focused on the mechanics of high-speed deformation with application to high-speed machining and grinding. To fully characterize the material behavior in such processes, the dynamic friction effects must be incorporated. As early as 1987, Ramesh and Clifton [70]

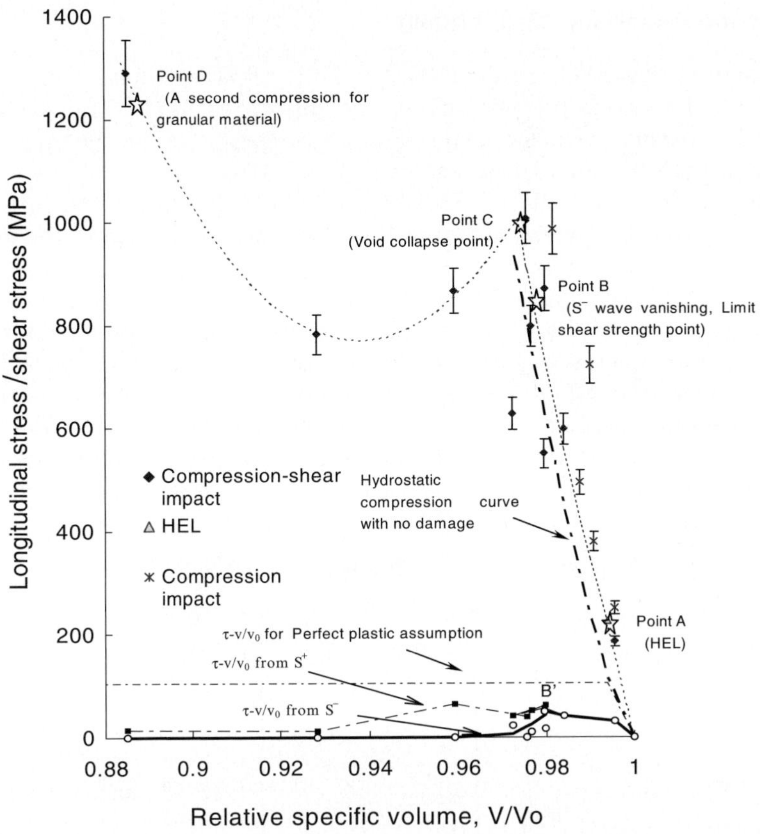

Fig. 3.35. Stress versus relative specific volume Hugoniot. 4 other curves are: Hydrostatic curve, $\tau\text{-}v/v_0$ for perfect plasticity, $\tau\text{-}v/v_0$ from S^+ and $\tau\text{-}v/v_0$ from S^-. A large drop of shear strength to zero at B′ on $\tau\text{-}v/v_0$ from S^- discloses the failure transition at B on the Hugoniot.

studied the dynamic properties of lubricants. Prakash and Clifton [71] suggested using the pressure-shear plate impact technique to investigate dynamic friction behaviors. The principal advantage of this technique is that it uses plane compression-shear waves to load the interface, resulting in a rapidly-attained steady state sliding condition with an effectively infinite loading stiffness [72]. Consequently, changes in kinetic friction can be measured for arbitrary loading history, and slip can be resolved to 30 nm. Because the total slip distance is approximately 40 μm, no significant wear particles are generated, making this experiment ideal for probing intrinsic kinetic friction values. However, this method requires that the constitutive response of the plates is known. Hence, it is most applicable in the loading range where the solid materials remain elastic. The analysis here assumes linear elastic response.

Experimental Setup and Theory

Irfan and Prakash [73] studied the dynamic friction behavior of CH tool steel/Ti-6Al-4V tribo-pair with this plate impact technique. The normal and transverse particle velocity histories of the free surface of the target were measured by NDI and TDI, as shown in Fig. 3.10a.

The wave propagation in the target and flyer is illustrated schematically in the time–distance diagram shown in Fig. 3.36 [72]. In the figure, the longitudinal wave fronts are represented with solid lines while the shear wave fronts are represented with dashed lines. It can be seen that the friction interface is sampled in three states during impact. State 1: the tribo-pair interface is under a compression stress σ_1 and a shear stress τ_1. State 1 allows the investigation of the dynamic sliding characteristics of the frictional interface under constant normal pressure. State 2: the longitudinal unloading wave from the free surface arrives at the frictional interface, reducing the compression from σ_1 to σ_2, and the shear from τ_1 to τ_2. It allows investigation of the dynamic

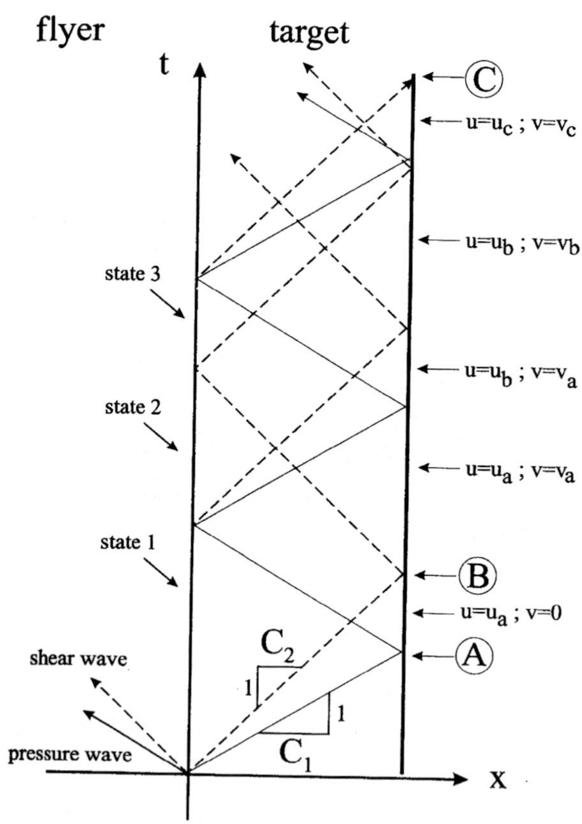

Fig. 3.36. Wave propagation in the flyer and target plates: time distance diagram showing the position of the stress wave fronts during the experiment.

sliding characteristics of a frictional interface subjected to a step change in normal stress. State 3: the shear wave reflects from the free surface and arrives at the frictional interface, changing the applied shear stress while maintaining the normal stress.

Using the method of characteristics for one-dimensional hyperbolic equations, the components of traction at the frictional interface, the slip velocity, and the accumulated displacement can each be related to the normal, $u_{\mathrm{fs}}(\mathrm{t})$, and transverse, $v_{\mathrm{fs}}(\mathrm{t})$, components of the measured free surface particle velocity histories of the target plate and the longitudinal, (ρc_1), and shear impedances, (ρc_2).
State 1:

$$\tau(t) = -(\rho c_2)_T \; v_{fs}(t)/2, \text{ and} \tag{3.61}$$

$$\sigma(t) = (\rho c_1)_T \; u_{fs}(t)/2, \tag{3.62}$$

where the subscript T denotes the target plate. When slipping occurs at the flyer-target interface, the measured free surface velocity of the target plate can be used to obtain the coefficient of kinetic friction,

$$\mu_k(t) = |\tau/\sigma| \,. \tag{3.63}$$

To calculate the interfacial slip velocity, the transverse particle velocity versus shear stress diagram is used, as shown in Fig. 3.37 [73]. For a no-slip condition, the state of the interface is denoted with the letter A. If the interface slip velocity is V_{slip}, the transmitted shear stress at the flyer-target interface reduces from τ_A to τ^*. Consequently, the transverse particle velocities on the target and flyer plates are represented with V_B and V_C, respectively. Then,

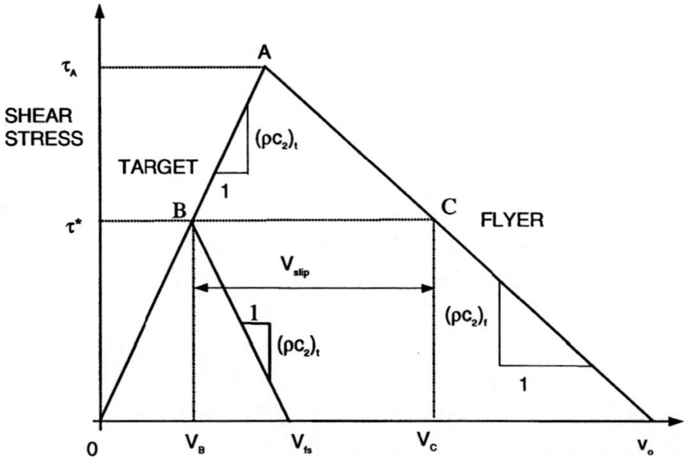

Fig. 3.37. Loci of all shear stress and transverse particle velocity states at the tribo-pair interface.

from the knowledge of the impact velocity V, the inclination angle α, the shear impedances of the flyer and target, and the measured free surface transverse particle velocity history, the slip velocity can be expressed as

$$V_{slip} = V_C - V_B = V \sin\alpha - \frac{(\rho c_2)_T + (\rho c_2)_F}{2(\rho c_2)_F} v_{fs}(t). \tag{3.64}$$

The accumulated slip distance can be obtained from (3.64) by integrating the slip velocity in time

$$\delta_{slip} = \int_0^t V_{slip}(t)dt. \tag{3.65}$$

State 2:

When the longitudinal compression wave reflects from the free surface of the target plate, it reduces the compressive stress at the interface from σ_1 to σ_2

$$\sigma_2 = \frac{[(\rho c_1)_F - (\rho c_1)_T]}{[(\rho c_1)_F + (\rho c_1)_T]}\sigma_1. \tag{3.66}$$

The corresponding expressions for the frictional stress and slip velocity can be obtained as

$$\tau_2(t) = -(\rho c_2)_T \, v_B(t)/2 \quad \text{and} \tag{3.67}$$

$$V_2^{slip} = V \sin\theta - \left[\frac{(\rho c_2)_T + (\rho c_2)_F}{(\rho c_2)_F}\right] v_B(t). \tag{3.68}$$

State 3:

$$\tau_3(t) = -(\rho c_2)_T (v_C(t) - v_A(t))/2, \tag{3.69}$$

$$\sigma_3 = \sigma_2, \quad \text{and} \tag{3.70}$$

$$V_3^{slip} = V \sin\theta + \left[\frac{(\rho c_2)_T - (\rho c_2)_F}{(\rho c_2)_F}\right] v_A(t) - \left[\frac{(\rho c_2)_T + (\rho c_2)_F}{(\rho c_2)_F}\right] v_C(t). \tag{3.71}$$

Experimental Results and Discussion

Table 3.2 summarizes the experimental results for the Ti-6Al-4V and CH tool steel tribo-pair. Ti-6Al-4V was used as the flyer plate and CH tool steel was used as the target plate. Table 3.3 lists the material parameters used for calculation.

Experiments 9,701 and 9,702 were designed to study the effect of different normal pressures on the interfacial friction resistance of sliding interfaces. Experiments 9,703 and 9,704 were used mainly to study the effect of normal pressure on interfacial frictional behavior when the surface roughness of the CH tool steel plate increased substantially. Experiment 9,801 was conducted to study the increase in surface roughness of the Ti-6Al-4V plate on the friction

Table 3.2. Summary of plate impact pressure shear friction experiments [73] .

Shot No	Skew angle θ (degree)	Impact velocity (m s^{-1})	Normal stress (GPa)	Ti-6Al-4V roughness (μm)	CH tool steel roughness (μm)
9701	35	76	1.017	0.1	0.02
9702	35	95	1.292	0.07	0.02
9703	35	102.7	1.374	0.09	0.12
9704	35	80.5	1.078	0.09	0.11
9801	35	79.2	1.059	0.18	0.10

Table 3.3. Properties of Ti-6Al-4V and CH tool steel [73].

Physical properties	Ti-6Al-4V	CH tool steel
Young's modulus (GPa)	110	207
Poisson's ratio	0.33	0.3
Longitudinal wave speed (km s^{-1})	6.255	5.98
Shear wave speed (km s^{-1})	3.151	3.264
Mass density (kg m^{-3})	4,400	7,860
Material hardness (HRC)	30	62
Thermal conductivity, k (W $(m\ K)^{-1}$)	6	42.3
Thermal diffusivity, α ($m^2\ s^{-1}$)	2.6×10^{-6}	12.2×10^{-6}
Specific heat, C_p (J $(Kg\ K)^{-1}$)	522	422

resistance of the tribo-pair. Here we just introduce the experimental results and corresponding discussion of Experiment 9,702.

Figure 3.38 shows the normal and transverse free surface particle velocity histories of the target plate of 9,702 [73]. The first arrival of the shear wave at the free surface of the target plate occurs at approximately 1.9 μs. Then there is a drop in the transverse velocity at about 3.6 μs, which is due to the step drop in normal pressure and indicates the beginning of State 2. The following rise at about 5.25 μs indicates the beginning of State 3. The dashed horizontal line in the figure represents the elastic prediction for the transverse particle velocity at the free surface of the target plate with no slip at the tribo-pair interface. It can be seen that the levels of the transmitted particle velocity in the three states are much lower than their corresponding no-slip predictions, so the frictional interface is understood to be in a state of dynamic slip.

Figure 3.39 shows the histories of the interfacial normal stress, friction stress, and interfacial slip velocity [73]. These quantities were calculated from the measured normal and transverse free surface particle velocity profiles shown in Fig. 3.38 using (3.61)–(3.71). Figure 3.40 shows the variation of the slip velocity and the coefficient of kinetic friction with the accumulated slip distance [73]. These figures demonstrate that during State 1, the normal pressure σ_1 at the frictional interface, the transmitted shear stress τ_1, the slip velocity V_1^{slip} and the coefficient of kinetic friction μ_{k1} are nearly

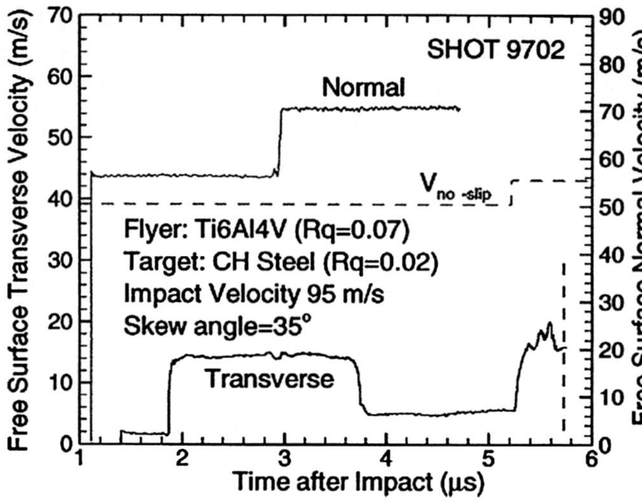

Fig. 3.38. Normal and transverse free surface particle velocity histories of the target plate of 9702.

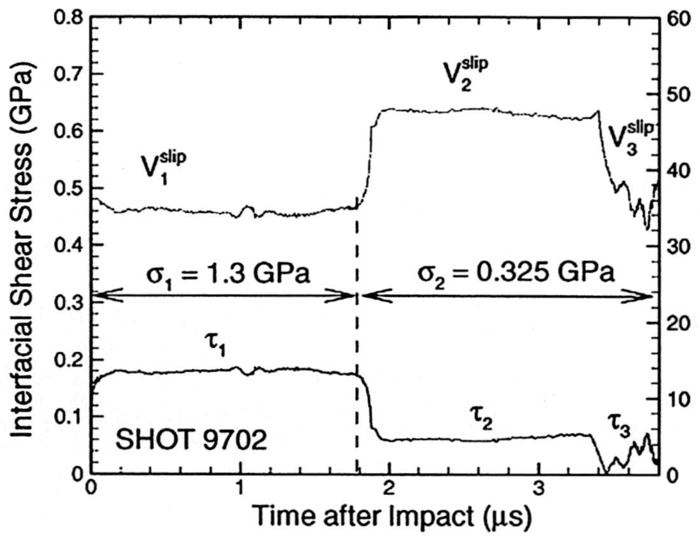

Fig. 3.39. Interfacial normal pressure, friction stress and slip velocity history of 9702.

steady at 1.3 GPa, 0.18 GPa, 34 m s^{-1} and 0.14, respectively. This value of μ_{k1} is relatively low as compared to the reported value of 0.36 for the coefficient of friction between ground steel and Ti-6Al-4V specimens under 200 g of load [74]. The possible reasons are the high normal pressure, high slip speed and small slip distance.

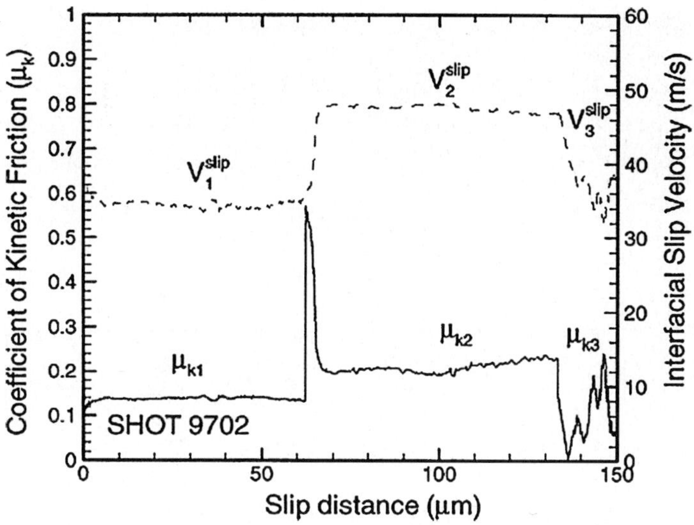

Fig. 3.40. Coefficient of kinetic friction and slip velocity as a function of interfacial slip distance of 9702.

During the transition from State 1 to State 2, the normal pressure reduces to $\sigma_2 = 0.325\,\text{GPa}$ in a very short time (about 5–10 ns). It is interesting to note that the corresponding friction stress τ_2 does not follow this step change in normal pressure. Instead, the friction stress gradually approaches a new steady state ($\tau_2 = 0.06\,\text{GPa}$, $V_2^{\text{slip}} = 48\,\text{m s}^{-1}$, $\mu_{k2} = 0.22$). This causes an instantaneous jump in the value of coefficient of friction as shown in Fig. 3.40. This evolution of friction stress in response to a step change in normal pressure provides insight into the evolution of the effective area of contact and introduces a critical relaxation time to the dynamic friction process. Additionally, the steady coefficient of kinetic friction μ_{k2} is much higher than μ_{k1}, which means that the steady-state friction stress may not decrease in proportion to the drop in normal pressure during high speed sliding. However, this result was not observed in the similar plate impact pressure-shear friction experiments for WC/Ti-6Al-4V tribo-pair [75]. In these experiments the level of the steady state coefficient of kinetic friction before and after the step decrease in compression were approximately the same, implying that the steady-state friction stress changes approximately linearly with the applied normal pressure. These results are important for understanding the mechanics of high speed machining of Ti-6Al-4V alloys.

The bulk temperature rise at the tribo-pair interface can be estimated by solving the one-dimensional transient heat conduction equations with the heat generated at the interface assuming that all the frictional work is converted to heat [73],

$$\dot{q}_{Ti-6Al-4V}(t) = \beta\tau(t)V_{slip}(t), \text{ and} \tag{3.72}$$

$$\dot{q}_{CHsteel}(t) = (1-\beta)\tau(t)V_{slip}(t), \tag{3.73}$$

where β governs the partitioning of heat in the tribo-pair materials. The fraction β is estimated by equating the temperatures at the tribo-pair interface to yield

$$\beta = \frac{k_1\sqrt{\alpha_2}}{k_2\sqrt{\alpha_1} + k_1\sqrt{\alpha_2}}, \tag{3.74}$$

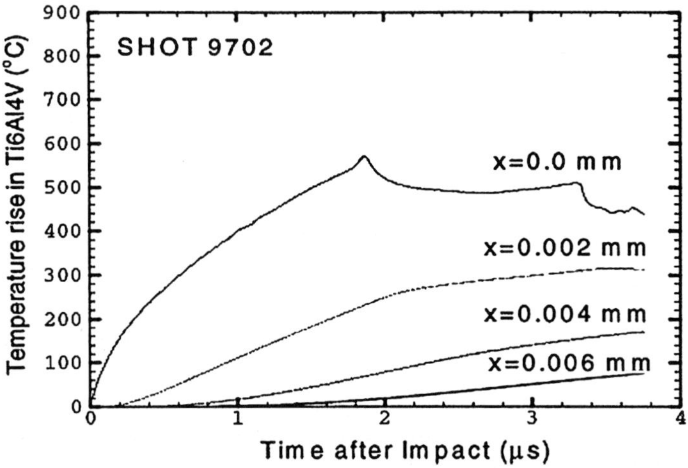

(a)

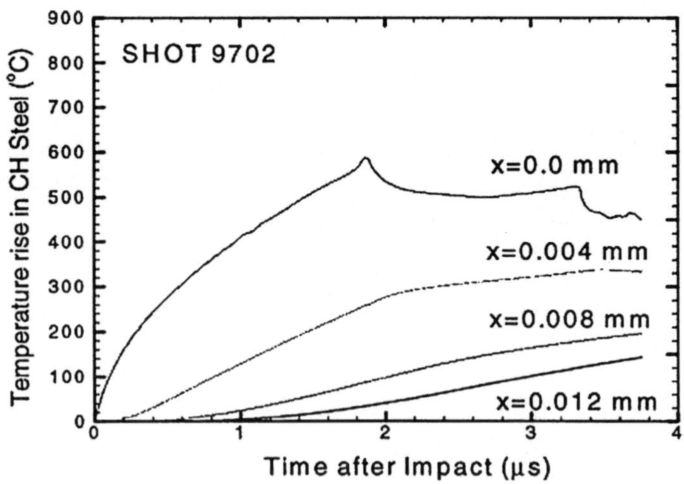

(b)

Fig. 3.41. Computed temperature rise for 9702, (**a**) in Ti-6Al-4V, (**b**) in CH tool steel.

where k_1, α_1, k_2 and α_2 are the thermal conductivity and thermal diffusivity of Ti-6Al-4V and CH steel, respectively, and are listed in Table 3.3. The calculated temperature rise and distribution with time for 9,702 are shown in Fig. 3.41 [73]. It can be seen that the maximum temperature rise is about 600°C with a very thin layer at the frictional interface, the thickness of which is of the order of the surface roughness of the tribo-pair materials.

Other experiments were conducted by varying the surface roughness and impact velocity. Moreover, by appropriate selection of the thickness of flyer and target plates, the tribo-pair interface was subjected to step changes in normal pressure and step changes in applied shear stress. The experimental results provide insight into time-resolved dry sliding characteristics of metal on metal at normal pressure of about 1.5 GPa, slip speeds up to 60 m s^{-1} and interfacial temperature as high as 800°C.

The compression-shear plate impact experimental technique was used for investigating transient friction behavior for other materials such as WC/4,340 tribo-pair [75, 76], and WC/4,340 steel tribo-pair at high temperature (300°C) [71]. The plate impact friction experiments represent a significant improvement over conventional dynamic friction experiments by allowing control of interfacial tractions with the use of combined pressure–shear loading waves instead of manipulating actuator motion. Also, by measuring the combined normal and transverse motion of the rear surface of the target plate, critical frictional parameters such as the applied normal pressure, the interfacial slip distance, and the interfacial slip speeds can be interpreted by using the framework of one-dimensional plane wave analysis. The results of these experiments provide new insights into the evolution of interfacial sliding resistance with accumulated interfacial slip, and its dependence on surface roughness, slip velocity, normal pressure and interfacial temperature.

3.5 Summary

Inclined parallel plate impact experiments, variously known as pressure-shear or compression-shear experiments, along with their associated time-resolved observation methods, have enabled a range of scientifically and technologically interesting investigations of dynamic material behavior. These experiments are of fundamental value for enabling more detailed characterization of the shocked state and for providing a controlled means of exercising the sample material along a dynamic loading path that is distinctly different than the principle Hugoniot. Only recently has the Isentropic Compression Experiment (ICE) [77, 78] provided another method of off-Hugoniot, controlled dynamic loading with good diagnostics.

While alternate methods of producing combined states of dynamic compression and shear were tried, only the compression-shear methods developed independently by Clifton and Gupta, and their respective coworkers, have found extensive application. The reasons for this are, first, only these methods provide controlled states of compression and shear in which the planar longitudinal and shear wave fronts are parallel, which greatly simplifies quantitative analysis. The second reason is the effective measurement techniques the two groups developed.

Both groups used slotted compressed gas guns and a key protruding from the sabot that slides down this slot, thereby maintaining the angular orientation of the inclined flyer plate. Clifton, Abou-Sayed, and Kim devised pressure-shear experiments in which interferometry was used to independently monitor the normal and transverse motion of the rear free surface of the target. This measurement method has been applied successfully to characterize dynamic metal plasticity, interfacial friction, and lubricant response. The method developed by Gupta and Keough instead used embedded particle velocity gauges to measure selected components of Lagrangian particle velocity in the sample. This approach has proven to be very useful for investigating the dynamic response of insulating materials, including the dynamic mean stress curve, solid state phase transformations, and high strain rate deformation of polymers.

These experimental developments were preceded by nearly two decades of theoretical study of combined stress wave loading, which have continued to the present. Theoretical descriptions exist for combined stress loading wave propagation in hyperelastic, elastic–plastic, and elastic–viscoplastic materials.

Together, the accumulated theory and experimental methods continue to offer opportunities for investigating and understanding additional aspects of dynamic material response under combined compression and shear loading. This is illustrated by Prakash's investigations of sliding interfaces and Tang's recent method for better characterizing damage in brittle solids. Additional areas in need of investigation include experimental investigation of the effects of pre-stress on combined compression–shear loading, which have already been addressed by theory, and detailed analysis of shear wave amplitude histories measured in non-metals to characterize their inelastic constitutive response [45, 55].

Acknowledgment

John H. Carpenter is gratefully acknowledged for his careful reading and critical comments on the manuscript. Sandia National Labortories is a multiprogram laboratory operated by Sandia Corporation, a Lockheed Martin Company, for the United States Department of Energy's National Nuclear Security Administration under contract DE-AC04-94AL85000.

References

1. Gupta, Y.M.: Shock wave experiments at different length scales: recent achievements and future challenges. In: Furnish, M.D., Chhabildas, L.C., Hixson, R.S. (eds.) Shock Compression of Condensed Matter – 1999, pp. 3–10. AIP, New York (2000)
2. Bridgman, P.W.: Effects of high shearing stress combined with high hydrostatic pressure. Phys. Rev. **48**, 825 (1935)
3. Bridgman, P.W.: Shearing phenomena at high pressure of possible importance for geology. J. Geol. **44**, 653 (1936)
4. Bridgman, P.W.: Shearing phenomena at high pressures, particularly in inorganic compounds. Proc. Am. Acad. Arts Sci. **71**, 387 (1937)
5. Courant, R., Hilbert, D.: Methods of Mathematical Physics, vol. 2. Interscience Publishers, New York (1962)
6. Jeffrey, A., Taniuti, T.: Nonlinear Wave Propagation with Application to Physics and Magnetohydrodynamics. Academic Press, New York (1964)
7. Rakhmatulin, Kh. A.: On the propagation of elastic–plastic waves owing to combined loadings. J. Appl. Math. Mech. (Prikladnaya Matematika I Mekhanika), **22**, 1079–1088 (1958)
8. Cristescu, N.: On the propagation of elastic–plastic waves for combined stresses. J. Appl. Math. Mech. (Prikladnaya Matematika I Mekhanika), **23**, 1605–1612 (1959)
9. Bleich, H.H., Nelson, I.: Plane waves in an elastic–plastic half-space due to combined surface pressure and shear. Trans. ASME J. Appl. Mech. **33**, 149 (1966)
10. Clifton, R.J.: An analysis of combined longitudinal and torsional plastic waves in a thin-walled tube. Fifth U.S. National Congress of Applied Mechanics, pp. 465–480, ASME (1966)
11. Ting, T.C.T., Nan, N.: Plane waves due to combined compressive and shear stresses in a half space. Trans. ASME J. Appl. Mech. **36**, 189–197 (1969)
12. Ting, T.C.T.: A unified theory on elastic–plastic wave propagation of combined stress. In: Sawczuk, A. (ed.) Proceedings of IUTAM Symposium on Foundations of Plasticity, pp. 301–316, Noordhoff Int. Pub., Leyden (1973)
13. Nowacki, W.K.: Stress Waves in Non-Elastic Solids. Pergamon Press, Oxford (1978)
14. Lipkin, J., Clifton, R.J.: An experimental study of combined longitudinal and torsional plastic waves in a thin-walled tube, pp. 292–304. 12th International Congress of Applied Mechanics, Stanford University (1968)
15. Lipkin, J., Clifton, R.J.: Elastic waves of combined stresses due to longitudinal impact of a pretorqued tube, Part 1: Experimental results and Part 2: Comparison of theory with experiment. J. Appl. Mech. **37**, 1107–1120 (1970)
16. Johnson, J.N.: Shock propagation produced by planar impact in linearly elastic anisotropic media. J. Appl. Phys. **42**, 5522 (1971)
17. Chhabildas, L.C., Swegle, J.W.: Dynamic pressure-shear loading of materials using anisotropic crystals. J. Appl. Phys. **51**, 4799 (1980)
18. Abou-Sayed, A.S., Clifton, R.J., Hermann, L.: The oblique-plate impact experiment. Exp. Mech. 127–132 (1976)
19. Gupta, Y.M.: Shear measurements in shock-loaded solids. Appl. Phys. Lett. **29**, 694 (1976)

20. Gupta, Y.M., Keough, D.D., Walter, D.F., Dao, K.C., Henley, D., Urweider, A.: Experimental facility to produce and measure compression and shear waves in impacted solids. Rev. Sci. Instrum. **51**(2), 183–210 (1980)
21. Kim, K.S., Clifton, R.J., Kumar, P.: A combined normal- and transverse-displacement interferometer with an application to impact of y-cut quartz. J. Appl. Phys. **48**(10), 4132–4139 (1977)
22. Karpp, R., Chou, P.C.: The method of characteristics, Chapter 6. In: Chou, P.C., Hopkins, A.K. (eds.) Dynamic response of materials to intense impulsive loading, (U.S. Government, 1973), pp. 283–362. Available from NTIS, order number AD768416. Library of Congress No. 73-600247 (1973)
23. Abou-Sayed, A.S., Clifton, R.J.: Pressure shear waves in fused silica. J. Appl. Phys. **47**, 1762 (1976)
24. Malvern, L.E.: Introduction to the mechanics of a continuous medium. Prentice-Hall, Englewood Cliffs, NJ (1969)
25. Ting, T.C.T.: The initiation of combined stress waves in a thin-walled tube due to impact loading. Int. J. Solids Struct. **8**, 269–293 (1972)
26. Abou-Sayed, A.S., Clifton, R.J.: Analysis of combined pressure–shear waves in an elastic/visco-plastic material. J. Appl. Mech. **44**, 79 (1977)
27. Hill, R.: The Mathematical Theory of Plasticity. Clarendon Press, Oxford (1950)
28. Ting, T.C.T.: Nonlinear Waves in Solids. Friendship Press, Beijing (1985, in Chinese)
29. Mashimo, T., Ozaki, S., Nagayama, K.: Keyed powder gun for the oblique-impact shock study of solids in several 10s of GPa region. Rev. Sci. Instrum. **55**, 226 (1984)
30. Koller, L.R., Fowles, G.R.: Compression-shear waves in Arkansas novaculite. In: Timmerhaus, K.D., Barber, M.S. (eds.) High Pressure Science and Technology, Proceedings of Sixth AIRAPT Conference, Boulder CO, 1977, vol. 2, p. 927. Plenum Press, New York (1979)
31. Young, C., Dubugnon, O.: A reflected shear-wave technique for determining dynamic rock strength. Int. J. Rock Mech. Min. Sci. Geomech. Abstr. **14**, 247–259 (1977)
32. Clifton, R.J., Klopp, R.W.: Pressure-shear plate impact testing. In: Newby, J.R., Davis, J.R., Refsnes, S.K. (eds.) American Society of Metals Handbook, vol. 8, pp. 230–239 (1985)
33. Tang, Z.P., Xu, S.L., Dai, X.Y., Hu, X.J., Liao, X.L., Cai, J.: S-wave tracing technique to investigate the damage and failure behavior of brittle materials subjected to shock loading. Int. J. Impact Eng. **31**(9), 1172–1191 (2005)
34. Clifton, R.J.: Shear waves and phase transformations. In: Dunn, J.E., Fosdick, R., Slemrod, M. (eds.) Shock Induced Transitions and Phase Structures in General Media, IMA, vol. **52**, p. 45. Springer-Verlag, Berlin New York Hiedelberg (1998)
35. Kolsky, H.: An investigation of the mechanical properties of materials at very high rates of loading. Proc. Phys. Soc. Lond., Series B **62**, 676–700 (1949)
36. Chhabildas, L.C.: Pressure-shear loading of materials. In: Schmidt, S.C., Holmes, N.C. (eds.) Shock Waves in Condensed Matter-1987, p. 579. Elsevier, Amsterdam (1988)
37. Swegle, J.W.: A suggested technique for determining in-material longitudinal and shear particle velocity histories in a single-inclined-plate impact experiment. J. Appl. Phys. **49**, 4280 (1978)

38. Aidun, J.B., Gupta, Y.M.: Simultaneous measurement of in-material longitudinal and transverse particle velocity histories in a compression-shear experiment. J. Appl. Phys. **65**, 1898 (1989)
39. Gupta, Y.M.: Measurement of compression and shear waves in an impact experiment role of gauge leads in particle-velocity measurements. J. Appl. Phys. **51**, 1835 (1980)
40. Gupta, Y.M.: Determination of impact response of PMMA using combined compression and shear loading. J. Appl. Phys. **51**, 5352–5361 (1980)
41. Barker, L.M., Hollenbach, R.E.: Laser interferometer for measuring high velocity of any reflecting surface. J. Appl. Phys. **43**, 4669–4675 (1972)
42. Kim, K.S., Clifton, R.J.: Pressure-shear impact of 6061-T6 aluminum. J. Appl. Mech. **47**, 11 (1980)
43. Espinosa, H.: Dynamic compression-shear loading with in-material interferometric measurements. Rev. Sci. Instrum. **67**(11), 3931–3939 (1996)
44. Mello, M., Prakash, V., Clifton, R.J.: Multi-point interferometer for monitoring two dimensional wave motion. In: Schmidt, S.C., Dick, R.D., Forbes, J.W., Tasker, D.G. (eds.) Shock Compression of Condensed Matter-1991, pp. 763–768. AIP, New York (1992)
45. Gupta, Y.M.: Development of a method for determining dynamic shear properties. Draft Final Technical Report to the Defense Nuclear Agency under contract No. DNA 001-76-C-0384, (Unpublished, available from: Librarian, Poulter Laboratory, SRI International, 333 Ravenswood Avenue. Menlo Park, CA 94025) (1978)
46. Abou-Sayed, A.S., Clifton, R.J.: Pressure-shear waves in 6061-T6 aluminum due to oblique-plate-impact. J. Appl. Mech. **44**, 85 (1977)
47. Gilat, A., Clifton, R.J.: Pressure-shear waves in 6061-T6 aluminum and alpha-titanium. J. Mech. Phys. Solids **33**, 263 (1985)
48. Gilat, A.: A viscoplastic theory with anisotropic hardening and its application to pressure-shear plate impact experiments. J. Appl. Mech. **52**, 629–633 (1985)
49. Frutschy, K.J., Clifton, R.J.: High-temperature pressure-shear plate impact experiments using pure tungsten carbide impactors. Exp. Mech. **38**, 116–125 (1998)
50. Frutschy, K.J., Clifton, R.J.: High-temperature pressure-shear plate impact experiments on OFHC copper. J. Mech. Phys. Solids **46**(10), 1721–1743 (1998)
51. Tong, W., Clifton, R.J., Huang, S.: Pressure-shear impact investigation of strain rate history effects in Oxygen-free high-conductivity copper. J. Mech. Phys. Solids **46**(6), 1251–1294 (1992)
52. Duvall, G.E., Graham, R.A.: Phase transitions under shock-wave loading. Rev. Mod. Phys. **49**, 523 (1977)
53. Aidun, J.B.: Study of shear and compressive waves in shocked calcium carbonate. Ph. D thesis, Washington State University, Pullman, WA (1989)
54. Aidun, J.B., Gupta, Y.M.: Shear wave measurements for improved characterization of shock-induced phase transformation in carrara marble. Geophys. Res. Lett. **16**, 191–195 (1989)
55. Aidun, J.B., Gupta, Y.M.: Shear and compression waves in shocked calcium carbonate. J. Geophys. Res. **100**, 1955 (1995)
56. Grady, D.E.: Compression wave studies in Oakhall limestone. Sandia National Laboratory Report SAND83-0370, Albuquerque, NM (1983)
57. Wang, C.-Y., Meltzer, M.: Propagation of sound waves in a rock undergoing phase transformations. J. Geophys. Res. **78**, 1293 (1973)

58. Gupta, Y.M.: Shear and compression wave measurements in shocked polycrystalline Al_2O_3. J. Geophys. Res. **88**, 4304–4312 (1983)
59. Mashimo, T., Nagayama, K.: Shock compression and shear properties of sapphire (AlO). Mat. Res. Soc. Proc. Elsevier, **22**, 143–146 (1984)
60. Mashimo, T.: Measurements of the combined compression–shear shock waves of over 20 GPa in sapphire. In: Schmidt, S.C., Holmes, N.C. (eds.) Shock Waves in Condensed Matter-1987, pp. 285–287. Elsevier Science, Amsterdam, The Netherlands (1988)
61. Yuan, G., Feng, R., Gupta, Y.M.: Compression and shear wave measurements to characterize the shocked state in silicon carbide. J. Appl. Phys. **89**, 5372–5380 (2001)
62. Conner, M.P.: Shear Wave Measurements to Determine the Nonlinear Elastic Response of Fused Silica Under Shock Loading. M. S. Thesis, Washington State University, Pullman (1988)
63. Walley, S.M., Balzer, J.E., Proud, W.G., Field, J.E.: Response of thermites to dynamic high pressure and shear. Proc. R. Soc. Lond. A **456**, 1483–1503 (2000)
64. Zavattierti, P.D., Espinosa, H.D.: An examination of the competition between bulk behavior and interfacial behavior of ceramics subjected to dynamic pressure-shear loading. J. Mech. Phys. Solids **51**, 607–635 (2003)
65. Li, T., Tang, Z.P., Cai, J.: Experimental investigation of the failure mechanism near the impact surface of nylon-66 subjected to combined compression and shear shock loading. Mat. Lett. **61**, 1436–1438 (2007)
66. Gupta, Y.M.: High strain-rate deformation of a polyurethane elastomer subjected to impact loading. Polym. Eng. Sci. **24**, 851–861 (1984)
67. Gupta, Y.M.: Compression and shear wave propagation in plane and filled elastomer. Final Report to the Office of Naval Research under contract No. N00014-81-K-0840, Washington State University, (unpublished) (1986)
68. Tang, Z.P., Gupta, Y.M.: Shock-induced phase transformation in cadmium sulfide dispersed in an elastomer. J. Appl. Phys. **64**(4), 1827–1837 (1988)
69. Boyle, V.M., Frey, R.B., Bines, A.L.: Parallel/oblique impact on thin explosive samples. In: Schmidt, S.C., Tao, W.C. (eds.) SCCM-1995, pp. 819–822. Shock Compression of Condensed Matter-1995, AIP Conference Proc. 370, AIP Press, New York (1996)
70. Ramesh, K.T., Clifton, R.J.: A pressure-shear plate impact experiment for studying the rheology of lubricants at high pressures. ASME J. Tribol., vol. **109**, p. 215 (1987)
71. Prakash, V., Clifton, R.J.: Time resolved dynamic friction measurements in pressure-shear. In: Ramesh, K.T. (ed.) Experimental Techniques in the Dynamics of Deformable Bodies, vol. 165, pp. 33–47. New York, ASME, AMD
72. Frutschy, K.J., Clifton, R.J.: Plate-impact technique for measuring dynamic friction at high temperatures. J. Tribol. ASME **119**, 590–593 (1997)
73. Irfan, M., Prakash, V.: Time resolved friction during dry sliding of metal on metal. Int. J. Solids Struct. **37**, 2859–2882 (2000)
74. Bhushan, B., Gupta, B.K.: Handbook of Tribology: Material Coatings and Surface Treatments. McGraw Hill, NY (1991)
75. Prakash, V.: Frictional response of sliding interfaces subjected to time varying normal pressures. J Tribol. **120**, 97–102 (1998)
76. Prakash, V.: A pressure-shear plate impact experiment for investigating transient friction. Exp. Mech. **35**(4), 329–336 (1995)

77. Hall, C.A., Asay, J.R., Knudson, M.D., Stygar, W.A., Spielman, R.B., Pointon, T.D.: Experimental configuration for isentropic compression of solids using pulsed magnetic loading. Rev. Sci. Instrum. **72**(9), 3587–3595 (2001)
78. Reisman, D.B., Toor, A., Cauble, R.C., Hall, C.A., Asay, J.R., Knudson, M.D., Furnish, M.D.: Magnetically driven isentropic compression experiments on the Z accelerator. J. Appl. Phys. **89**(3), 1625–1633 (2001)

4

Dynamic Fragmentation of Solids

D. Grady

The topic pursued in the present chapter is that of dynamic fragmentation. Dynamic fragmentation is identified with both the process and the out-come of an event in which a monolithic structure or body is subjected to intense and abrupt forces causing it to break up into a number of component parts (fragments). Usually the concern is with the fragmentation of an entity composed of condensed matter; i.e., ductile or brittle solids, or a liquid. The concept of dynamic fragmentation is not alien to gaseous matter, however. One of the more thought provoking examples is that of the early universe theorized as initially composed of a nearly homogenous gas of elementary particles before density perturbations and instabilities in the gravitational attraction precipitate fragmentation evolving into the distribution and fabric of galaxies and galactic clusters observed today.

The more mundane topic to be addressed here, however, will be the dynamic fragmentation of condensed matter – more specifically, solid condensed matter. The dynamic fragmentation of liquids has been treated elsewhere (e.g., [34]).

Even within this restricted consideration of dynamic fragmentation, the subject is quite broad; both in material response, and in the kinematic conditions leading to fragmentation. The shatter of an impacted sheet of glass differs starkly from the rupture of an explosive-driven metal shell. Impact-induced spall bears no apparent similarity to the breakup of a stretching metal jet. The strain energy resident in a stone or concrete pillar subject to an excessive compressive load will cause the pillar to fail explosively and experience extensive dynamic fragmentation.

In a given dynamic fragmentation event there are a number of features that might be of concern. How many pieces did the body break into? What was the largest piece? What was the smallest piece? What was the disposition of the fragment debris in terms of momentum and kinetic energy? What was the stress, impulse or energy requirement needed to produce the fragmentation event? These and other issues are explored in the present chapter.

Section 4.1 will address probabilistic issues of dynamic fragmentation, including both historic, and more recent, representations of fragment distributions along with the arguments supporting the respective distributions. In Sect. 4.2 methods used to predict fragment size in a fragmentation event are surveyed. These include empirical methods as well as techniques that are based on stress, impulse, and energy driving the fragmentation process. Size prediction methods are also considered that are founded on the inherent flaw structure of engineering solids. Section 4.3 focuses on issues specific to the fragmentation of brittle solids. Some applications to fragmentation in the impact spall process are treated in Sect. 4.4.

4.1 Size Distributions and Probabilistic Methods in Dynamic Fragmentation

The violent disruption of a formerly contiguous body into a diverging collection of discrete fragments is a consequence of an intriguing mix of physics and statistics that is not well understood, even with the considerable progress of the past few decades. Dynamic fragmentation events range from the common place, such as explosive blasting of rock in the mining and quarry industry, to the exciting and esoteric issues of understanding the size distribution and clustering of the galaxies resulting from the big bang genesis of the universe.

The statistical nature of the sizes and velocities of fragments resulting from an explosive or impact event is self-evident. Less evident, however, is the mathematical and statistical framework most appropriate for the statistical prediction and description of a dynamic fragmentation event. The various approaches that have been pursued are considerable and no clear consensus has yet been achieved. Empiricism governed the statistical description of the size distribution of fragmentation events over most of the past century. More recently, a range of approaches has been pursued including geometric statistics, fractal physics, maximum entropy and information theory ideas, percolation theoretical methods and classical statistical mechanics.

Statistical issues in the dynamic fragmentation event are the principal concern in many engineering and scientific applications. Properties of the distribution in size and shape of the fragment debris resulting from such an event are commonly the foremost features sought. In the detonation of exploding munitions what is the number, the mass, and the velocity of the most lethal fragments? In the accidental or intentional detonation of a device containing hazardous material, what fraction of the disrupted material results in particulate of a size small enough to remain airborne?

Statistics of the fragmentation event will permeate all sections of the present article. This section reviews methods that pursue strictly probabilistic approaches to the fragmentation event. Physical issues regarding the propagation and interaction of fracture is largely ignored. It will be found in numerous cases that this probabilistic approach is quite adequate. In other very pertinent problems, this approach will be found lacking, and physics that is more detailed needs to be pursued.

It should be noted in the outset, however, that a generally accepted theory of statistical fragmentation has not yet emerged. Consequently, a wide number of empirical and semi-theoretical approaches are currently in active use in addressing the numerous applications involving the statistical disruption and fragmentation of condensed matter. There are a number of reasons for this.

First, physical and engineering applications of dynamic fragmentation are extremely diverse and an extensive literature on the subject going well back into the last century already exists. A number of empirical and semi-empirical theories for describing the consequences of fragmentation events are already comfortably entrenched in the various fields concerned with dynamic fragmentation. Portions of this literature have been reviewed by Grady and Kipp [42].

Second, it is difficult to perform experiments to select among different theories that meet the rather stringent statistical conditions assumed in the development of the theories. Methods exist for bringing the theories into accord with the more complex statistical conditions of the experiments, but uniqueness is lost and experiment no longer serves as a method for discrimination among theories.

Statistical complexities arise in experiments for several reasons: Fragmentation experiments are invariably statistically inhomogeneous. That is, the fragmentation intensity or average fragment size varies as a function of position within the fragmenting body because of spatial variations of the forces and energies causing fragmentation. An example is provided by the fragmentation caused when a stone strikes an untempered glass windshield of a sufficiently fast-moving automobile. Fracture intensity is severe near the point of impact where dynamic forces are high but gradually lessens in regions away from the impact point as the stresses driving fracture attenuate with distance.

Further complications arise when the size of the fragmented body is finite. If the number of fragments is few the distribution in fragment size will depend on the total mass of the body. In addition, in certain applications of interest there is a fundamental minimum fragment size brought about by specific material conditions which will skew the predicted statistical fragment distribution. This might occur for example when a strong aggregate is weakly bonded into a competent body. The aggregate size would then represent the minimum fragment size into which the body could be broken. The nucleus of an atom might represent such an aggregate in that the forces bonding nucleons together would be considered weak compared to the forces necessary to disrupt an individual nucleon.

Here we will pursue some of the probabilistic methods that have been pursued to address the statistical consequences of fragmentation events including some of the complications noted above.

4.1.1 Early Applications and Empirical Distributions

The reduction of a monolithic body to particulate of the same material through the application of some set of mechanical forces has been of interest in a wide range of both industrial and scientific concern. The blasting of rock in mines and quarries; the periodic collision of planetary bodies resulting in the distribution of objects throughout the asteroid belt and other regions of the solar system; the dynamic processes which occur where land and water meet, producing the sand, silt and gravel beaches on this earth, are just a few representative examples of applications that have tested scientific thought over the past several centuries.

We will identify, in the present section, some of the more prominent scientific attempts over about the last century and a half to provide predictive order and understanding to such fragmentation processes. These efforts have generally focused on one or the other of two general issues. Namely, what are the forces and energies required to cause the fragmentation events? And, what are the size and the distribution of particulate resulting from these breakup processes?

Over the time span considered here most of the attention has focused on the spread, or the distribution, in particulate size resulting from the fragmentation event. Historical efforts undertaken to address the former issue of the fragment size are assessed in Sect. 4.2.

Logarithmic-Normal Fragment Distribution

Analytic representations of the distributions of fragments resulting from various breakage events have historically been of two general types. One is of the normal distribution form (actually logarithmic-normal). The other form, which has stronger empirical roots, carries a measure of the fragment dimension in the argument of an exponential function. Many forms of the latter distribution have been constructed along with attempts to experimentally or theoretically justify them.

The logarithmic-normal fragment distribution has generally been considered inappropriate for dynamic single-event fragmentation processes such as through impact or explosive loading. Actually, the logarithmic-normal fragment distribution has remarkable analytic flexibility, and can probably adequately represent most of the fragment distributions generated both in the laboratory and in nature easily within the accuracy with which they can be measured. Theoretical developments of the logarithmic-normal fragment distribution (e.g., [22, 65]) are based on the application of the asymptotic limit of the central limit theorem to continued repetitive breakage process such as might occur in crushing and grinding, or perhaps in repeated impact events. Applicability is stated to require that the continued partitioning of a body occur in a self-similar manner independent of particle size. Thus, crusher loads, geological deposits, and the size frequency of meteors are representative of distributions that are expected to be represented by a logarithmic-normal distribution. The functional form of the logarithmic-normal fragment distribution is,

$$f(x) = \frac{1}{\sqrt{2\pi}\sigma x} e^{-\frac{1}{2\sigma^2}(\ln x - \mu)^2}, \tag{4.1}$$

where the probability density function $f(x)$ provides the number distribution of fragments of size x in terms of the scale parameter μ and shape parameter σ. The expected value of the distribution is,

$$<x> = e^{\mu + \sigma^2/2}, \tag{4.2}$$

and the standard deviation, normalized by the expected value,

$$s.d./<x> = (e^{\sigma^2} - 1)^{1/2}. \tag{4.3}$$

The distribution provided in (4.1) is conveniently normalized by the expected value through $X = x/ < x >$ yielding,

$$f(X) = \frac{1}{\sqrt{2\pi}\sigma X} e^{-\frac{1}{2\sigma^2}(\ln X + \sigma^2/2)^2}. \tag{4.4}$$

Logarithmic-normal distributions from (4.4) for various values of the shape parameter σ are illustrated in Fig. 4.1. The distribution can range from the

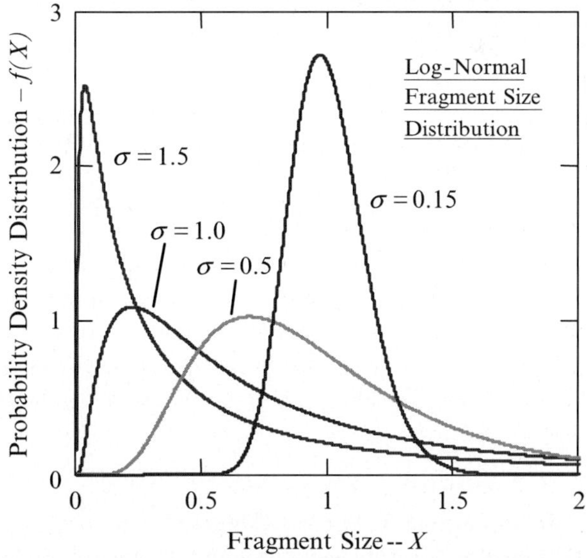

Fig. 4.1. Probability density functions for logarithmic-normal fragment distribution. The expected value is unity and curves are for various values of the shape parameter

highly skewed distributions characteristic of the shatter of brittle solids such as glass and ceramics to a strongly centered distribution representative of the fragmentation of expanding shells or rings of ductile metals.

Rosin–Rammler and Related Fragment Distributions

A framework for representation of the distribution of fragments, which has found wide application, appears to have had its genesis within the field of coal and ore crushing during the 1930s. Fragment distributions were determined in that field by sieving methods where fragments within a given size range are separated through a collection of variously sized sieving screens. Fragments within each size range are weighed providing a measure of the mass fraction of fragments within each size range and consequently the fragment size distribution.

The most frequently quoted fragment distribution from this period is attributed to Rosin and Rammler [86]. The Rosin–Rammler distribution is most commonly written as a cumulative mass fraction greater than a size x,

$$\frac{M(x)}{M_o} = 1 - e^{-(x/\lambda)^n}, \tag{4.5}$$

where λ is the size scale of the distribution and the shape parameter n typically ranges between about 0.5 and 1.5. The Rosin–Rammler fragment mass distribution is of the Weibull extreme value statistical distribution form [52]. The limiting form of the Rosin–Rammler distribution,

$$\frac{M(x)}{M_o} = (x/\lambda)^n, \tag{4.6}$$

is usually attributed to Schuhmann [87], and is quite appropriate in most applications (see also Gaudin [24]).

Attempts have been made to theoretically derive the distribution of Rosin and Rammler, and most are based on the elementary fragmentation algorithm explored extensively be Lienau [70]. Lienau investigated the one-dimensional problem of the distribution in fragment lengths resulting from the random placement of breaks on a line. Provided the average fragment length λ is small compared with the total initial length L, it is readily shown that the fragment length probability density distribution is of the Poisson form,

$$f(x) = \frac{1}{\lambda} e^{-x/\lambda}. \tag{4.7}$$

If relatively few breaks occur on the line length L it can also be shown that the density distribution in fragment lengths corresponding to (4.7) is binomial (e.g., [33, 38]).

Bennett [3] used the Poisson relation in (4.7) in an effort to justify the Rosin–Rammler distribution during an extensive investigation of crushing and shatter of hard coals. He assumed that the Poisson process applied to the random placement of fracture planes partitioning the length, width, and breadth of a fractured body, and arrived at the Rosin–Rammler relation in (4.5) with a value of the shape parameter (his distribution index) of $n = 1$. He offered arguments of non-ideality in the breakage process to justify values of n both less than and greater than unity.

Gilvarry [27] noted certain inconsistencies in the derivation of Bennett [3] and undertook an elaborate physics-based statistical development of fragment distributions resulting from single fracture events in brittle solids. Gilvarry postulated the existence of independent edge, surface and volume flaws underlying the breakage process in brittle bodies. By ascribing a Poisson process to each type of fracture-initiating defect he arrived at the Gilvarry distribution,

$$\frac{M(x)}{M_o} = 1 - e^{-\left(\frac{x}{\lambda_1}+\left(\frac{x}{\lambda_2}\right)^2+\left(\frac{x}{\lambda_3}\right)^3\right)}, \tag{4.8}$$

where, again, $M(x)/M_o$ is the cumulative mass fraction of fragments of size measure x. Gilvarry noted that in most instances edge flaws (first term in the exponential) dominated, resulting in the Rosin–Rammler equation with $n = 1$. Variations from $n = 1$ occurred, according to Gilvarry, when other flaw types contributed, or in repeated breakage processes such as crushing where the single fracture condition was not met.

The present author has noted oddities in the Gilvarry [27] development; particularly in his method of mass normalization ((16) in his paper) of the derived statistical relation. It seems to this writer that Gilvarry replicates the inconsistencies noted in the analysis of Bennett [3]. Consequently, the Rosin–Rammler fragment distribution should at present probably remain noted as a historically interesting and experimentally convenient empirical relation for characterization of fragment debris.

Gaudin and Meloy [25] have provided one further interesting alternative distribution in the same vein. They worked with the binomial distribution as the basis for randomization of the fragment size, rather than with the Poisson distribution, and arrive at the cumulative mass fraction distribution,

$$\frac{M(x)}{M_o} = 1 - \left(1 - \frac{x}{\lambda}\right)^r, \tag{4.9}$$

where λ is regarded as a characteristic feed size (initial body size) and r is a measure of the reduction ratio in the breakage process. Again, the Gaudin–Meloy development appears to suffer from mass normalization inconsistencies similar to the Gilvarry efforts.

Several of the distributions described here and used in the coal and ore industry are illustrated in Fig. 4.2. The distributions have been normalized to a mass average fragment size of unity. A shape index of $n = 1$ is used in the Schuhmann and the Rosin–Rammler relations. A relatively modest value of $r = 3$ is used in the Gaudin–Meloy equation. As r increases the Gaudin–Meloy and the Rosin–Rammler relations become coincident.

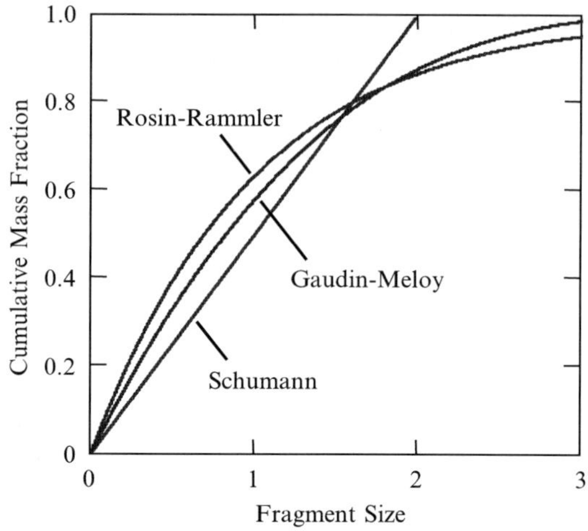

Fig. 4.2. Cumulative fragment distributions used to describe consequences of shatter and crushing in the coal and ore industry

The Mott–Linfoot Fragment Distribution

To those researchers working in the field of exploding munitions fragmentation, and in a number of areas closely akin to munitions fragmentation, no theoretical work has had more profound impact than that of Mott in a series of several reports during the World War II years [76–78,81], and in several papers written shortly after [79, 80]. The Mott distribution, or the Mott–Linfoot distribution as it will be referred to here, has been used widely to characterize fragment distributions from the breakup of exploding bombs and shells [81]. Mott and Linfoot also used the one-dimensional Poisson distribution derived by Lienau [70] as the statistical basis for their development.

Their acceptance of the Lienau distribution was bolstered by fragmenting munitions data available to them at the time, which were found to plot reasonably linear in a log number versus cube root of the fragment mass representation. Since $m^{1/3}$ is proportional to a length measure of the fragment they reasoned that the same random variable considered in the Lienau one-dimensional development also applied in the multidimensional fragmentation event. Further, in examining fragments from the available data, they observed that a substantial portion retained inner and outer surfaces of the original munitions case. This suggested that in the fragmentation of a plate or area the appropriate length scale would be proportional to $m^{1/2}$. Thus, a plot of log number versus $m^{1/2}$ should by the reasoning given, provide a better fit to the fragment distribution data. Thus, the fragment cumulative number probability distribution proposed by Mott and Linfoot [81] would be,

$$F(m) = 1 - e^{-(m/\mu)^{1/2}}, \tag{4.10}$$

where the characteristic mass μ is the distribution scale parameter. The corresponding probability density distribution is then,

$$f(m) = \frac{1}{2\mu}\left(\frac{m}{\mu}\right)^{-1/2} e^{-(m/\mu)^{1/2}}. \tag{4.11}$$

Numerous researchers have successfully used this distribution in various forms over the past six decades to organize and compare vast amounts of exploding munitions fragmentation data. Mott expended considerable subsequent effort in a quest to justify the functional form assumed in the equations above. In general, he did not succeed. However, his efforts resulted in some of the strongest physics based analysis of the dynamic fragmentation process that exists to the present day.

4.1.2 Poisson Processes and Geometric Methods

The idea of randomly placing points within a body (a Poisson process) and using these points to randomly partition the body through some algorithm is natural and was pursued by early workers either to provide a rational for observed fragment distributions, or to suggest a distribution when experiments were not available. The body of interest could be a line (one-dimension), a surface (two-dimensions), or a volume (three-dimensions). As will be shown, flexibility in choosing a geometric fragmentation method increases with the higher dimensions.

As noted earlier, the seminal effort within this approach to random fragmentation is probably the study of Lienau [70] on the fragmentation of a line. Later work attempted to extend Lienau's method to the fragmentation of higher dimensional bodies (surfaces and volumes). The efforts of Bennett [3], Gilvarry [27], and Gaudin and Meloy [25], in attempting to justify observed fragment distributions in brittle solids, relied heavily on the one-dimensional Lienau distribution.

Mott and Linfoot [81] also explored geometric fragmentation in the breakup of exploding munitions. Finally, the present discussion would be remiss if application of the classic tessellation methods of Voronoi and Dirichlet to fragmentation were not addressed.

The geometric methods that will be considered here do not exhaust those that have been explored. In addition, it will become apparent that these geometric fragmentation approaches, divorced from the fragmentation physics, do not in-of-themselves; explain the diverse observations in the random fragmentation of bodies. When used with care, the geometric methods do have useful application in selected practical fragmentation processes, however.

Lienau Distribution

Fundamental to the geometric fragmentation of a body are the theoretical efforts of Lienau [70]. He considered the elementary problem of an extended

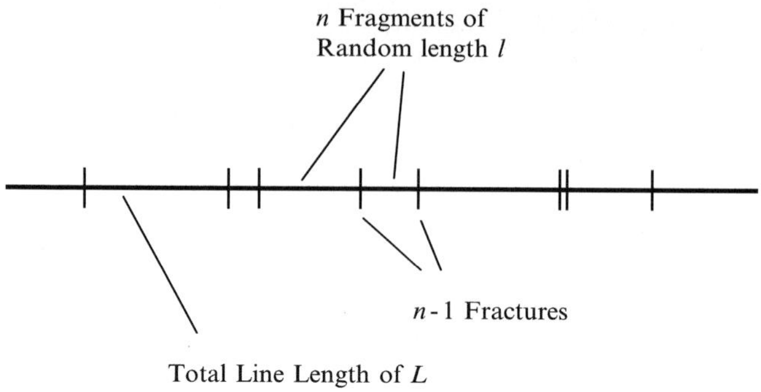

Fig. 4.3. Line of total length L broken at random into fragments of variable length l by $n-1$ fractures

body such as a glass rod or a stretching wire subjected to forces resulting in the multiple fracturing of the body. If any point on the body is as likely as another to fracture the problem is statistically well posed. The problem is modeled as that of an infinite one-dimensional body, or line, in which breaks are introduced with equal probability at any point on the line as illustrated in Fig. 4.3.

Thus, as stated, the random geometric fragmentation of a one-dimensional body appears decidedly unambiguous. An analytic solution requires only a proper probabilistic description of the random breaks, and the lengths of the segments delineated by these breaks. We shall show later that even this prescription for the statistical fragmentation of a one-dimensional body is arguable. At this point, however, we proceed with the solution leading to the one-dimensional Lienau fragment size distribution.

Consider a line of length L in which breaks on the line are introduced at random [33]. Since we are initially interested in partitioning the line into a large number of fragments (the average length is very small compared to the total length L), the finite length of the line is not of consequence and can effectively be considered infinite. The average spacing between breaks λ, or equivalently the frequency of breaks per unit length $h_o = 1/\lambda$ characterizes the statistical distribution. The random distribution of points on a line is described by Poisson statistics. If an arbitrary length l of the line is examined then the probability of finding n points (fractures) within the length l is given by,

$$P(n,l) = \frac{(l/\lambda)^n e^{-l/\lambda}}{n!}. \tag{4.12}$$

The probability of occurrence of fragments of length l within a tolerance dl is then,

$$f(l)\,dl = P(0,l)\,P(1,dl) = (1/\lambda)\,e^{-l/\lambda}dl. \tag{4.13}$$

Thus, the one-dimensional probability density distribution is,

$$f(l) = (1/\lambda)\,e^{-l/\lambda}, \tag{4.14}$$

and the cumulative fragment distribution,

$$F(l) = 1 - e^{-l/\lambda}. \tag{4.15}$$

Binomial Distribution

If, in Fig. 4.3, the line length L is broken into relatively few pieces, then the random placement of fractures is no longer governed by Poisson statistics. Probabilistic aspects of the fragmentation are instead governed by the binomial probability function [33],

$$P_{j,k}(p) = \frac{k!}{j!\,(k-j)!} p^j (1-p)^{k-j}. \tag{4.16}$$

Given n fragments ($n-1$ fractures) the probable occurrence of a fragment of length l within the interval dl is the product of the probability of no fractures in length l and one fracture within length dl where, respectively, $p = l/L$ and $dp = dl/(L-l)$,

$$f(l)\,dl = P_{0,n-1}(p)P_{1,n-1}(dp) = \frac{n-1}{L}\left(1 - \frac{l}{L}\right)^{n-2} dl, \tag{4.17}$$

with $f(l)$ providing the requisite probability density function. The cumulative distribution function is then,

$$F(l) = 1 - (1 - l/L)^{n-1}, \tag{4.18}$$

and the average fragment size is $\lambda = L/n$. Binomial fragment size probability density functions are shown in Fig. 4.4, and compared with the corresponding Poisson distribution for $n = 5$. The two of course converge as n becomes large.

Discrete Fracture Sites Distribution

A further interesting constraint occurs when the body has a limiting grain size which determines the minimum spacing between fractures. Setting this minimum size to δ, the number of fracture sites per unit length is $N = 1/\delta$. If fractures occur at $N_f < N$ sites then the probability of fracture at any specific site is $p = N_f/N = N_f\delta$. The probability of finding a fragment of length $l = n\delta$ is then,

$$P(n\delta) = (1-p)^{n-1}p. \tag{4.19}$$

The average fragment length is $\lambda = 1/N_f = \delta/p$ and the discrete fragment probability relation becomes,

$$P(l) = \frac{\delta}{\lambda}\left(1 - \frac{\delta}{\lambda}\right)^{(l/\delta)-1}. \tag{4.20}$$

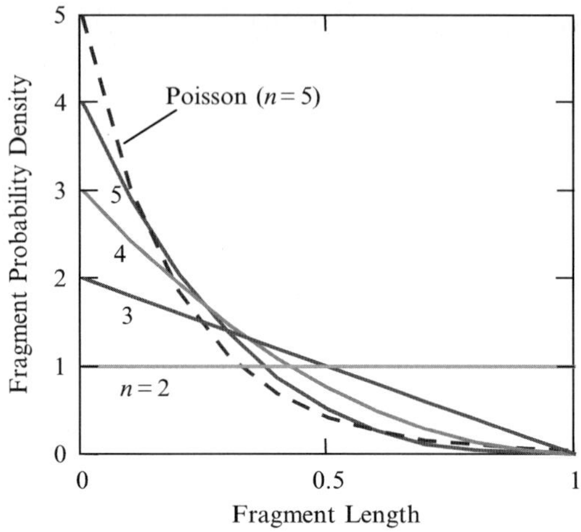

Fig. 4.4. Illustrates fragment probability distributions for fragmentation of unit length body into $n = 2$, 3, 4 and 5 fragments. Dashed line shows the Poisson distribution approximation for the case of $n = 5$ fragments

The cumulative fraction of fragments of size $l > n\delta$ is provided by,

$$\sum_{i=n}^{\infty} \frac{\delta}{\lambda}\left(1 - \frac{\delta}{\lambda}\right)^{n}. \tag{4.21}$$

The relation $\sum\limits_{n=r}^{\infty} x^n = x^r/(1-x)$ then provides for the cumulative fragment distribution,

$$F(l) = 1 - \left(1 - \frac{\delta}{\lambda}\right)^{l/\delta}. \tag{4.22}$$

The above relations converge to the exponential Lienau distribution as δ/λ becomes small.

Probability fragment size relations can also be developed for a one-dimensional body of finite length and with discrete fracture sites [33]. The relations are somewhat more complex but tractable.

4.1.3 Geometric Fragmentation of a Surface

The random fragmentation of a one-dimensional body through a Poisson process as was pursued by Lienau [70] seemed to various later workers to provide the basis for rationalizing the statistical fragmentation of more complex breakup events. Bennett [3] considered fragmentation of a brittle solid through the random distribution of mutually orthogonal planes distributed according to the Lienau method. Gilvarry [27] pursued the similar, but more complex, admixture of one-, two- and three-dimensional Poisson fragmentation processes. The approach of Gaudin and Meloy [25] was similar to that of Bennett except they chose the one-dimensional binomial distribution as the basis for a more general statistical development.

Mott and Linfoot [81] also considered the Lienau distribution in exploring fragmentation statistics in the entirely different application of the explosive breakup of rapidly expanding cylindrical metal shells. Thus, the fragmentation of bodies through random geometric methods had appeal to these earlier workers. Pursuing some of these efforts is worthwhile. The nature of the fragmentation application of concern to Mott led naturally to consideration of the geometric breakup of an area, or two-dimensional body. Mott and Linfoot showed that certain two-dimensional geometric statistical fragmentation methods are amenable to rigorous analytic solutions.

Mott Random Lines Fragmentation

In Fig. 4.5 selected geometric fragmentation methods are illustrated which have been pursued by others to rationalize observed fragment distributions. The first illustrates the partitioning of a plane with vertical and horizontal parallel lines, while the second, with randomly oriented straight lines. Both statistical fragmentation algorithms were considered in the report provided by Mott and Linfoot [81]. The first is seen to be the two-dimensional version of the Poisson process pursued by Bennett [3] and others.

The solution of the random horizontal and vertical lines fragmentation geometry put forth by Mott and Linfoot proceeds as follows (e.g., [38]). The probability density distribution over fragment lengths and widths is provided by the juxtaposition of the one-dimensional Lienau distribution over both the x (horizontal) and y (vertical) directions,

$$f(x, y) = \frac{1}{x_o y_o} e^{-x/x_o - y/y_o} . \tag{4.23}$$

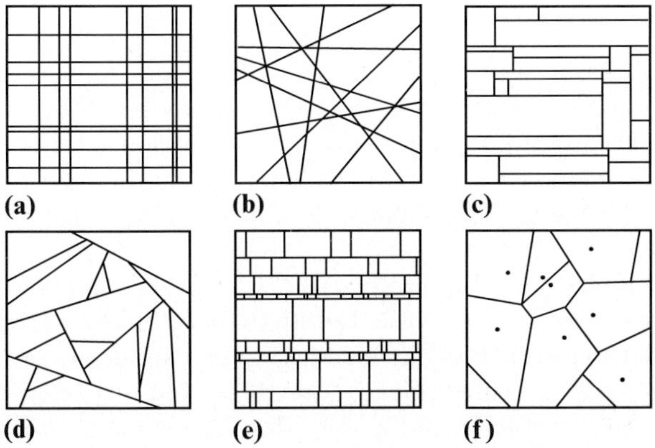

Fig. 4.5. Various geometric random fragmentation algorithms explored by Mott and Linfoot [81] and others

The cumulative distribution $F(z)$ over all fragments of size greater than $z = \sqrt{xy}$ is then provided by the integral,

$$1 - F(z) = \iint\limits_{xy > z^2} \frac{1}{x_o y_o} e^{-x/x_o - y/y_o} dx dy. \tag{4.24}$$

The integral is readily manipulated, and when integrated over the y coordinate yields,

$$\frac{1}{x_o} \int_0^\infty e^{-\frac{1}{x_o}\left(x + \frac{x_o}{y_o}\frac{z^2}{x}\right)} dx. \tag{4.25}$$

Not immediately obvious is the transformation,

$$x = z\sqrt{\frac{x_o}{y_o}} e^\theta, \tag{4.26}$$

which leads in turn to the integral,

$$2\frac{z}{z_o} \int_0^\infty e^{-2\frac{z}{z_o}\cosh\theta} \cosh\theta d\theta, \tag{4.27}$$

where $z_o = \sqrt{x_o y_o}$. Mott and Linfoot recognized this as one of a series of integral solutions for the modified Bessel functions of integer order [1] and arrived at the cumulative fragment size probability distribution,

$$F(z) = 1 - 2\frac{z}{z_o} K_1(2z/z_o), \tag{4.28}$$

and probability density distribution,

$$f(z) = 4\frac{z}{z_o^2} K_o(2z/z_o). \tag{4.29}$$

The intent of the exercise by Mott and Linfoot was to rationalize experimentally observed distributions of fragments resulting from exploding munitions, and a more intuitive analytic distribution (the well-known Mott distribution) developed by them in which they predicted the fragment size dependence $\ln f(z) \sim z$. They were not fully satisfied when this expectation was compared with (4.29).

Mott and Linfoot [81] also briefly explored the second algorithm in Fig. 4.5 in which partitioning lines are oriented at random. They were unable to find an analytic solution except at the small fragment limit where they suggested that $\ln f(z) \sim z$ was satisfied. Grady and Kipp [42] explored this algorithm through computer methods and found that the distribution did not differ markedly from the Bessel solution obtained for the vertical and horizontal lines fragmentation algorithm.

A useful and interesting feature of the vertical and horizontal lines fragmentation method is the ability to vary the density of lines in the two orientations, and assess the distribution in fragment aspect ratio as well as size. This distribution has application, for example, in the breakup of biaxial stretching metal shells where stretching rates in orthogonal directions differ. Pursuing the previous solution the distribution,

$$k(r) = \frac{1}{r_o} \frac{1}{(1 + r/r_o)^2}, \tag{4.30}$$

over the fragment aspect ratio $r = x/y$ is readily obtained.

Voronoi–Dirichlet Fragmentation

The present discussions of random geometric fragmentation would be remiss without consideration of the Voronoi–Dirichlet construction (e.g., [7]). This method for the random partitioning of space has received by far the lion's share of attention in a much broader spectrum of literature. The resulting distributions have been proposed for such applications as the distribution of galactic matter throughout the universe [59] and the formation of geologic columnar structures such as the Giant's Causeway in Northern Ireland [97].

The construction algorithm is illustrated in (f) of Fig. 4.5. The method begins with a random (statistically homogeneous) distribution of points on the surface (or within the volume, if three-dimensional space is considered). Space is then randomly partitioned by construction of perpendicular bisecting lines (or surfaces) as illustrated. On a regular (periodic) lattice of points the same process creates the Wigner–Seitz cells used, for example, in the construction of Brillouin zones in solid state physics [64]. The space is also randomly partitioned through the reciprocal, or dual, Delauney construction [95] created through the joining, with lines (or surfaces), the points in each Voronoi–Dirichlet cell.

Analytic relations for the fragment size distributions resulting from the Voronoi–Dirichlet construction have not been directly determined. A computational determination of the resulting fragment size distributions has been widely pursued, however (e.g., [18]), and analytic expressions which successfully reproduce the computational distributions have been arrived at by inductive methods [60].

Both the analytic distributions and their inductive development are of interest to the present pursuit of statistical fracture through geometric means. Kiang [60] first considered the one-dimensional Voronoi–Dirichlet construction where points are distributed at random on a line (a Poisson process) and then the degenerate (one-dimensional) Voronoi cell from the perpendicular bisector (the midpoint) of each point pair is determined. Thus, the Voronoi–Dirichlet distribution on a line is the dual of the Lienau distribution considered earlier (or the degenerate Delauney distribution). Whereas, in the Lienau distribution random points on the line were considered as breaks or fractures, in the present one-dimensional Voronoi–Dirichlet distribution these same random points constitute in some sense the centroid of fragments with fractures occurring at the bisector points.

The fragment size distribution for the one-dimensional Voronoi–Dirichlet distribution can be determined directly as follows [38]. The probability of finding a length l_i between a Poisson point pair is given by the Lienau expression,

$$f(l_i)dl_i = \frac{1}{\lambda}e^{-l_i/\lambda}. \tag{4.31}$$

The probability of finding a point pair of length l_1 adjacent to a point pair of length l_2 is then the product,

$$f(l_1)f(l_2)dl_1dl_2 = \frac{1}{\lambda^2}e^{-(l_1+l_2)}dl_1dl_2. \tag{4.32}$$

Implementing the transformation,

$$l = (l_1 + l_2)/2, \tag{4.33}$$

$$\xi = (l_1 - l_2)/2, \tag{4.34}$$

and integrating over the variable ξ leads to the one-dimensional Voronoi–Dirichlet distribution,

$$f(l) = \frac{2}{\lambda}\left(\frac{2l}{\lambda}\right)e^{-2l/\lambda}. \tag{4.35}$$

The distribution in (4.35) is a gamma function of order $n = 2$. Kiang [60] offered without proof that symmetrically higher order gamma functions would provide analytic fragment distributions for Voronoi–Dirichlet partitioning of an area or a volume. Following Kiang we will write the general expression for fragment distributions over fragment mass,

$$f(m) = \frac{1}{\mu}\frac{n}{\Gamma(n)}\left(\frac{nm}{\mu}\right)^{n-1}e^{-nm/\mu} \tag{4.36}$$

where $n = 2$, 4 or 6 is for line, surface or volume fragmentation, respectively.

Computational distributions from Voronoi–Dirichlet constructions on an area performed by Kiang [60] were in acceptable agreement with (4.36) for $n = 4$. A degree of controversy was generated by Kiang's proposal among subsequent authors as to the adequacy of (4.36); both for and against. Apparently, the construction of computer algorithms to generate Voronoi–Dirichlet fragment distributions is not a trivial exercise. In any case, for the present geometric fragmentation investigations, (4.36) is an adequate analytic representation of Voronoi–Dirichlet distributions in line, area or volume fragmentation.

Fragment size (area) distributions resulting from both the Voronoi–Dirichlet algorithm and the horizontal and vertical lines algorithm (Bessel distribution) are compared in Fig. 4.6. The two density distributions are normalized to unit expected value. The comparison reveals the stark differences resulting from differing randomization algorithms and certainly raises questions as to the applicability of statistical geometric methods to fragmentation.

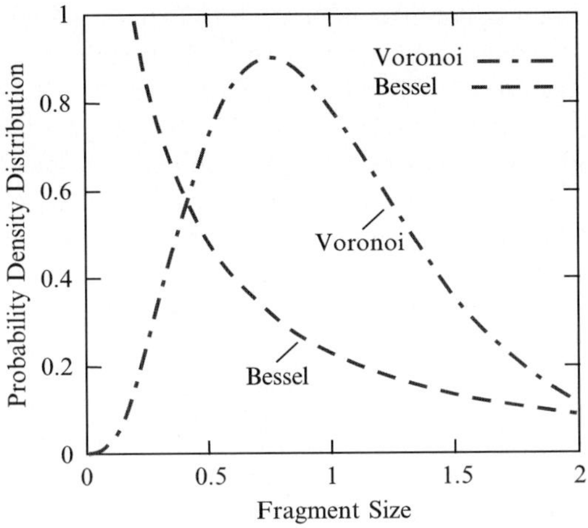

Fig. 4.6. Comparison of fragment size distributions resulting from the random horizontal and vertical lines algorithm, and the Voronoi–Dirichlet tessellation in the geometric fragmentation of a surface

Mott Cylinder Segmentation

Mott was clearly not satisfied with the results from the fragment distributions resulting from the random placement of infinite lines on a surface. It was becoming apparent to him that such methods did not lead to unique distributions, and was very much dependent on the algorithm selected. Nevertheless, he made one last attempt at the use of a random geometric method before abandoning the approach. Perhaps he reasoned that if the method was algorithm dependent, then if a method was selected which more reasonable approximated the fracture event of interest, that more credence could be attributed to the resulting distribution.

From the elongated and sliver-shaped fragments recovered from exploding munitions tests Mott surmised that fracture in an end-detonated metal shell with cylindrical symmetry would occur through longitudinal running cracks with occasional crack branching and crack intersection resulting in the observed fragments. Thus he proposed the statistical algorithm illustrated in (e) of Fig. 4.5. For the random distribution of longitudinal and transverse lines on the plane, Mott made the interesting selection of the one-dimensional Voronoi–Dirichlet distribution discussed in the previous subsection rather than the Lienau distribution used in his earlier geometric pursuits. He may have made this choice for one of several reasons. It is possible that he carried through the analysis using the Lienau distribution and found, as can be shown, that the distribution does not converge in the small fragment limit. Alternatively, ideas to emerge in his later work, and to be discussed in a later section, may have influenced this selection: Namely, that the physics of fracture interaction precludes the close proximity of parallel fractures and thus limits the number of smaller fragments. The Voronoi–Dirichlet distribution is observed to better provide a statistical constraint limiting the number of the close parallel fractures and hence the number of smaller fragments.

The method, shown in Fig. 4.5 consists of first inscribing randomly positioned horizontal lines according to the one-dimensional Voronoi–Dirichlet distribution from (4.35), and then segmenting each horizontal strip with randomly positioned vertical lines where the average spacing within any strip is proportional to the width of that strip, $x_o = py$.

Following methods similar to the solution resulting from the horizontal and vertical random lines algorithm, one arrives at the fragment density distribution as a function of the fragment area a,

$$f(a) = \frac{2}{a_o} \frac{1}{\sqrt{4a/a_o}} \int_0^\infty \left(\xi\sqrt{4a/a_o} - 1\right) \left(1 + 1/\xi^2\right) e^{-\xi\sqrt{4a/a_o} - 1/\xi^2} d\xi. \quad (4.37)$$

This distribution is provided in Mott's third internal report [77], and differs only in the distribution variable $\lambda = \sqrt{a/a_o}$ used by him. The scale parameter is $a_o = py_o^2$ where y_o is the average spacing of the axial strips. The constant p is the aspect ratio of the average length of fragments in a strip to the width of the strip. Mott suggested $p \approx 5$.

Mott found that the present geometric distribution was again close to the Mott–Linfoot distribution ($\ln N \sim \sqrt{a}$) but not markedly better than the earlier attempt with random infinite lines. The two geometric distributions are compared with the Mott–Linfoot distribution in Fig. 4.7. It is interesting that

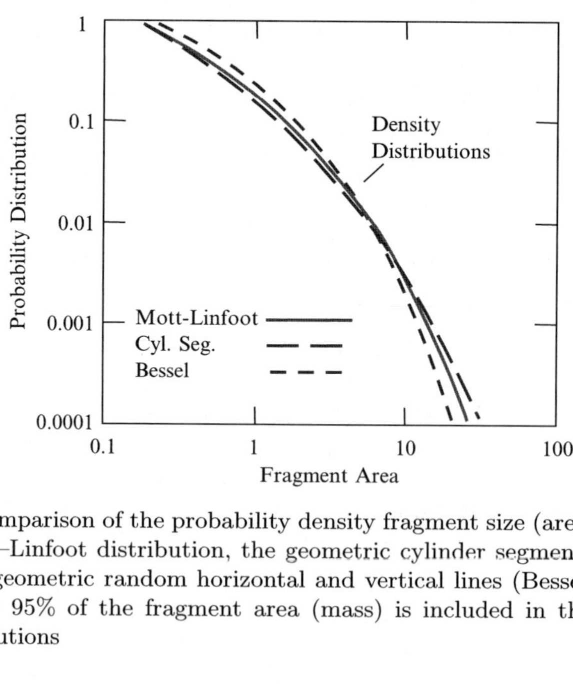

Fig. 4.7. A comparison of the probability density fragment size (area) distributions from the Mott–Linfoot distribution, the geometric cylinder segmentation distribution, and the geometric random horizontal and vertical lines (Bessel) distribution. Approximately 95% of the fragment area (mass) is included in the range of the plotted distributions

the distributions from the two algorithms tend to straddle the Mott–Linfoot proposed distribution, each with respectively larger and smaller variance. It is likely that little else can be said.

Other Geometric Algorithms

Other algorithms for randomly partitioning a surface are readily constructed (e.g., [42]). Several alternative algorithms are illustrated in Fig. 4.5. Mott and Linfoot [81] also considered the algorithm shown in (b). They were not able to generate the resulting fragment distribution except in the small fragment limit where the fragment probability proportional to the square root of the fragment area was demonstrated. Fragmentation algorithms shown in (c) and (d) were considered by Grady and Kipp [42], and both result in an exponential fragment distribution. Further random construction algorithms have also been explored [42].

At this point, a degree of healthy suspicion should be growing as to the applicability of random geometric fragmentation algorithms to actual physical fragmentation phenomena. Later it will be shown that some utility of these methods can be made use of in modeling the statistical fragmentation phenomena, but they should be employed only with a sensible understanding of the underlying physics.

4.1.4 Alternative Fragment Size Statistics Methods

Geometric fragmentation methods explored in the previous subsection pursued techniques for randomly distributing the boundaries delineating the fragments in the breakup of a competent body. Although the results are useful in some applications, one finds that there are many ways to randomly partition a body, and that the resulting fragment size distributions are not unique. Approaches other than geometric methods have been pursued in attempting to develop fragment size distributions appropriate to observation. Some of these methods are considered here.

The Mott–Linfoot Distribution

The original distribution developed by Mott and Linfoot [81], as was previously described in the historical efforts, was not based on geometric arguments. Existing fragmenting munitions data and some impetus from the work of Lienau [70] led them to propose that a measure of the fragment size $x \sim \sqrt{a}$ was a random variable distributed according to a Poisson process. The fragment area is a and is proportional to the mass m since they were interested in the breakup of thin shells. This led immediately to the Mott–Linfoot distribution,

$$f(m) = \frac{1}{2\mu}\left(\frac{m}{\mu}\right)^{-1/2} e^{-(m/\mu)^{1/2}}, \tag{4.38}$$

in fragment mass where μ is the distribution scale parameter. The corresponding cumulative fragment distribution is then,

$$F(m) = 1 - e^{-(m/\mu)^{1/2}}. \tag{4.39}$$

The Mott–Linfoot distribution has been remarkably successful in both correlating munitions fragmentation data, and in developing predictive fragmentation relations, over the intervening six decades. Mott and Linfoot tacitly suggested that if the preponderance of fragments were of a size smaller than the thickness of the exploding shell, then the fractional power in the exponent should be 1/3 rather than 1/2. This argument is certainly consistent with their assumption above that a length measure of the fragment was a random Poisson variable.

The Exponential Distribution

Grady and Kipp [42] have offered an alternative development and explanation for the distributions in fragment size observed in munitions fragmentation. They first suggest that if such fragmentation can be represented by a mechanism-independent statistical description, then perhaps the fragment mass m, as opposed to fragment size (either $m^{1/2}$ or $m^{1/3}$ in the Mott–Linfoot development), is the more appropriate random variable. They then propose that the mass of the fragment is distributed over fragment number according to a Poisson (or binomial if the fragment number is small) process which parallels the development of Lienau outlined earlier.

Thus, if the fragment mass is viewed as a random scalar variable, then the random fragmentation of the mass is analogous to the one-dimensional Lienau problem. Fragmentation is determined by breaks distributed randomly over the scalar measure of mass. The breaks determine a Poisson variate and lead to a cumulative fragment probability distribution,

$$F(m) = 1 - e^{-m/\mu}, \tag{4.40}$$

and density distribution,

$$f(m) = \frac{1}{\mu} e^{-m/\mu}. \tag{4.41}$$

In contrast to the Mott distribution, the present distribution keeps the same linear exponential functional form for both area and volume fragmentation. Some support for application of an exponential distribution in the form of (4.41) to munitions fragmentation is provided by the experimental study of Mock and Holt [75], and in later observations of Mott [78].

The Hazard Function Distribution

The fragment distribution of Mott and Linfoot, and the exponential distribution suggested by Grady and Kipp, can be generalized through standard methods of hazard (or survival) statistics (e.g., [52]). Consider a collection of fragments with an occurrence frequency, or hazard function, $h(m)$ where m is the fragment mass. Then according to the tenets of survival statistics the probability of finding a fragment of size m within interval dm is just the product of finding no fragments smaller than m and that of finding one fragment in the interval m to $m + dm$, or,

$$f\left(m\right) dm = e^{-\int\limits_0^m h(m)dm} h\left(m\right) dm. \tag{4.42}$$

Thus the fragment probability density distribution is,

$$f\left(m\right) = h\left(m\right) e^{-\int\limits_0^m h(m)dm}, \tag{4.43}$$

and the cumulative fragment distribution is,

$$F\left(m\right) = 1 - e^{-\int\limits_0^m h(m)dm}. \tag{4.44}$$

This derivation of the fragment distribution is fully rigorous. The unknown of course is the functional form of the hazard function $h(m)$. This function is determined either by exploring physics specific to the fragmentation process of interest, or else by hypothesizing a functional form and validating through experiment.

For example, given no additional insight into the fragmentation process there is little reason to assume a bias of $h(m)$ toward either the large or the small masses and the simplest assumption is $h(m) = h_o$ a constant (a Poisson process). Thus $f(m)$ in (4.43) is,

$$f\left(m\right) = h_o e^{-h_o m}, \tag{4.45}$$

and the cumulative fragment number distribution is,

$$F\left(m\right) = 1 - e^{-m/\mu}, \tag{4.46}$$

where $\mu = 1/h_o$ is the distribution scale parameter and the average fragment mass. Grady and Kipp [42] suggested applicability of this distribution to dynamic fragmentation.

Mott and Linfoot [81], following the theoretical work of Lienau [70], and inspection of some exploding munitions fragmentation data, were led to assume that a linear size measure x of the fragment dimension was a random variable and hence $h_o dx$ provided the chance of fracture determining a fragment of mass m. Mott was interested in the breakup of thin-walled cylinders where fragment dimensions were large compared to wall thickness. Hence $x \sim m^{1/2}$ and $dx \sim (1/2)m^{-1/2}dm$. Thus, an appropriate hazard function,

$$h\left(m\right) = \frac{1}{2\mu}\left(\frac{m}{\mu}\right)^{-1/2}, \tag{4.47}$$

follows from Mott's assumption, and (4.43) becomes,

$$f(m) = \frac{1}{2\mu}\left(\frac{m}{\mu}\right)^{-1/2} e^{-(m/\mu)^{1/2}}, \tag{4.48}$$

and the corresponding cumulative fragment distribution is,

$$F(m) = 1 - e^{-(m/\mu)^{1/2}}. \tag{4.49}$$

It is reasonable to consider limiting the functional dependence of $h(m)$ to the power law,

$$h(m) = \frac{\beta}{\mu}\left(\frac{m}{\mu}\right)^{\beta-1}, \tag{4.50}$$

which leads to the cumulative fragment size distribution relation of the Weibull form,

$$F(m) = 1 - e^{-(m/\mu)^{\beta}}, \tag{4.51}$$

where μ and β are the scale and shape parameters of the distribution, respectively. This relation clearly encompasses the previous examples. Namely, $\beta = 1/2$ corresponds to the distribution arrived at by Mott and Linfoot [81] while $\beta = 1$ corresponds to that suggested by Grady and Kipp [42].

Generality of the parameter β for a munitions-specific scaling equation is warranted for several reasons. Mott and Linfoot argued that when the fragment distribution is dominated by fragments of a size smaller than the case thickness, then $\beta = 1/3$ is probably more appropriate. Although apparently observed under certain testing conditions, such statistical behavior was not found in the cylinder fragmentation tests of Mock and Holt [75] for example. For specific munitions systems a range of expansion strain rates will lead to statistical heterogeneity (a different size parameter at different positions along the munitions case). This breakup feature will broaden the distribution leading to smaller effective values of β. On the other hand a degree of case scoring, or other processing, with the intention of biasing the distribution toward a unique size has the effect of increasing the distribution shape parameter β. (Note that as β approaches infinity (4.51) approaches a Heaviside function with all fragments the same size.)

Mott Statistical Fracture Distribution

As noted earlier, several technical reports published within the year of 1943 revealed the maturing of Mott's understanding of the dynamic fragmentation process and a statistical theory of fragmentation emerged which is still one of the leading theories available. The theory was published several years later in the open literature [79]. This development is summarized here and leads to statistical fragment distributions that differ starkly from the earlier distribution proposed by Mott and Linfoot [81], and others.

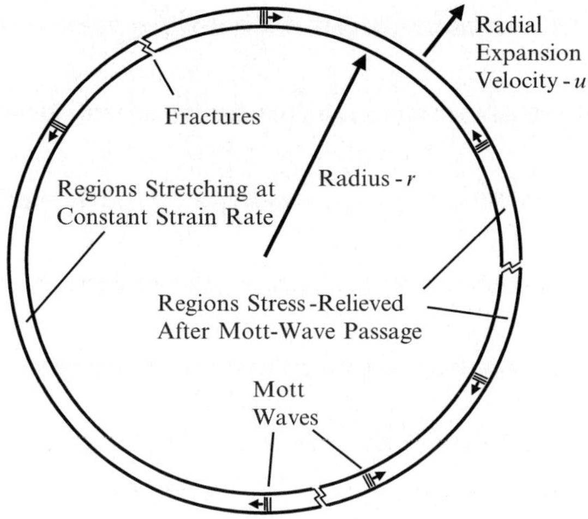

Fig. 4.8. Mott cylinder illustrating the one-dimensional activation and interaction of fractures leading to the statistical fragmentation of the body

The Mott theory of fragmentation is most readily conceptualized by considering the Mott cylinder (or ring) as illustrated in Fig. 4.8. The Mott cylinder is an idealization of an expanding cylindrical shell whose outward motion is imparted by some radial impulse. Mott in particular focused on the natural fragmentation of exploding pipe bombs. The model is applicable to other test conditions such as the radial acceleration and fragmentation of metal rings through explosive or magnetic methods.

An explosively driven expanding metal cylinder is a decidedly multidimensional fragmentation event. Fragmentation of the Mott cylinder is only an approximation to this event. The theory attempts to capture the characteristic circumferential spacing of fractures and the statistical distribution in that spacing. It is not intended to account for the axial propagation and interaction of cracks within a finite length cylinder. The Mott cylinder is an expanding metal body with outwar radial velocity u and radius r at the time when multiple fracture and break up of the cylinder proceeds. Just preceding break up, the cylinder body is in circumferential tension and undergoing uniform circumferential stretching at a rate given by the ratio $\dot{\varepsilon} = u/r$.

Mott proposed that fragmentation proceeded through the random spatial and temporal occurrence of fractures resulting in a distribution in fracture spacing. Release waves propagate away from sites of fracture relieving the tension and precluding the need for further fracture within the regions encompassed by tension release waves. Fragmentation is complete when fracture-induced release waves circumferentially subsume the entire cylinder (see Fig. 4.9).

Thus, within the model for dynamic fragmentation proposed by Mott, two physical issues require attention. First, is the issue of when and where

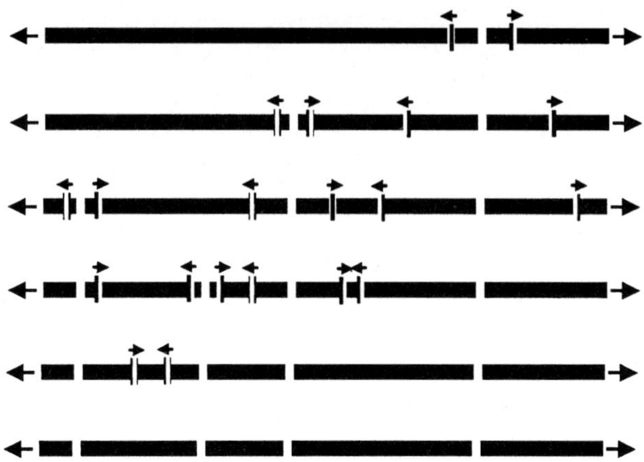

Fig. 4.9. The graphical statistical fragmentation undertaken by Mott is illustrated from top to bottom as fractures randomly activate and tensile release waves subsume the body

fractures activate on the Mott cylinder. Second, is the nature of propagation of tensile release waves (Mott waves) from the sites of fracture.

Mott Fracture Activation: Mott put forth arguments that energy dissipation was not of consequence in the fracture process and proposed instead a statistical strain-to-fracture criterion. Mott assumed that fractures occurred with equal probability at any point around the circumference of the cylinder at a frequency governed by a strain dependent hazard function $\lambda(\varepsilon)$ (e.g., [52]) such that $\lambda(\varepsilon)d\varepsilon$ provides the statistical number of fractures occurring within a unit length of the cylinder circumference within the strain interval $d\varepsilon$.

Mott expected $\lambda(\varepsilon)$ to be a strongly increasing function of strain and suggested both an exponential hazard function,

$$\lambda(\varepsilon) = Ae^{\gamma\varepsilon}, \tag{4.52}$$

and a power-law hazard function,

$$\lambda(\varepsilon) = \frac{n}{\sigma}\left(\frac{\varepsilon}{\sigma}\right)^{n-1}. \tag{4.53}$$

The former relation leads to Gumbel extreme value statistics, while the latter leads to Weibull statistics. Mott pursued in detail the exponential hazard function. Here the two-parameter power-law hazard function is explored. For sensibly large n the parameter σ is the expected value of the strain to fracture while σ/n is proportional to the standard deviation.

Mott Tension Release: The region of the material that has experienced tensile release at strain ε is given by,

$$D_x(\varepsilon) = \int_0^{\varepsilon} 2g(\varepsilon - \eta)\lambda(\eta)d\eta, \tag{4.54}$$

where $\lambda(\eta)d\eta$ is the statistical number of fractures activated on the Mott cylinder within a strain η within interval $d\eta$. The function $g(\varepsilon - \eta)$ is the distance traveled by a tensile stress release wave over the strain interval $\varepsilon - \eta$ for one fracture. The factor of two accounts for right and left facing release waves. (Since strain rate is assumed to be constant over the duration of the fracture process, strain and time are synonymous through $\varepsilon = \dot{\varepsilon}t$.)

In (4.54) $D_x(\varepsilon)$ is seen to provide the fraction of the Mott cylinder which has been encompassed by stress release waves emanating from sites of fracture at a current strain ε. The equation also determines the fraction of the cylinder in which further fracture is precluded. A form of (4.54) was derived by Mott in the original 1943 reports.

Correspondingly, the number of fractures activated at strain ε is given by,

$$N(\varepsilon) = \int_0^{\varepsilon} (1 - D_x(\eta))\lambda(\eta)d\eta, \tag{4.55}$$

where the factor $D_x(\eta)$ accounts for the fraction of the Mott cylinder no longer deforming in tension.

The functional form of the stress release function $g(\varepsilon)$ must be specified. There are several possibilities. If the expanding Mott cylinder is elastic at the time of fracture then a constant elastic release wave velocity governed by the elastic modulus is sensible. Mott, however, considered an expanding ductile metal cylinder and assumed a material on the tensile yield surface governed by a constant flow stress Y. Instantaneous fracture and rigid-ideally-plastic constitutive response leads to the stress release function (derived later in this chapter),

$$g(\varepsilon) = \sqrt{\frac{2Y}{\rho\dot{\varepsilon}^2}}\varepsilon. \tag{4.56}$$

It is readily demonstrated that stress wave diffusion rather than wave propagation governs stress release under the assumed plastic deformation conditions.

The physical and statistical principles outlined were then used by Mott to determine a distribution in fragment length (or fracture spacing) resulting from the plastic fracture of the expanding Mott ring. It should be emphasized that this one-dimensional distribution bears no relationship to the earlier two-dimensional Mott distribution arrived at intuitively by Mott and Linfoot.

The Mott Graphical Solution: Mott was not able to develop an analytic solution and proceeded with a graphical method. It is likely that with a modest amount of additional time Mott would have developed an analytic solution, which was his nature. It is interesting that nearly concurrently other workers [54] were pursuing transformation reaction kinetics in metals and developed analytic tools ideally suited to Mott's statistical fragmentation theory.

The graphical solution to the statistical fragment size distribution derived by Mott is illustrated in Fig. 4.9 and proceeds as follows: A parameter D is defined which is a function of time t (or strain). At any time $D(t)$ is the fraction of the stretching body that has been subsumed by the Mott release waves propagating from points of fracture. Since further fractures are assumed to occur only in the fraction of the body not yet encompassed by these release waves, $1-D$, the number of fractures which will appear in the time increment t to $t+dt$ is,

$$dN = (1 - D)\,\lambda\,(\dot{\varepsilon}t)\,\dot{\varepsilon}dt, \tag{4.57}$$

where, as noted previously, Mott used the exponential fracture frequency relation $\lambda\,(\dot{\varepsilon}t) = A\exp\,(\gamma\dot{\varepsilon}t)$.

The tension release fraction $D(t)$ of the plastic stretching body is determined by the collective Mott waves emanating from fractures initiating prior to time t. Each Mott wave propagates according to (see (4.56)),

$$x_i = \sqrt{2Y/\rho\dot{\varepsilon}}\,(t - t_i)^{1/2}\,, \tag{4.58}$$

where t_i is the initiation time of the ith fracture. Introducing a dimensionless time through the expression $\xi = \gamma\varepsilon = \gamma\dot{\varepsilon}t$, (4.58) becomes,

$$x_i = \sqrt{2Y/\rho\dot{\varepsilon}^2\gamma}\,(\xi - \xi_i)^{1/2}\,, \tag{4.59}$$

where Mott recognized the leading expression on the right hand side as the normalizing length scale x_o (proportional to the average fragment length) for the distribution in fragment lengths. The constant γ is the Mott fragmentation parameter commonly used by those who work with arena fragmentation data through the Mott method.

To determine the fragment distribution, Mott worked with (4.58) and (4.59), using a deck of cards and graph paper, and arrived at the solution shown in Fig. 4.10. The statistical sequence shown in Fig. 4.9 is illustrative of the method performed numerous times by Mott to acquire the statistical distribution. Mott noted that the average fragment size was about 1.5 x_o. Again, this distribution bears no relation to the earlier distribution inferred by Mott and Linfoot for fragmenting munitions data.

Wesenberg and Sagartz [98] performed fragmentation experiments through impulsive magnetic inductive expansion of aluminum cylindrical shells. Using computer methods and an appropriate random number generator they produced fragment size distributions by solving the same pair of equations as Mott. Wesenberg and Sagartz displayed their calculated distributions as the average of the individual results of 10 rings, 100 rings and 1,000 rings, respectively, and concluded that a reasonably large number of calculations was required to achieve sensible convergence. Their distribution resulting from the average of 1,000 rings is compared with the Mott distribution in Fig. 4.10.

Analytic Solution of the Mott Distribution: Analytic solutions of the Mott fragmentation model have been pursued by Grady [29, 30, 38]. An inspection

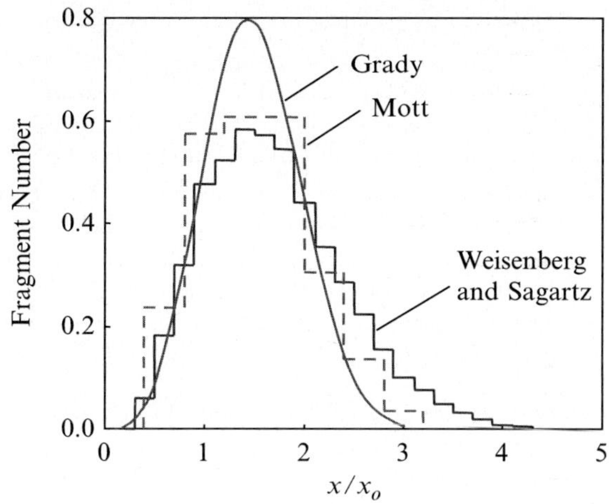

Fig. 4.10. Mott distribution for the spacing of fractures on a fragmenting expanding cylinder. Comparisons are shown with fragment distributions calculated by Wesenberg and Sagartz [98] and by analytic methods of Grady [29, 30]

of (4.54) reveals that the function $D_x(\varepsilon)$ will exceed unity at sufficiently large strain. This non-physical result is a consequence of not accounting for two factors in the fracture activation and stress wave process. First, the fracture activation function $\lambda(\varepsilon)$ does not exclude the activation of further fractures within regions previously stress relieved. Second, the stress release function $g(\varepsilon)$ does not account for the impingement and the overlap of opposing release waves from separate neighboring fractures. Thus, (4.54) is only applicable for a dilute number of fractures early in the fracture and release process. To account for fracture exclusion and wave impingement in the statistically random Mott model a statistical method introduced by Johnson and Mehl [54] in their pursuit of reaction kinetics in metals is appropriate. Exclusion and impingement is accounted for through the relation,

$$D(\varepsilon) = 1 - e^{-D_x(\varepsilon)}, \tag{4.60}$$

which provides the fraction of the Mott cylinder encompassed by fracture stress release waves at any strain ε. $D(\varepsilon)$ and $D_x(\varepsilon)$ are equivalent at early times as they should. The function $D(\varepsilon)$, however, approaches unity as ε becomes large. The relation corresponding to (4.55) then becomes,

$$N(\varepsilon) = \int_0^\varepsilon (1 - D(\eta))\lambda(\eta)d\eta = \int_0^\varepsilon e^{-D_x(\eta)}\lambda(\eta)d\eta. \tag{4.61}$$

With this contribution to the statistical fragmentation theory, it is then possible to develop analytic solutions to the distribution in fracture spacing for fragmentation of the Mott cylinder. This analytic solution has been derived, and details of the solution method are provided elsewhere [29, 30, 38]. The calculated analytic distribution in fracture spacing by this analytic method is,

$$f(x) = \frac{\beta^2}{4}\frac{1}{x_o}\left(\frac{x}{x_o}\right)^3 e^{-\frac{1}{4}(x/x_o)^3} \int_0^1 \left(1-y^2\right) e^{-\frac{3}{4}(x/x_o)^3 y^2}\, dy, \tag{4.62}$$

where $\beta = 3\Gamma(2/3)$ and x_o is the corresponding distribution scale parameter. The present analytic fragment length distribution is also compared with the Mott distribution in Fig. 4.10.

4.1.5 Entropy and Energy Approaches to Fragment Distributions

Concepts from classical statistical mechanics have found infrequent application to fragment size statistics. These approaches have been strictly exploratory. Validation of these theories through experiment or through broader community consideration has not been largely pursued. Such methods are nonetheless intriguing and may lend insight into the topic of dynamic fragmentation statistics. Several examples pursuing both entropy maximum methods and energy-based statistical mechanical methods are explored in this subsection.

Entropy Methods

An alternative method of developing a fragment size distribution relation is based on the proposition of maximizing the information entropy associated with a fragmentation event. Efforts to constrain fragment distributions through entropy methods have been pursued in several studies (e.g., [14, 21, 42]). Following the approach undertaken by Grady and Kipp, a fragmentation event is considered in which a body of mass M is broken into N fragments. In the development, energy is ignored and the proposition is made that any distinct fragment distribution is equally likely subjected to the constrain of N fragments of total mass M.

The problem is formulated by partitioning the mass M into equal intervals $\delta m = M/\eta$. If n_s fragments have mass within m_s to $\mathrm{m_s}+\delta m$, then the number of ways that the N fragments can be distributed is,

$$P(n_1, n_2, \ldots n_\eta) = N! \prod_{s=1}^{\eta} 1/n_s!. \tag{4.63}$$

The entropy is written as,

$$\ln P = \ln N! - \sum_{s=1}^{\eta} \ln n_s!, \tag{4.64}$$

or in the limit $\delta m \ll M$,

$$\ln P = \ln N! - \int_0^\infty \ln n(m)!\, dm, \tag{4.65}$$

where $n(m)$ provides the fragment number distribution over mass. The entropy $\ln P$ is maximized subject to constraints,

$$N = \int_0^\infty n(m) dm, \tag{4.66}$$

and,

$$M = \int_0^\infty m n(m) dm. \tag{4.67}$$

Variations provide,

$$\begin{aligned} \delta \ln\ P &= -\int_0^\infty \delta \ln n!\ dm = 0, \\ \delta\ N &= \int_0^\infty \delta\ n dm = 0, \\ \delta\ M &= \int_0^\infty m\delta\ n dm = 0. \end{aligned} \tag{4.68}$$

Introduction of undetermined multipliers α and β, and making use of the approximation $\delta \ln n! \simeq \ln \delta n$ provides,

$$\ln\ n(m) + \alpha + \beta\ m = 0, \tag{4.69}$$

or,

$$n(m) = e^{-\alpha - \beta m}. \tag{4.70}$$

Constraints provide $N = e^{-\alpha}/\beta$ and $M = N/\beta$ yielding,

$$n(m) = \frac{N^2}{M} e^{-Nm/M}. \tag{4.71}$$

Finally, identifying the average fragment mass $\mu = M/N$ provides the probability density distribution over fragment mass,

$$p(m) = \frac{1}{\mu} e^{-m/\mu}. \tag{4.72}$$

It is observed that the present maximum entropy distribution provides the same exponential distribution as the earlier Poisson process with a uniform hazard function.

Englman et al. [21] have also pursued a maximum entropy formulation in which energy dependence in the fragmentation event is retained, and included as a further constraint. This development leads to a more complex distribution that is power law over a limited range of fragment size. Cohen [14] develops a fragmentation entropy expression and provides a method, through the second law of thermodynamics, for examining the applicability of commonly used empirical fragment distribution relations.

Energy and Statistical Fragment Distributions

A number of the statistical formulations that provide relations for the distributions in fragment size pay little or no attention to energetics in the breakup event. These include the various geometric fragmentation algorithms, the hazard function methods discussed earlier, fractal considerations, as well as the previous entropy maximum approach. Even the Mott strain-to-failure model, while accounting for key physical details of the fracture interaction process, does not address energy consumed in the fragmentation event.

There is evidence that fracture energy governs the characteristic fragment size in the breakup event as attested to, for example, by the repeated consistency with experimental data of the Rittinger [85] fragment size relation, or the more recent energy-based fragmentation models (described in later sections). It is also reasonable to consider energy principles in the processes governing the statistical distribution of fragments.

Griffith [51] put forth an interesting energy-based statistical formulation leading to functional expressions for fragment size distributions. (Note that this is not the same Griffith whose seminal works were instrumental in the development of classical fracture mechanics.) The approach proposed by Griffith was investigated briefly by Grady and Kipp [43], and pursued further in Grady et al. [48] and Grady and Winfree [46]. The resulting distributions will be found to differ starkly from the exponential distribution over fragment mass provided by the previous entropy maximum method as well as several of the earlier developments.

The approach is as follows: Consider a body of total mass M. Conceptualize the partitioning of that body into arbitrary but equal sized mass elements (referred to as molecules by Griffith). Now consider the random partitioning of a total energy E among the mass elements. If the energy content of an element is viewed as discrete and indexed by an integer j then classical statistical mechanics provides the statistical number of mass elements n_j having an energy e_j as,

$$n_j = Ae^{-\beta e_j}, \tag{4.73}$$

where A and β are normalizing parameters constrained by the total mass M and energy E. Recognizing that n_j is proportional to the mass fraction of the body with energy e_j, and going to the limit of a continuous function, one arrives at,

$$m(e) = Ae^{-\beta e}, \tag{4.74}$$

where $m(e)$ is the mass distribution over energy. Constraining A and β yields,

$$\lambda(e) = \frac{1}{\bar{e}} e^{-e/\bar{e}}, \tag{4.75}$$

where $\lambda(e) = m(e)/M$ is the spectral mass fraction and $\bar{e} = E/M$ is the average specific energy.

The argument is then made that the specific energy e associated with a mass element can be related to the size x of the fragment within which it resides. The issue of whether the element resides on the surface of, or within the interior of, the fragment is not raised. A fragment of diameter x has surface energy Γx^2 and mass ρx^3 (numerical coefficients are ignored). Thus the specific energy and size are related through,

$$e = \frac{\Gamma}{\rho x}, \tag{4.76}$$

where the surface energy Γ is assumed to be independent of the size of the fragment.

Identifying a size scale through $\sigma = \Gamma/\rho\bar{e}$ (4.75) transforms to a mass fraction distribution over fragment size through the relation $\lambda(e(x))dx = \lambda(e)de$ yielding,

$$\lambda(x) = \frac{\sigma}{x^2} e^{-\sigma/x}, \tag{4.77}$$

where $\lambda(x)$ is a different functional form and $\sigma \ll L$, the size of the unbroken body, is assumed.

Equation (4.77) is a well-behaved distribution that nicely captures the statistical size cutoff in the small particle limit. Equally important (4.77) leads to a finite integrated surface area (and energy), a feature not found in several more prominent empirical size distribution relations. The transformation from (4.75) to (4.77) seems to be the correct analytic method but it differs importantly from the same step presented by Griffith [51]. The significance is that the resulting relations of Griffith are more in line with experimental fragment distributions of brittle solids than is the derivation continued here. Those interested in pursuing energetic approaches to fragment sizes and distributions are encouraged to inspect in detail the original analysis of Griffith [51].

Equation (4.77) can be related to the empirical Rosin–Rammler [86] mass versus fragment size density distribution relation,

$$m(x) = \frac{M}{\sigma} \left(\frac{x}{\sigma}\right)^{n-1} e^{-(x/\sigma)^n}. \tag{4.78}$$

The energy-based distribution from (4.77) is equivalent to (4.78) for the special case of $n = -1$ and provides, for this special case, a physical statistical mechanical basis for the above empirical distribution. It is observed, however, that $n = -1$ falls outside the range of about $0.5 \leq n \leq 1.5$ for this parameter commonly associated with the Rosin and Rammler distribution. More recently

Brown and Wohletz [10] have also defended a Weibull (Rosin and Rammler) form for fragment size distributions on physical grounds.

For $n = -1$ (energy-based statistics) (4.78) provides,

$$m(x) = \frac{M}{\sigma}\left(\frac{\sigma}{x}\right)^2 e^{-(\sigma/x)}. \tag{4.79}$$

In the limit of large fragments $x \gg \sigma$ this distribution becomes,

$$m(x) = M\frac{\sigma}{x^2}, \tag{4.80}$$

and the number density,

$$n(x) = \frac{m(x)}{\rho x^3} = \frac{M}{\rho}\frac{\sigma}{x^5}. \tag{4.81}$$

Equation (4.81) is an energy-based statistical relation intended for providing probabilistic predictions of the large fragments produced in an impulsive fragmentation event. Applications, for example, would be hypervelocity impacts on space debris shielding where concerns are with most lethal (probably largest) fragments generated in such events [46, 48]. It is also of interest that the energy-based relation of (4.88) provides a power law dependence between number and size (hence fractal) in contrast to the Rosin–Rammler relation which is exponential over the same size range. Fractal processes in fragmentation are discussed further in a later section.

Justification for the applicability of an equilibrium energy-based statistical mechanical approach to size distribution in mechanical breakage is problematic. Numerous theoretical efforts have been pursued based on various randomization processes not associated with energetics and have made sensible correlations with experimental data lending some credence to these methods. None is capable of satisfactorily describing the broad range of experimental fragment size distributions that have been observed, however. A number of approaches including geometric statistics and percolation theories lead in the limit to exponential fragment distributions and appear to reasonably describe the results of very simple fragmentation experiments such as the breakup of expanding rings and shells. They do not adequately represent fragment size distributions resulting from more complex breakup events, however. It has been argued [42] that the additional complexity is a consequence of a spectrum of strain rates participating in the fragmentation event and that a superposition of exponentials (Poisson mixtures) could adequately characterize the distribution. This conjecture, however, does not lead to a predictive theory.

Empirical distributions continue to be used in application with the Rosin–Rammler (Weibull) distribution noted above one of the more common. A comparison of the energy-based statistical mechanical distribution with the Rosin–Rammler distribution is shown in Fig. 4.11. Several curves for

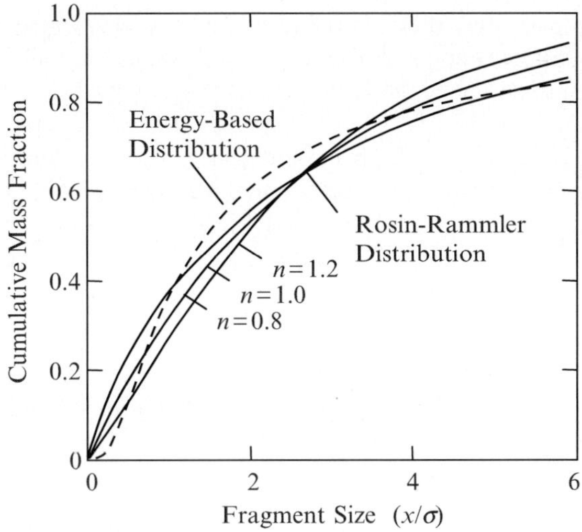

Fig. 4.11. Cumulative fragment mass versus fragment size comparing the energy-based statistical mechanical distribution with the empirical Rosin–Rammler distribution with $n = 0.8$, 1.0, 1.2

the latter are shown covering observed values of the distribution index n. Although quite adequate within the range of most data the empirical distribution has problems in the tails of the distribution. In the small particle limit, the Rosin–Rammler relation can be shown to integrate to infinite fragment number and, depending on the value of n, infinite surface area as fragment size approaches zero. Also, as noted above, in the large particle limit the expected length-scale-invariant power-law behavior is not provided. Discrepancies between the statistical mechanical and the empirical distributions in the large fragment size limit are normally not of concern because of the relatively few large fragments constituting this end of the distribution. This portion of the distribution is important, however, to applications where statistical prediction of the large fragments produced in the breakup process is a key objective. As observed in the figure, the statistical mechanical distribution predicts more frequent occurrence of large fragments than does the empirical Rosin–Rammler distribution.

Distribution Extremes

Issues of the statistical distribution of fragment size in a dynamic fragmentation event becomes critical when concern is with the distribution extremes; either at the small end, or the large end, of the fragment distribution. For example, as briefly discussed in the previous subsection, concern may be with the few largest fragments produced in the breakup event. Large fragments are a problem in quarry and mine blasting where oversize material requires secondary fragmentation efforts. Successful application of shielding with spaced armor can depend on the size of the largest fragments produced.

If a collection of fragments is reasonably described by a probability density distribution $f(m)$, with m the fragment mass, then a common and expedient method for estimating the statistical largest fragment follows from the complementary cumulative number distribution function,

$$N(m) = \int_{m}^{\infty} N_o f(m) dm, \tag{4.82}$$

with N_o the total number of fragments in the distribution. An estimate of the mass of the largest fragment $m_{\max}$ follows from solution of the relation,

$$N(m_{\max}) = \int_{m_{\max}}^{\infty} N_o f(m) dm = 1. \tag{4.83}$$

For example let $f(m) = (1/\mu)\exp(-m/\mu)$, the Poisson distribution, and,

$$N(m) = N_o e^{-m/\mu}. \tag{4.84}$$

Set $N(m_{\max}) = 1$ and the solution,

$$m_{\max} = \mu \ln N_o, \tag{4.85}$$

provides an estimate of the largest fragment in the distribution. For $N_o = 10$ the largest fragment is about 2.3 times the mean; for $N_o = 100$ the largest fragment is approximately 4.6 μ.

A more careful statistical assessment of the largest fragments is possible, however. The field of extreme value statistics is concerned with the largest and smallest (or extreme) values in a distribution (e.g., [2]). Let the cumulative probability function be,

$$F(m) = \int_{0}^{m} f(m) dm, \tag{4.86}$$

then, in the statistical treatment, the mass $\bar{m}$ of the largest fragment in a sample of N_o fragments will have a cumulative probability distribution function given by,

$$F_{N_o}(\bar{m}) = (F(\bar{m}))^{N_o}, \tag{4.87}$$

where $\bar{m}$ spans the same range as m. The corresponding probability density function for the largest fragment is then provided by,

$$f_{N_o}(\bar{m}) = \frac{d}{d\bar{m}} \left(F_{N_o}(\bar{m})\right). \tag{4.88}$$

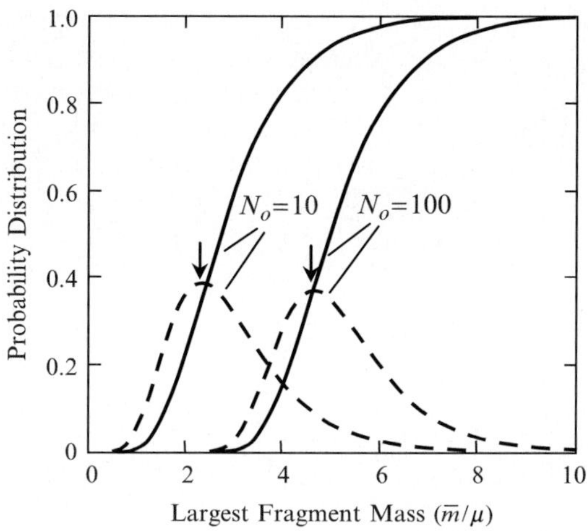

Fig. 4.12. Cumulative probability and probability density functions for largest fragment mass are illustrated for Poisson fragment distribution of 10 and 100 fragments. The arrows identify the distribution mode and corresponds estimate base on (4.85)

Again, consider the Poisson distribution $f(m) = (1/\mu)\exp(-m/\mu)$, for which the largest fragment cumulative distribution for N_o fragments is,

$$F_{N_o}(\bar{m}) = \left(1 - e^{-\bar{m}/\mu}\right)^{N_o} . \tag{4.89}$$

The Poisson largest fragment cumulative distribution from (4.89) and the corresponding probability density distribution derived from (4.88) are shown in Fig. 4.12 for a collection of $N_o = 10$ and $N_o = 100$ fragments. It is interesting that the estimate of largest fragment size provided by (4.85) corresponds to the distribution mode (identified by the arrows in the figure) and approaches a cumulative probability $F_{N_o}(m_{\max}) = 1/e \simeq 0.368$ for sensibly large N_o.

As the number of fragments in the distribution becomes large (4.87) approaches,

$$F_{N_o}(\bar{m}) = e^{-N_o(1-F(\bar{m}))} . \tag{4.90}$$

Again, for the Poisson distribution, (4.90) yields,

$$F_{N_o}(\bar{m}) = e^{-N_o e^{\bar{m}/\mu}} , \tag{4.91}$$

the Gumbel extreme value distribution. For $N_o = 100$ (4.91) reproduces the curve in Fig. 4.12 and even for as low as $N_o = 10$ the equation is a close approximation to the corresponding curve in the figure.

Fragment Velocity Statistics

The topics addressed in the present section up to this point have focused strictly on the statistics of the size (or the mass) of fragments produced in

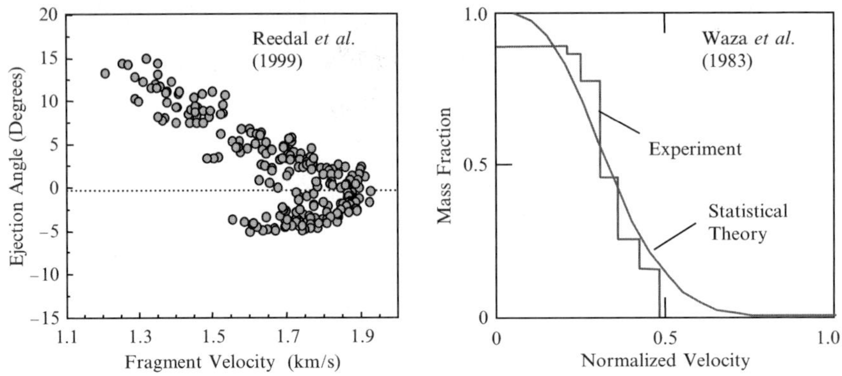

Fig. 4.13. Experimental observations of statistical variations in fragmentation ejecta velocity. Explosive fragmentation of a tungsten cylinder [84]. Impact fragmentation of a basalt rock [96]

a dynamic fragmentation event. Such events, the explosive fragmentation of a metal shell, or the spall debris resulting from a high-velocity impact, eject fragments at substantial velocity. There is observed to be a statistical spread in the ejecta velocity resulting from these events. Very little theoretical effort has been focused on the statistical aspect of ejecta velocity in dynamic fragmentation.

Two examples are illustrated in Fig. 4.13. The first shows measured experimental fragment velocities resulting from the natural fragmentation of an exploding tungsten cylindrical shell [84]. The ordinate is the angle of the ejection trajectory relative to the equator of the cylinder. Of interest here is the significant spread in velocities at any specific ejection angle. The second example is from a laboratory study attempting to replicate planetesimal collision events through the high-speed impact of steel projectiles onto spheres of basaltic rock [96]. The graph provides a cumulative distribution by mass of the velocity of the ejected fragments in a center of momentum coordinate system.

After a dynamic fragmentation event, expansion kinetic energy resides in the outward directed motion relative to some center of momentum associated with the fragmentation event. For an expanding cylindrical shell, for example, it would be relative to the average terminal case velocity.

The present development will consider an easily visualized example. Consider a spherical solid at the center of which an explosion leads to the fragmentation of the solid and outward expulsion of the resulting fragments. Depending on the energy of the explosion, the resulting collection may consist of 2, 3 4, 5 or any larger number of fragments. The event is illustrated for three fragments in Fig. 4.14. An impulsive fragmentation event has led to the expulsion of N (with $N = 3$) fragments resulting from the break up of a body of mass M in to individual fragment masses m_i. Each fragment

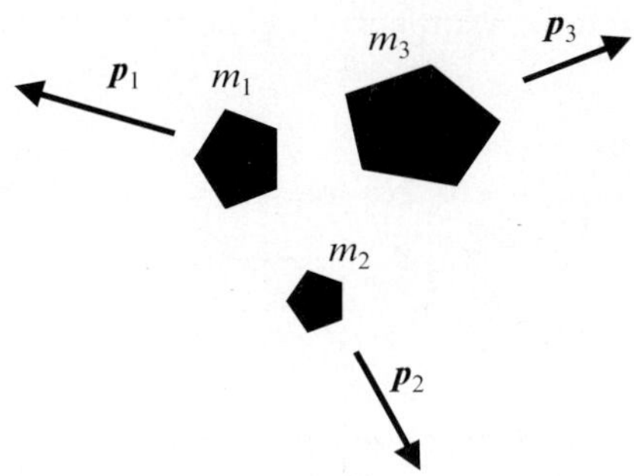

Fig. 4.14. Fragmentation event results in three fragments with statistically determined masses and momenta

emerges from a stationary center-of-momentum point with velocity u_i. The total kinetic energy of the fragment collection is E. The issue of interest is the probabilistic distribution of velocities (or momenta) over the collection of fragments subject to the following constraints on mass, momentum and energy (bold type indicates a vector quantity whereas normal type provides the scalar value of the same property),

$$\sum_{i=1}^{N} m_i = M, \tag{4.92}$$

$$\sum_{i=1}^{N} \boldsymbol{p}_i = \sum_{i=1}^{N} m_i \boldsymbol{u}_i = 0, \tag{4.93}$$

$$\sum_{i=1}^{N} e_i = E. \tag{4.94}$$

The quantities e_i and $\boldsymbol{p}_i$ are the individual fragment kinetic energies and vector momenta related through,

$$e_i = \frac{\boldsymbol{p}_i^2}{2m_i} = \frac{m_i \boldsymbol{u}_i^2}{2}. \tag{4.95}$$

In certain applications, physical conditions resulting from details of the fragmentation event will constrain the distribution of velocities over the resulting fragments. Here the premise is that no knowledge is available as to how the energy E is distributed over the N fragments and one simply seeks the most probable distribution given the specified constraints. The problem as posed is amenable to classical statistical mechanical. The methods are described in such classical papers as those of Maxwell [74] and Jeans [53]. The present development follows closely the derivation of Levesey [69] on the statistical treatment of disintegration in nuclear reactions.

The analysis is most easily followed by considering a limited number of fragments. The solution is pursued for $N = 3$ and then generalized to arbitrary fragment number. Momentum conservation from (4.93) yields the vector condition,

$$\boldsymbol{p}_1 + \boldsymbol{p}_2 + \boldsymbol{p}_3 = 0. \tag{4.96}$$

The rather obscure substitutions are made,

$$\boldsymbol{q}_1 = \sqrt{\frac{M}{M - m_1}}\boldsymbol{p}_1, \tag{4.97}$$

$$\boldsymbol{Q} = \sqrt{\frac{m_1 (m_2 + m_3)}{m_2 m_3}} \left(\frac{m_2}{m_2 + m_3}\boldsymbol{p}_1 + \boldsymbol{p}_2 \right), \tag{4.98}$$

that, when combined with energy conservation from (4.94), yield the expression,

$$2m_1 E = q_{1x}^2 + q_{1y}^2 + q_{1z}^2 + Q_x^2 + Q_y^2 + Q_z^2. \tag{4.99}$$

From this point the subscript on m_1 will be dropped, as the intention was to isolate attention to one fragment as constrained by the remaining fragments through the relations provided in (4.92)–(4.94).

Equation (4.99) is a six-dimensional hyper-sphere in the space of the six independent momentum components with radius R given by,

$$R = \sqrt{2mE}. \tag{4.100}$$

It is the classic functional form aspired to before ascribing equal statistical weights to equal volumes in momentum phase space. The statistical distribution of states for a fragment of mass m is then obtained by integrating over the hyper-sphere the remaining momentum variables. The process results in a probability distribution for kinetic energy of a fragment of mass m of,

$$f(e) \propto e^{1/2} \left(E - \frac{eM}{M - m} \right)^{1/2}. \tag{4.101}$$

Equation (4.101) provides the distribution for a mass M fragmented into $N = 3$ fragments with total kinetic energy E. From the equation the maximum kinetic energy (hence velocity) available to a fragment of mass m is,

$$e_{\max} = \frac{M - m}{M} E. \tag{4.102}$$

The analysis can be carried through for an arbitrary number of fragments N, and leads to a kinetic energy probability distribution for fragments [69],

$$f(e) \propto e^{1/2} \left(E - \frac{eM}{M - m} \right)^{\frac{3N-8}{2}}. \tag{4.103}$$

The transition to a Maxwell energy distribution follows when N becomes large and $M >> m$. Equation (4.103) converges to,

$$f(e) \propto e^{1/2} (E - e)^{\frac{3N}{2}}, \qquad (4.104)$$

and in turn for large N becomes the exponential Maxwell kinetic energy distribution,

$$f(e) \propto e^{1/2} \exp(-3e/2\bar{e}), \qquad (4.105)$$

where $\bar{e} = E/N$ is the average kinetic energy per fragment. For large N the energy distribution becomes independent of the fragment mass m. The corresponding velocity distribution resulting from the appropriate substitution in (4.105) does depend on the fragment mass, weighting the distribution for heavier fragments toward correspondingly lower velocities.

The distribution from (4.103) is illustrated in Fig. 4.15 for selected values of the fragment number. In the left hand plot the fragment size remains the same at $m = M/4$ while in the right hand plot all fragments are equal size $m = M/N$ in each distribution. Also illustrated in the right hand plot is the limiting Maxwell distribution from (4.105) for an average energy of $\bar{e} = E/5$ showing that for five fragments the distributions are already close.

There are many impulsive fragmentation events for which the statistical velocity distributions developed in this section would not apply. Detailed geometry and impact conditions would lead to a spectrum of velocities which may differ markedly from the Maxwell representation of (4.105). In events where initial conditions are not as precisely determined, or in the average response of many similar events, the present approach may provide useful statistical predictions. An application of the statistical theory has been applied to the data of Waza et al. [96] and is shown with the data in Fig. 4.13 [46,48].

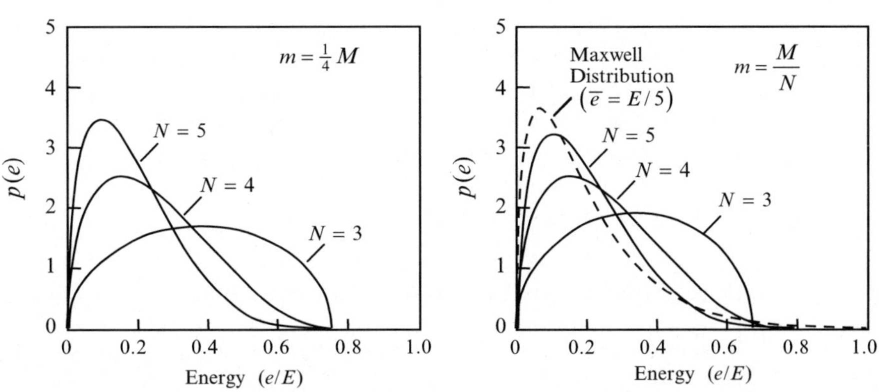

Fig. 4.15. Kinetic energy probability for 3, 4 and 5 fragments. Fragment mass is $M/4$ on the left and all fragments are equal size, M/N, on the right. Maxwell distribution for average mass of $E/5$ is shown for comparison

4.2 Fragment Size in Dynamic Fragmentation

In a dynamic fragmentation, the foremost issue by far is characterization of the intensity of the fragmentation. That is, in the simplest sense, the number of fragments the disrupted object breaks into, or, conversely, the average fragment size. This query can take a number of forms with differing levels of scrutiny. In some applications, prediction of the amount of new surface area created in the event may be adequate. In others, how this surface area distributes over fragment size is paramount. The concern may be with the few largest fragments created, or it may be with finest airborne particulate. A further complicating factor is that fragmentation intensity in a complex event varies continuously with position in the body. This is important where velocity and directionality of the resulting debris, as well as size, is an issue. Some of the issues as they relate to statistical fragment size distributions were pursued in the previous section.

Conditions driving the fragmentation intensity are many and have been characterized in various ways by analysts exploring the problem. Energy imparted to the body through the loading process leading to disruption, and attempts to correlate energy with the fragmentation intensity, was common in earlier approaches. Stress amplitude and impulse criteria have also been pursued. A kinematic criterion based on the rate of strain carrying the body to disruption has had some success. Geometry of the fragmenting body also plays an important role.

The material properties governing the resistance (or lack thereof) of the structure or body to the fracture processes leading to fragmentation are also quite varied. In some materials, continuum strength properties, such as fracture toughness, yield strength, and surface tension in liquids, appear to correlate with the intensity of fragmentation. In others, characterization of microscopic defect structure is apparently necessary to prediction of the resulting fragmentation. Mott's seminal theory of fragmentation is unique in requiring a determination of the statistical strain to failure of the sites responsible for fracture activation.

In the present section, some of the theoretical efforts that have been pursued to predict size (or new surface area) in dynamic fragmentation events are described. This survey is not exhaustive, but does provide a perspective for some of the directions that have been considered.

4.2.1 Historical Theories of Fragment Size Prediction

Much of the historical work in attempting to quantify fragment size in fragmentation events was stimulated by needs in the mine and quarry blasting, and the rock crushing, industries. In the arena of military application, the studies of Neville Mott during the World War II years are seminal.

Fragment Size Laws of Rittinger, Kick and Bond

Early efforts in ascertaining the reduction in size of a body subjected to a fragmentation event focused on the energy input to the system in the breakage process. The workers in this area were concerned with brittle and semi-brittle materials, and processes such as mine and quarry blasting, rock and mineral crushing, and grinding were applications of interest.

Apparently first, and most widely recognized, is the size reduction law of Rittinger [85] relating the proportionality between the specific energy input E to the increase in new fragment surface area,

$$dA \sim dE. \tag{4.106}$$

A characteristic fragment size x is proportional to the fragment distribution length scale and, in turn, $x \sim 1/A$ where A is the fragment surface area per unit volume. Consequently, the Rittinger law leads to a reduction in fragment size with a dependence on energy of the form,

$$x = \frac{x_o}{1 + \alpha E}, \tag{4.107}$$

where x_o is the initial size of the body and α an efficiency index of the fragmentation process.

An alternative theory of fragmentation was proposed by Kick [61] whereby equivalent amounts of energy resulted in equivalent geometric reductions in the sizes of fragments. The Kick hypothesis leads to,

$$E \sim \ln(x_o/x), \tag{4.108}$$

or,

$$x = x_o e^{-\alpha E}. \tag{4.109}$$

Some time later, Bond [6] recognized that neither Rittinger's law nor Kick's law fully worked in the engineering applications of interest and proposed a relation that turns out to be a compromise between the two previous relations resulting in,

$$x = \frac{x_o}{(1 + \alpha E)^2}. \tag{4.110}$$

Summarization of Charles

The fragment size relations of Rittinger, Kick, and Bond are nicely summarized in a study of Charles [11] where he generalized the earlier relations into the differential expression,

$$dE = -C dx / x^{\eta}, \tag{4.111}$$

where dE is the increment of the specific energy required to achieve a size reduction dx. Values of $\eta = 1$, $3/2$ and 2 retrieve the relations of Kick, Bond and Rittinger, respectively.

There is ample evidence over the diverse applications of interest in materials fragmentation to support any of the three fragment size laws under certain circumstances. In testing any of the relations, however, the difficulty is first, quantifying the energy input responsible for fragmentation, and second, characterizing the appropriate fragment size. Size was commonly characterized as the modulus or length scale k of a measured size distribution such as the frequently used Schuhmann [87] cumulative weight versus fragment size relation,

$$M(x) = (x/k)^n. \tag{4.112}$$

Energy characterization usually reduces to quantifying the proportional work in the breakage event, such as the relative time in a particular crusher, or the relative height in a drop-tower fragmentation apparatus. Efforts that more carefully quantify the strain energy fueling fracture in the brittle fragmentation process appear few, but one extensive effort by Bergstrom et al. [5] is discussed in a later section on brittle fragmentation.

Charles [11] examined diverse fragmentation data and found that the parameter η in his (4.111) depended on the breakage method applied, but ranged within the bounds provided by the Rittinger and the Kick fragment-size-reduction laws. Combining the Schuhmann distribution in (4.112) with his (4.111) provided a size reduction relation of the form,

$$k = \frac{x_o}{(1 + \alpha E)^{1/(\eta-1)}}. \tag{4.113}$$

A relation for α in terms of the energy reduction constants η and C and the Schuhmann index n is provided, however Charles identifies α as the appropriate empirical comminution constant. Although largely empirical, (4.112) and (4.113) provided a means of systemizing a broad body of fragmentation data on brittle materials and were eminently useful in application.

Mott Statistical Strain-to-Fracture Theory

Mott in 1943 encountered an entirely different application in dynamic fragmentation and his theoretical approach to the problem differs starkly from the previous authors. Mott addressed the issue of fragmenting munitions and undertook an effort to understand the influences of case metal, explosive properties, and system geometry, on the processes of natural explosive-driven fragmentation. His brief efforts within the early months of 1943 resulted in a remarkably sophisticated theory governing the size and statistics of the fragmentation process in exploding warheads. His seminal efforts continue to have influence in this area of dynamic fragmentation to the present day. In the present section the early theoretical ideas of Mott are briefly described. More on Mott's work is pursued in other sections of this chapter.

Mott was interested in the explosion-driven fragmentation of tubular munitions and his theoretical efforts primarily focus on the statistical frequency of axial fractures around the circumference of the cylindrical shell. The steel case metals investigated by Mott were observed to undergo extensive plastic expansion prior to fracture onset and fragmentation. Mott concluded that an ultimate tensile stress Y, and not the initial yield stress, was a governing fragmentation property and showed that when a fracture occurred, as previously illustrated in Fig. 4.8, regions away from the fracture were relieved of the driving stress by a diffusion-like wave which propagated according to the relation,

$$x \sim \sqrt{\frac{Yt}{\rho\dot{\varepsilon}}} \sim \sqrt{\frac{Y}{\rho\dot{\varepsilon}^2}}\varepsilon^{1/2}, \tag{4.114}$$

where x is the distance propagated after a time t from fracture initiation. The metal density is ρ, and $\dot{\varepsilon} \simeq u/r$, where u is the radial velocity of the case and $\dot{\varepsilon}$ is the circumferential stretching rate.

Although (4.114) is probably a reasonable approximation of the physics considered by Mott for the fracture of a plastically deforming metal shell, it is instructive in the present survey to also consider elastic release from points of fracture for comparative purposes. This would perhaps be more appropriate for brittle fracture where the elastic limit is not achieved before failure. The relation corresponding to (4.114) would then be,

$$x \sim ct \sim \frac{c}{\dot{\varepsilon}}\varepsilon, \tag{4.115}$$

where c is a measure of the elastic wave speed in the material. In (4.114) or (4.115) the length x is the nominal distance that the release wave will propagate before interacting with an adjacent fracture and is, consequently, proportional to the average circumferential spacing of the fractures. The time t and the strain $\varepsilon = \dot{\varepsilon}t$ are the time and corresponding cumulated strain over which the dynamic fracture process proceeds.

Mott proposed that fracture in the expanding shell was dependent on the strain and could be characterized by a function $\lambda(\varepsilon)$ that provided the statistical rate of fracture activation per unit length of the circumference. As discussed in the previous section, Mott supposed that $\lambda(\varepsilon)$ should be a rapidly increasing function of strain and explored both a power law,

$$\lambda(\varepsilon) = \frac{n}{\sigma}\left(\frac{\varepsilon}{\sigma}\right)^{n-1}, \tag{4.116}$$

and an exponential law,

$$\lambda(\varepsilon) = \frac{1}{\sigma}e^{(\varepsilon-\mu)/\sigma}. \tag{4.117}$$

The power law activation function in (4.116), or the exponential function in (4.117), lead to extreme value statistics of the Weibull and the Gumbel form, respectively.

The strain interval in (4.114) or (4.115) is determined by the statistical spread in strain allowed by either the Weibull or the Gumbel distribution and is proportional to the standard deviation in activation strain provided by either distribution. Thus, for an exponential activation function and the Gumbel distribution, the standard deviation is of order σ and,

$$\varepsilon \sim \sigma. \tag{4.118}$$

Thus, (4.114) provides for the characteristic fragment size,

$$x \sim \sqrt{\frac{Y\sigma}{\rho\dot{\varepsilon}^2}}, \tag{4.119}$$

which is functionally the same as the expression provided by Mott [79], where $\gamma = 1/\sigma$ is the well known gamma parameter introduced by Mott. Alternatively, (4.115) for elastic release leads to,

$$x \sim \frac{c\sigma}{\dot{\varepsilon}}. \tag{4.120}$$

In either case (4.119) or (4.120) the characteristic fragment size is dependent on the expansion rate $\dot{\varepsilon}$ to the inverse first power.

Use of the power-law activation function in (4.116), and the Weibull distribution is more subtle as was pointed out by Mott. The distribution moments for the Weibull distribution are dependent on the length scale of interest. Consequently, the strain interval corresponding to (4.118) for the Gumbel distribution is,

$$\varepsilon \sim \frac{\sigma}{x^{1/n}}. \tag{4.121}$$

Combining (4.114) and (4.121) leads to,

$$x \sim \left(\frac{Y\sigma}{\rho\dot{\varepsilon}^2}\right)^{\frac{n}{2n+1}}, \tag{4.122}$$

which provides a fragment size dependence on $\dot{\varepsilon}$ ranging between an inverse first and an inverse two-thirds power as reported by Mott. Equation (4.115) for elastic fracture yields,

$$x \sim \left(\frac{c\sigma}{\dot{\varepsilon}}\right)^{\frac{n}{n+1}}, \tag{4.123}$$

with an inverse first to inverse one-half power dependence on strain rate.

4.2.2 Weibull Critical Flaw Fragmentation

A model of dynamic fracture and fragmentation for brittle solids has been pursue that provides a method for predicting average and distributed fragment size in impact or explosive loading applications [Grady and Kipp, 1980]. The model has been successfully implemented into computational codes, and used to predict consequences of blasting in hard rock, explosive fragmentation of oil shale, and impact breakup of ceramics, among others. The model is seen to include features of the Weibull critical state theory for the stress-load induced fracture of solid bodies. This relationship between the theories offers alternative methods for assessing the material fracture parameters essential to the model. A strong similarity of the present model to the statistical activation and interaction fragmentation theory of Mott [77] is also apparent.

Central to the model is a statistical population of fracture-producing flaws characterized by the tensile elastic strain ε (or stress) at which fracture is activated. In computational applications this parameter is usually the mean tensile strain, or that collinear with the maximum principal tensile stress. Fracture-producing flaws are characterized through a statistical cumulative number density expression of the power-law form,

$$N(\varepsilon) = (\varepsilon/\eta)^m. \tag{4.124}$$

Note that the parameter η has dimensions of $[\mathrm{L}]^{3/m}$, and consequently does not determine an elastic strain (or stress) scale but rather the product of a strain and a length to some power.

At a site of fracture activation, release of the tensile stress is considered to propagate spherically outward with a velocity c_g. A limiting constant fracture growth velocity in the brittle material is assumed for c_g. The region of stress release about a site after fracture activation is then the volume,

$$\upsilon(t) = (c_g t)^3. \tag{4.125}$$

Kinematics of the tensile fracture event is characterized by an expansion strain rate $\dot{\varepsilon}$ such that,

$$\varepsilon = \dot{\varepsilon}t. \tag{4.126}$$

Again, in computer implementation the expansion rate is either an average of the principal strain rates, or that corresponding to the maximum principal tension.

Material properties for the model thus require the flaw structure parameters, η and m, the fracture stress release velocity c_g, and an elastic modulus and density, E and ρ. The model relations laid out in (4.124), (4.125) and (4.126) combine to provide the fraction of stress relieved material as dynamic fracture proceeds,

$$D(t) = \int_0^t \upsilon(t-\tau)dN = \int_0^t \upsilon(t-\tau)N'(\varepsilon)\dot{\varepsilon}\, d\tau. \tag{4.127}$$

In the present formulation of the brittle fracture model $D(t)$ is identified as the fracture damage, and is also the volume fraction of the fractured and stress-released material, with $0 \leq D(t) \leq 1$. The damage parameter D is used to determine both the fracture activity and the continuum stress state as a function of time through the fracture process. A simple linear energy function, with the damage D identified as an isotropic internal state variable,

$$U = \frac{1}{2} E\varepsilon^2 (1 - D), \tag{4.128}$$

provides both the fracture strain energy release,

$$\Gamma = - \left. \frac{\partial U}{\partial D} \right|_{\varepsilon} = \frac{1}{2} E\varepsilon^2, \tag{4.129}$$

and the continuum stress,

$$\sigma = \left. \frac{\partial U}{\partial \varepsilon} \right|_{D} = E(1 - D)\varepsilon, \tag{4.130}$$

with $E(1 - D)$ the fracture-damage reduced elastic modulus of the material.

Fractures activated in the dynamic event are provided through the relation,

$$N_g(t) = \int_0^t (1 - D) N'(\varepsilon) \dot{\varepsilon} d\tau = \int_0^t e^{-D} N'(\varepsilon) \dot{\varepsilon} d\tau. \tag{4.131}$$

In the first equation $1 - D$ approaches zero through (4.127) rather abruptly, and is probably not inappropriate for brittle failure. In the second, the exponential statistical convergence of Johnson and Mehl [54] theory is suggested which, when used, softens the rate at which damage approaches one. The spectrum of fracture number activation over time is provided by the derivative of (4.131),

$$n_g(t) = (1 - D) N'(\varepsilon) \dot{\varepsilon}. \tag{4.132}$$

A sensible measure of the fragment size distribution is provided by relating fragment size to the damage growth time within the vicinity of the fracture,

$$x \sim c_g (t_D - t), \tag{4.133}$$

and eliminating time between (4.132) and (4.133). The time t_D is provided by $D(t_D) = 1$ in (4.127). This approach assumes that fractures activate statistically over time, but arrest approximately together during the final catastrophic growth phase occurring at time t_D. Although this solution does not boast the rigor of the one-dimensional statistical Mott distribution, the arguments are nonetheless reasonable. Also like the Mott solution, however, is the prediction of a rather narrow spread in fragment size. The theory does not predict the Schuhmann like power law distribution observed in experimental brittle fracture, or the fractal behavior discussed in a later section. Selection of the mean, or the mode, of the distribution provides a sensible measure of the average fragment size.

It is further shown [Grady and Kipp, 1980] that under uniform strain rate loading a maximum stress (the fracture stress) is achieved,

$$\sigma_M = \alpha(m,\eta)E\dot{\varepsilon}^{3/(m+3)}, \tag{4.134}$$

with a characteristic fragment size,

$$x_m = \beta(m,\eta)c_g\dot{\varepsilon}^{-m/(m+3)}, \tag{4.135}$$

where $\alpha(m,\eta)$ and $\beta(m,\eta)$ are calculable functions of the brittle material flaw structure parameters m and η.

Close correspondence of the present dynamic fracture model with Weibull strength of materials relations is observed by noting that the derivative of the cumulative number density in (4.124),

$$N'(\varepsilon) = \frac{m}{\eta}\left(\frac{\varepsilon}{\eta}\right)^{m-1} = h(\varepsilon), \tag{4.136}$$

provides the power-law statistical failure hazard function $h(\varepsilon)$ integral to the Weibull theory. Cumulative probability of failure $P(\varepsilon)$ of a body of volume V, subjected to a stress load $\sigma = E\varepsilon$, is derived from,

$$dP = (1-P)Vh(\varepsilon)d\varepsilon, \tag{4.137}$$

or,

$$P(\varepsilon) = 1 - e^{-V(\varepsilon/\eta)^m}. \tag{4.138}$$

Scale dependence of the Weibull strength is noted where equivalent probability of failure requires,

$$V(\varepsilon/\eta)^m = \text{constant}. \tag{4.139}$$

4.2.3 Impulse and Energy Criteria for Fragment Size

A nominal fragment size resulting from a dynamic fragmentation event can be estimated from one of several reasonably basic physical arguments. Here methods are described in which elementary momentum and energy constraints are applied to estimate a characteristic fragment size in a dynamic fragmentation event.

Consider the one-dimensional fracture event pictured in Fig. 4.16 in which a fragment of length x_o is in the process of forming. A coordinate x is identified from the center of the fragment. The body before fracture proceeds is stretching at a uniform strain rate $\dot{\varepsilon}$ so that the right hand region from center

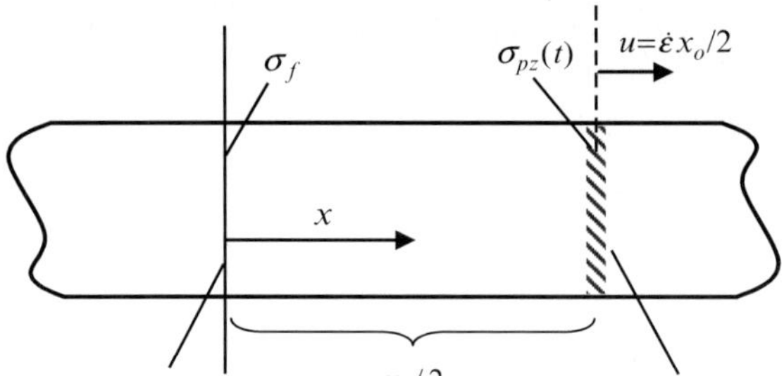

Fig. 4.16. Illustrates a fragment in the formation process for consideration of momentum and energy balance estimates of fragment size

to the fracture zone contains a momentum $\rho x_o^2\dot{\varepsilon}/8$. Through the fracture event momentum is reduce to zero when,

$$\int_0^{t_f} (\sigma_f - \sigma_{pz}(t))dt = \rho x_o^2\dot{\varepsilon}/8. \tag{4.140}$$

Assuming stress $\sigma_{pz}(t)$ in the fracture zone reduces rapidly to zero, (4.140) yields,

$$\sigma_f t_f = \sigma_f x_o/2c = \rho x_o^2\dot{\varepsilon}/8, \tag{4.141}$$

where c is the elastic wave speed determining the time over which the momentum is reduced to zero and σ_f is the tensile fracture stress. Solving for the length x_o gives,

$$x_o = 4\sigma_f/\rho c\,\dot{\varepsilon}, \tag{4.142}$$

for a measure of the nominal fragment size based on the premise that residual individual fragment momentum is brought to rest upon completion of fragment formation in the fracture event.

An alternative fragment size criterion is based on the assumption that the kinetic energy in the same region identified in the figure is fully dissipated in the fracture zone. Identifying this dissipation with a fracture energy Γ, energy balance yields the relation,

$$\frac{1}{24}\rho\dot{\varepsilon}^2(x_o/2)^3 = \Gamma, \tag{4.143}$$

providing a nominal fragment length of,

$$x_o = 2\left(24\Gamma/\rho\dot{\varepsilon}^2\right)^{1/3}. \tag{4.144}$$

Relating Γ to a more familiar *fracture toughness* through $\Gamma = K_c^2/2E$ yields,

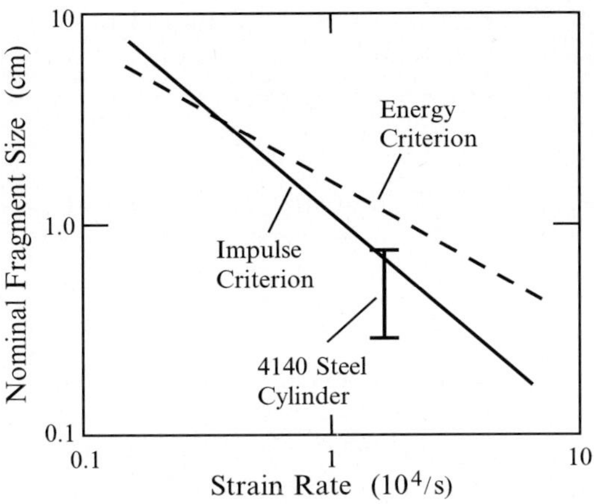

Fig. 4.17. Comparison of impulse and energy based fragment size prediction and data from 4140 steel cylinder fragmentation test

$$x_o = 2\left(\sqrt{12}K_c/\rho c\dot{\varepsilon}\right)^{2/3} . \tag{4.145}$$

Considering the relatively elementary momentum and energy principles used to estimate a nominal fragment size in (4.142) and (4.145), it is instructive to apply the relations to a reasonably well characterized material under representative fragmentation conditions. A 4140 steel heat treated to a Rockwell hardness of 40 has been used to investigate exploding case fragmentation [40]. Properties for the material include a density of 7,870 kg m^{-1}, an elastic modulus of 200 GPa, a yield stress of 1.1 GPa and a fracture toughness of 80 MPa $m^{1/2}$. Using the yield stress as a measure of the fracture stress (probably somewhat low due to the hardening properties of the steel) the predicted average fragment dimension (fracture spacing) at a strain rate in the neighborhood of $10^4\,s^{-1}$ from the (4.142) impulse-based ($\sigma_f = 1.1$ GPa) and the (4.145) energy-based ($K_c = 80$ MPa $m^{1/2}$) predictions are shown in Fig. 4.17. Curves are compared with the span of fragment widths covering about 80% of the mass of the fragments in one expanding cylinder fragmentation test on the 4140 steel. Quantitative agreement between the two predictive curves is remarkable considering the use of markedly different static properties in the functionally different dynamic relations. The different strain rate functionality ($\dot{\varepsilon}^{-1}$ for impulse and $\dot{\varepsilon}^{-2/3}$ for energy) is found to span the range of strain rate dependences arrived at in later more sophisticated fragment size analysis.

4.2.4 Mott Fracture with Dissipation

As discussed earlier in this chapter, Mott [77,79]) developed a fragmentation model for the breakup of an expanding ductile metal shell. He proposed a statistical strain to fracture model and calculated that the average fracture spacing was approximately,

$$x \sim \sqrt{\frac{Y\sigma}{\rho\dot{\varepsilon}^2}}, \tag{4.146}$$

when he assumed an exponential strain-to-fracture function (Gumbel statistics) or,

$$x \sim \left(\frac{Y\sigma}{\rho\dot{\varepsilon}^2}\right)^{\frac{n}{2n+1}}, \tag{4.147}$$

when he assumed a power law function (Weibull statistics). In both expressions, the strength property Y is the tensile flow stress at fracture, ρ the metal density, and $\dot{\varepsilon}$ the cylinder expansion rate. Also, in both cases, σ is proportional to the standard deviation in strain to fracture, and is the key material parameter governing fragmentation in the Mott statistical theory. As noted earlier $\sigma \sim 1/\gamma$ where γ is the Mott fragmentation parameter resulting from his development of the theory.

An alternative application of the Mott fragmentation model has led to markedly different physics governing the characteristic fracture spacing [47, 62]. Mott assumed that upon achieving the strain to failure criterion that fracture was instantaneous, and that the release of tensile stress propagates away from the point of fracture according to (solutions to propagation of Mott waves will be developed shortly),

$$\rho\dot{\varepsilon}\xi\frac{d\xi}{dt} = Y, \tag{4.148}$$

where ξ is the distance the Mott stress release wave has propagated from the point of fracture. The combination of (4.148) with the statistical strain to failure relations lead to the Mott predictions of characteristic fragment size provide in (4.146) or (4.147).

An alternative approach to the Mott problem assumes that fracture is not instantaneous, but that stress reduces continuously with some measure of the opening of the fracture [62]. A simple approach is to assume a crack opening model in which stress is reduced linearly from Y to zero with some fracture opening displacement parameter $0 \leq \eta \leq \eta_c$. In this case, a work of fracture $W = Y\eta_c/2$ is performed in completing the fracture process. An alternative relation, corresponding to (4.148), governing the motion of the stress release wave,

$$\rho\dot{\varepsilon}\xi\frac{d\xi}{dt} = \frac{Y^2}{2W}\eta, \tag{4.149}$$

is obtained where,

$$\frac{d\eta}{dt} = \dot{\varepsilon}\xi, \tag{4.150}$$

determines the crack opening displacement η. Solutions to this coupled system of ordinary differential equations are,

$$\xi(t) = \frac{1}{12} \frac{Y^2}{\rho W} t^2, \tag{4.151}$$

for the motion of the tensile release wave and,

$$\eta(t) = \frac{1}{36} \frac{\dot{\varepsilon} Y^2}{\rho W} t^3, \tag{4.152}$$

for the crack opening displacement over the interval $0 \leq \eta \leq \eta_c$. Rather than instantaneous fracture as assumed by Mott, a fracture duration of,

$$t_f = \left(\frac{72 \rho W^2}{Y^3 \dot{\varepsilon}} \right)^{1/3}, \tag{4.153}$$

is required for fracture completion.

Through substitution of this characteristic fracture time into (4.154) a distance of propagation of tensile release $\xi(t_f)$ is calculated. It was proposed [62] that twice this distance provides a reasonable measure of the nominal fracture spacing,

$$x = \left(\frac{24W}{\rho \dot{\varepsilon}^2} \right)^{1/3}. \tag{4.154}$$

Relating W to the Γ in the previous more elementary energy-based criterion provided by (4.144), finds that the present fragment size prediction is functionally the same (dimensional arguments would yield this conclusion) and a factor of two smaller.

The present relation for fragment size and that put forth by Mott in (4.147) differ starkly in the underlying physics, the functional dependence, and the governing material parameters. In the present (4.154) fracture energy controls the fragment size. In Mott's relation, a statistical strain-to-fracture parameter is the governing property.

Reconciliation of the two outwardly different theories for characteristic fragment size in the present one-dimensional ductile fracture model has been pursued [45]. The differences in the predicted strain rate dependence are readily understood, at least in part. Mott chose to use a Gumbel distribution to represent statistical strain to failure. The Gumbel distribution lacks a distribution shape parameter and leads to a unique inverse first power strain-rate dependence. Use of the more flexible Weibull distribution, however, which has a distribution shape parameter, leads instead to a nonunique strain rate dependence ranging between an inverse two-thirds and first power dependence, encompassing the energy-based prediction in (4.154). Intriguing, however, is that identical strain rate dependence between (4.147) and (4.154) requires $n = 1$ in the Weibull distribution, or a constant (uniform) fracture activation function.

Experimental data appear to support the predicted strain rate dependence over the range of nominal fragment sizes achieved in the fragmentation of ductile metals. Experimental examples supporting either bound (two-thirds or first power) can be found, however. For example, fragmentation of ductile uranium rings exhibits very nearly an inverse two-thirds power dependence on strain rate [45], whereas selected exploding steel cylinders show an inverse first power behavior [49].

Concerning the characteristic fragment size, Mott's statistical fragmentation theory and the energy-based theory can be sensibly reconciled if Mott's statistical fracture activation function is not necessarily equated to a fracture completion function. If, at some loading strain rate $\dot{\varepsilon}_o$, Mott's theory predicts an average fragment size smaller than that predicted by the energy-based theory, then it can be shown that all fractures activated by the Mott function cannot reach completion because of energy limitations. It is also readily shown that, for an exponential or strong powered law ($n \gg 1$) fracture activation function, a characteristic strain rate is determined from,

$$\dot{\varepsilon}_o \sim \sqrt{\frac{Y^3 \sigma^3}{\rho W^2}}. \tag{4.155}$$

At strain rates greater than $\dot{\varepsilon}_o$ the fracture hazard function, provided by (4.116) for fracture activation, is constant, or nearly constant, and of the form $\lambda(\varepsilon) = \lambda_o = 1/\sigma$. The statistical scale parameter σ is a continuum material length scale of order, $\sigma \sim W/Y$ with Y the flow stress at breakup and W the fracture energy.

A one-dimensional computational simulation of the Mott fragmentation process performed by Kipp and Grady [63] provides support for this interpretation of the two theories. At that time it was recognized that interplay between dynamics of the fragmentation event and the population of flaws seeding the multiple fracture process could lead to conditions in which flaw structure controlled the extent of fragmentation on one hand, while energy limitations controlled fragmentation on the other.

Computer simulations of dynamic fragmentation were performed to support experimental fragmenting ring studies performed at that time [47]. A one-dimensional finite difference wave code was used to calculate the response of an aluminum rod or wire 0.1 m in length stretching plastically at a flow stress $Y = 100\,\mathrm{MPa}$ and at a uniform stretching rate $\dot{\varepsilon} = 10^4\,\mathrm{s}^{-1}$. Fracture sites were introduced randomly in time according to a constant nucleation rate parameter $\lambda(\varepsilon) = \lambda_o$, and randomly positioned within the length of the rod. The nucleation rate λ_o was the only parameter varied over the series of calculations. When fracture was nucleated at a computational cell, stress in that cell was relaxed from the flow stress Y to zero as the cell distended, such that the plastic fracture energy within that cell of $W = 2 \times 10^4\,\mathrm{J\,ms}^{-2}$ was dissipated. The number of fragments produced as the nucleation rate λ_o was varied over approximately one order of magnitude was determined from the simulations and is shown in Fig. 4.18.

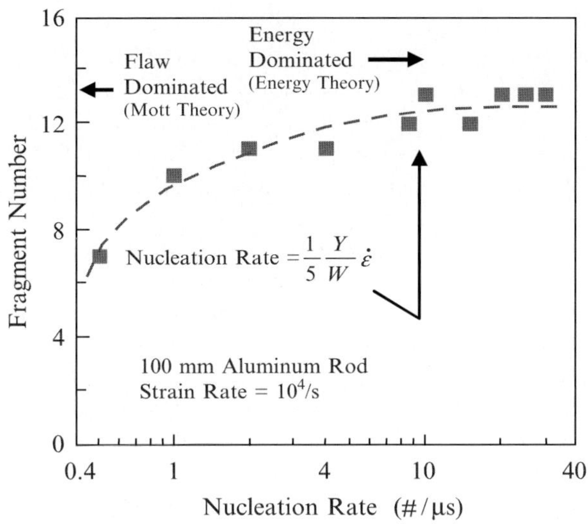

Fig. 4.18. Fragment number from computational simulations

At the lower nucleation rates every fracture nucleation site was computed to grow to full fracture. The number of fractures and the corresponding characteristic fragment size is then governed fully by the flaw structure and the fracture activation (seeding) function. As the nucleation rate is increased the number of activated fractures that fail to grow to completion, correspondingly increases. At the highest nucleation rates, the number of fragments becomes independent of the number of fracture sites nucleated and is determined strictly by the fracture energy W resisting fracture growth. The predicted energy-governed constant fracture survival rate, $\lambda_o \cong Y\dot{\varepsilon}/5W$, identified in the figure, is sensibly consistent with the computed transition from flaw-limited to energy-limited fragmentation. Distributions in fragment lengths from the computer simulations were also found to be consistent with predictions of the Mott statistical theory and with experiment.

4.2.5 Mott-Wave Solutions

In Mott's seminal study of the explosive-driven fragmentation of metal shells, an analysis of the propagation of stress release waves away from sites of fracture was integral to the model. The analysis was undertaken by Mott to assess the interacting of multiple fractures and to establish the characteristic spacing between fractures. Some of these features were introduced earlier. Here we focus on the solutions of such waves (Mott waves), including the original plastic wave solution derived by Mott, as well as on several extensions of this solution. The initial calculation of Mott assumed a rigid constant-flow-stress plastic body. Lee [68] extended this calculation to an elastic–plastic

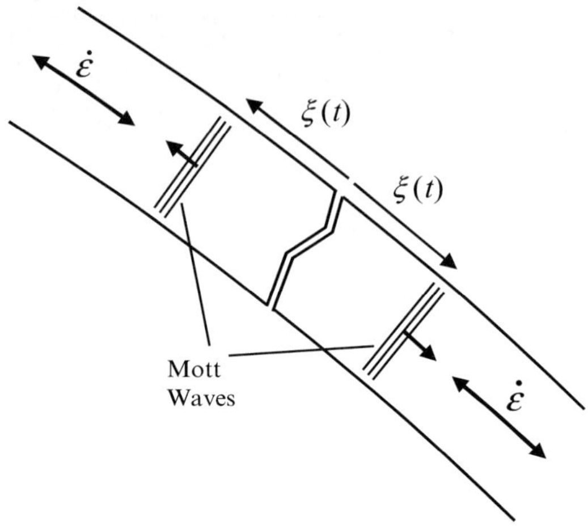

Fig. 4.19. Illustrates diffusion-like Mott waves propagating from point of fracture into regions stretching at a rate governed by the expansion velocity and at a flow stress Y

body showing more clearly the range of validity of the Mott solution. Comparison of the Mott solution with heat-like diffusion equations demonstrates the decidedly diffusive character of Mott-wave propagation. Hardening and softening plasticity are readily introduced into the Mott calculation. Fracture dissipation in the Mott wave analysis was described in the previous subsection.

The dynamic fracture analysis undertaken by Mott is one-dimensional. It is best visualized as that of a uniformly stretching rod or expanding ring such as previously illustrated in Fig. 4.8 and as illustrated in more detail in Fig. 4.19. The model can be usefully applied to fragmentation problems such as the fragmentation of a rapidly expanding cylinder in which the circumferential stretching rate substantially exceeds the axial. Similar wave propagation occurs in expanding spherical shells where principal strains are closer to equiaxial, although the Mott-like analysis becomes more complex. Here, for clarity, the analysis focuses on a stretching one-dimensional filament or wire. Prior to fracture the body is uniformly stretched to an axial instability strain ε and continues to increase at a constant stretching rate $\dot{\varepsilon}$.

Mott Rigid-Plastic Solution

Mott considered the body rigid-perfectly plastic and straining in tension under a constant flow stress Y. A more realistic model of an expanding and thinning shell (rigid plastic) might be as illustrated in Fig. 4.20, where the equivalent tension T (equivalent tensile stress times shell thickness) initially increases due to deformation hardening, and subsequently decreases when thinning overcomes hardening. The onset of tension instability can also be enhanced by adiabatic heating or cavitation damage during the dynamic deformation.

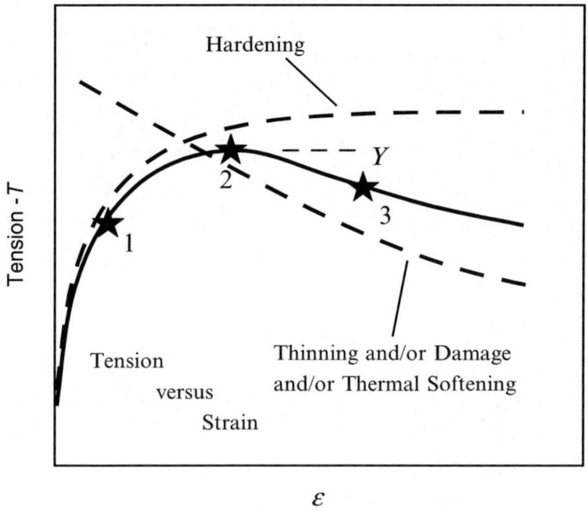

Fig. 4.20. Illustrates the in-plane tension as a function of thinning strain for expansion of a deformation hardening metal shell

Mott assumed that fracture occurred at or near point 2 (load maximum) in the figure.

At onset of breakup through-the-thickness fractures are considered to occur at random in both time (or equivalently strain) and in spatial location on the stretching body. Mott assumed that fractures occur instantaneously, relieving the tensile stress at the points of fracture to zero. Thus, fracture resistance and corresponding fracture energy during the breakage process is ignored.

Accordingly, Mott considered a one-dimensional body stretching plastically under a tensile stress Y, and uniformly at a constant stretching rate $\dot{\varepsilon}$. Fracture was initiated by setting the tensile stress to zero at time $t = 0$ at some Lagrangian position $h = 0$. Regions of the rod experiencing tensile stress less than Y were considered rigid. A moving boundary (the Mott wave) separates material stretching uniformly at stress Y in front of the boundary from rigid material moving at a uniform velocity behind the boundary. This Mott wave propagates outward from the point of fracture at $h = 0$. Features of the stress and velocity associated with the Mott wave at some time $t > 0$ are illustrated in Fig. 4.21. Position of the Mott wave is $\xi(t)$ while crack opening is identified by the coordinate $\eta(t)$. The velocity field is then,

$$u(h,t) = \begin{cases} \dot{\varepsilon}\xi(t) & 0 \le h < \xi(t) \\ \dot{\varepsilon}h & \xi(t) \le h \le h_o \end{cases}, \tag{4.156}$$

where h_o is some arbitrary distance greater than the propagation distance of interest.

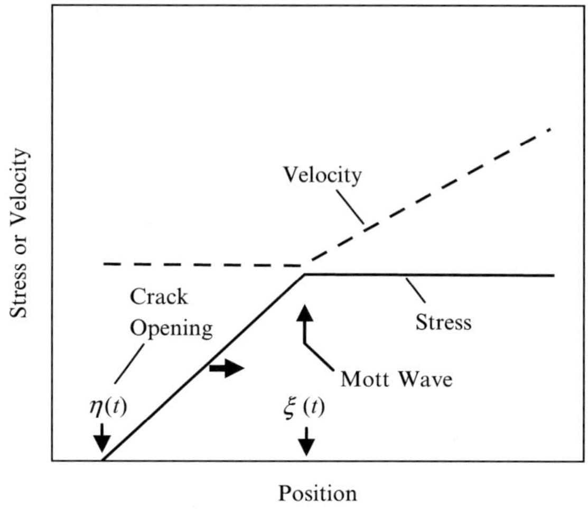

Fig. 4.21. Stress and velocity field at time t after fracture at position $h = 0$ drops tensile stress from Y to zero

The corresponding stress field is equally apparent. The total momentum of the rod within the region $0 \leq h \leq h_o$ is given by,

$$\rho\dot{\varepsilon}\xi^2 + \int_{\xi}^{h_o} \rho\dot{\varepsilon}h dh = \frac{1}{2}\rho\dot{\varepsilon}(\xi^2 + h_o^2). \tag{4.157}$$

Equating the time rate of change of momentum to the imbalance in tensile stress yields the differential equation,

$$\rho\dot{\varepsilon}\xi\frac{d\xi}{dt} = Y, \tag{4.158}$$

for the position $\xi(t)$ of the Mott wave at time t. Integration yields,

$$\xi(t) = \sqrt{\frac{2Yt}{\rho\dot{\varepsilon}}}. \tag{4.159}$$

Thus, fractures occurring in the stretching body lead to the propagation of waves away from points of fracture, which unload the tensile stress. The time dependent motions of these Mott waves are governed by both material properties and kinematic conditions according to the relations above. Within regions subsumed by Mott waves further fracture will not occur. Subsequent fracture will only occur in regions not yet reached by the unloading Mott waves, which continue to stretch unimpeded at a rate $\dot{\varepsilon}$ and flow stress Y.

Lee Elastic–Plastic Solution

Mott recognized that excessively high velocities of the interface $\xi(t)$ at early times was a consequence of the rigid-plastic assumption and inconsistent with a more rigorous elastic–plastic treatment of the problem. Mott acknowledged in his war years reports of an analysis due to E.H. Lee, which was published some years later [68]. Lee considered the same initial and boundary conditions posed by Mott, but treated material response behind the interface as elastic. It is shown that the initial drop in stress from $\sigma = Y$ to $\sigma = 0$ at the origin $h = 0$ propagates as a decaying shock discontinuity in stress and particle velocity at an elastic wave speed c. This shock discontinuity decays to zero at a distance of,

$$h = \frac{2Y}{\rho c \dot{\varepsilon}} = \lambda, \tag{4.160}$$

and at a time of,

$$\tau = \lambda / c. \tag{4.161}$$

Subsequent reflected elastic waves and the interface $\xi(t)$ are acceleration discontinuities (discontinuities in the slopes of stress and particle velocities). Continued solution reveals that the interface $\xi(t)$ is a polygon in the h vs. t domain with vertices,

$$h = n\lambda, \quad t = n^2\lambda/c, \quad n = 1, 2, 3, \ldots, \tag{4.162}$$

where each segment propagates at a velocity of,

$$c_n = \frac{c}{2n-1}. \tag{4.163}$$

The rigid-plastic solution of Mott [79] and the elastic–plastic solution of Lee [68] are compared in Fig. 4.22. The rigid-plastic solution of Mott is found

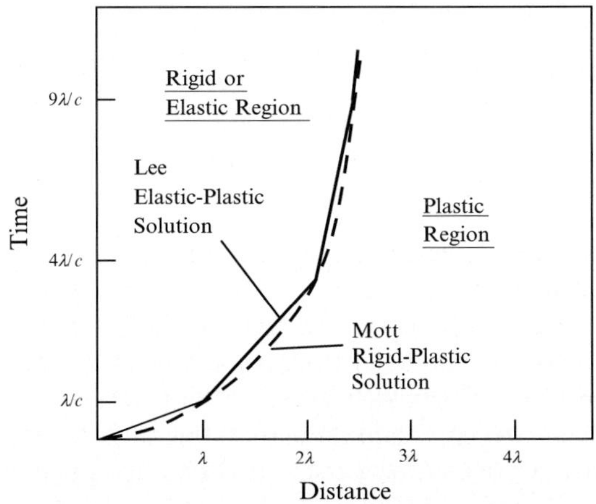

Fig. 4.22. Comparisons of interface separating plastic region and either the rigid or the elastic region according to solutions of Mott [79] and Lee [68]

to envelop the elastic–plastic solution touching at the vertices. Within several characteristic distances λ the rigid-plastic solution is found to be a very good approximation to both the position of the interface and to the stress and velocity field behind the interface.

Stress Diffusion Solution

It is the simplicity of Mott's analysis that reveals the underlying physics. The Mott solution is intended to apply to the point at which hardening in the stretching rod saturates and the tension versus strain loses its hyperbolic character. At this point of stationary tension the governing equations transition to parabolic and the diffusive nature implicit in the motion of the Mott wave is revealed.

The diffusive character of the stress release process is readily demonstrated by writing the linear diffusion equation,

$$\frac{\partial^2 \sigma}{\partial h^2} - \frac{1}{\kappa}\frac{\partial \sigma}{\partial t} = 0, \tag{4.164}$$

with a diffusion constant of the form,

$$\kappa = Y/2\rho\dot{\varepsilon}. \tag{4.165}$$

Consider the same problem treated by Mott in which fracture at $t = 0$ and $h = 0$ instantly decreases the tensile stress from $\sigma = Y$ to $\sigma = 0$. This classic solution (e.g., [73]) can be immediately written down for the stress,

$$\sigma/Y = erf\,(\zeta)\,. \tag{4.166}$$

The velocity is determined through the momentum relation,

$$u/\sqrt{4\kappa t} = \sqrt{\frac{4}{\pi}}\exp\left(-\zeta^2\right) + 2\zeta erf\,(\zeta) - \zeta. \tag{4.167}$$

The functional dependence in (4.166) and (4.166) is on the similarity parameter,

$$\zeta = h/\sqrt{4\kappa t}. \tag{4.168}$$

The present diffusion equation solution and the rigid-plastic solution of Mott are compared in Fig. 4.23.

The rigid-plastic solution of Mott with or without fracture dissipation, the elastic–plastic solution of Lee, and the solution to the diffusion equation are, of course, only models of the actual processes of fracture and stress unloading occurring in the rupture of a rapidly stretching ductile ring or shell. Which model most accurately depicts the material behavior probably cannot be answered. All, however, reveal physics of the fracture process and point to the decidedly diffusive nature of stress wave propagation.

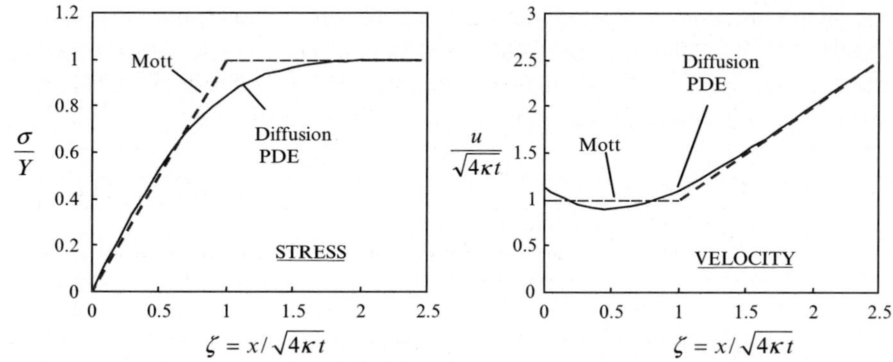

Fig. 4.23. Comparisons of solutions for Mott's rigid-plastic model and a linear stress-diffusion model of fracture in a stretching plastic rod

Hardening and Softening in the Mott Wave Solution

Mott introduced a model of, and a solution for, the rate of propagation of tensile stress unloading in expanding thin sheets (or wires) of metal undergoing uniform plastic deformation at a constant stretching rate. Mott chose to examine the case where the tension was approximately constant $T = Y$ over the problem time of interest (point 2 on the stress versus expansion strain in Fig. 4.20). It is not difficult to imagine fracture-inducing defects such that fracture and subsequent stress release occurs during the hardening phase of the deformation (point 1 in Fig. 4.20). Similarly, fracture delayed until well into the softening phase (point 3) is also possible. In such cases the equation corresponding to (4.158) governing propagation of the Mott wave can be generalized to,

$$\rho\dot{\varepsilon}\xi\frac{d\xi}{dt} = T(\varepsilon) = Y + H\dot{\varepsilon}t. \tag{4.169}$$

Here we solve the special case provided by the factor on the right where Y is the initial tension and H is a linear strain hardening (or softening) constant. Equation (4.1169) readily integrates to,

$$\xi(t) = \sqrt{\frac{2Yt}{\rho\dot{\varepsilon}}}\left(1 + \frac{1}{2}\frac{H\dot{\varepsilon}}{Y}t\right)^{1/2}, \tag{4.170}$$

for the position of the Mott wave at any time t.

Motion of the Mott wave is apparent. For positive H (hardening), the time dependent motion obtained by Mott for constant Y is observed at early times. At late time the wave propagates at a constant velocity of $\sqrt{H/\rho}$. For negative H (softening) the wave arrests after a time $t = -Y/H\dot{\varepsilon}$ and travels a distance $\xi = (-Y^2/\rho\dot{\varepsilon}^2 H)^{1/2}$.

Models for Mott wave propagation in other than axial one dimension are also possible. For example, equal stretching rates in orthogonal in-plane directions (an expanding spherical shell) where tension at a point is reduced from Y to zero (initial rupture is a hole in the metal membrane) can also be solved. Continued dissipation in material behind the circular Mott wave due to further hoop expansion must be accounted for, but the solution provides further demonstration of the diffusive nature of the stress release propagation process.

4.2.6 Energy-Based Criteria for Fragment Size

Fracture of solids occurs at sites of weakness inherent to all engineering materials. It follows that theories of dynamic fracture and fragmentation attempt to capture this inherent flaw structure as an intimate part of the theory. Characterization of the flaw structure for specific materials is difficult in practice however, and usually parameter fitting to indirect data is the only available solution. Second, fracture and fragmentation models based on inherent flaw activation commonly ignore energy consumed in fracture activation and growth through the creation of new surface area, plastic work, or viscous dissipation. Such energy must come from the dynamic load driving fracture, and it is reasonable to expect that this energy consumption process may control in some way the extent of fragmentation.

Consequently, efforts were pursual to develop more robust theories of fragment size based on energy principles, with less attention attributed to the underlying microstructure involved in the fracture process. Toward this goal Grady [31, 32], and Glenn and Chudnovsky [26], pursued methods in which the work from the dynamic load driving fracture is balanced against energy consumed in the fracture process to provide first order estimates of fragment size in the dynamic event. More recently, Drugan [20] and Zhou et al. [99] have pursued approaches where attention to details of the brittle fracture dissipation and interaction processes have provided increased structure to the predicted dependence of fragment size on dynamic loading parameters.

Kinetic and Strain Energy Criteria

The energy-based theory first found to provide sensible predictions of characteristic fracture spacing and fragment size in a fragmentation event focused on the kinetic energy associated with the velocity divergence within the spall zone [31]. By considering the fragment size as a thermodynamic variable, a local kinetic energy relative to the center of mass of the fragment is calculated. Assuming some model for fracture energy dissipation in the spall process, and further assuming that the local kinetic energy fuels this dissipation in the fragmentation event, one can strike a balance between the driving energy and the fracture energy, and derive a prediction of the characteristic fragment size. The energy balance concept leading to an optimum amount of fragment surface area or, equivalently, an optimum fragment size is illustrated in Fig. 4.24. (Fragment surface area and fragment size are related through the fragment distribution and fragment geometry. In the simplest estimate $A \approx 1/s$ where

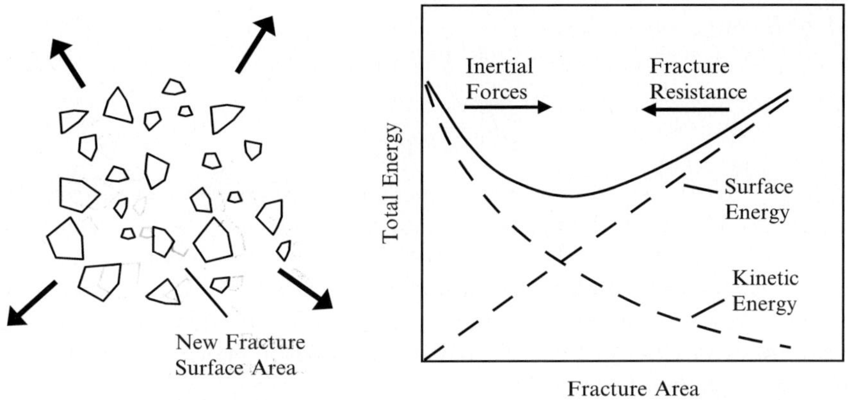

Fig. 4.24. Thermodynamic balance of local kinetic energy and fragment surface energy provides new fracture surface area or average fragment size in dynamic fragmentation event

A is the fragment surface area per unit volume and s is the characteristic fragment size.) Different assumptions about the geometric shapes for the fragments (spheres, cubes, etc.) lead to small differences in the magnitude of the local kinetic energy, and different implementations of the energy balance in the theory lead to variations in predicted average fragment sizes of as much as a factor of two.

In particular, for uniform elastic volumetric expansion within the interior of a solid (or a liquid) body, a spherical region of radius a will have a kinetic energy referenced to its mass center of,

$$T = \frac{2\pi}{45}\frac{\dot{\rho}^2}{\rho}a^5. \tag{4.171}$$

This kinetic energy, in terms of a volume measure of the expansion strain rate $\dot{\varepsilon} = \dot{\rho}/\rho$, and a diameter of the insipient fragment $s = 2a$ becomes,

$$T = \frac{\pi}{720}\rho\dot{\varepsilon}^2\, s^5. \tag{4.172}$$

Without specifying a particular fracture mechanism, if Γ is identified as the energy per unit area required to produce new fragment surface, then the work required to create a spherical fragment of diameter s is,

$$U = \pi\, s^2\Gamma. \tag{4.173}$$

By assuming a complete transfer of kinetic energy T into work of fragmentation U, an expression for the characteristic fragment size is obtained,

$$s = 2\left(\frac{90\Gamma}{\rho\dot{\varepsilon}^2}\right)^{1/3}. \tag{4.174}$$

Glenn and Chudnovsky [26] extended the present energy balance by including an elastic energy of dilation,

$$E = \frac{\pi}{12\rho c^2} \frac{P_s^2}{} s^3, \tag{4.175}$$

in addition to the kinetic energy in fueling the fragmentation process. This extension of the theory brings in an additional material property, dilatational tensile spall strength P_s. The Glen and Chudnovsky model would be appropriate, for example, if the material flaw structure included a strongly increasing number of fracture defects that limited increase in tensile elastic stress (and energy) beyond this tensile spall level. An energy balance equivalent to the previous approach yields,

$$\frac{\pi}{720}\rho\dot{\varepsilon}^2 s^5 + \frac{\pi}{12\rho c^2} \frac{P_s^2}{} s^3 = \pi\ s^2\Gamma. \tag{4.176}$$

The resulting cubic can be solved for the nominal fragment size s. At high loading rates the solution converges to that of the kinetic energy balance of (4.174), while at lower loading rates the predicted fragment size approaches a constant governed by the tensile spall strength P_s. At this lower limit, the relation of Glenn and Chudnovsky approaches the energy-based expression of Rittinger [85] discussed earlier.

In the present development of the energy-based fragmentation model, where an interior region undergoes dilatational expansion until fracture occurs through spall cavitation, the model applies equally well to a brittle or a ductile solid, or to a liquid. The nature of the fracture resistance and dissipation may differ. The present energy balance also applies to expanding rings (or stretching jets) and to expanding shells (or stretching sheets) to predict characteristic fragment size. Here ductile metals are more commonly of interest, and deformation is well into the plastic region when fragmentation occurs. In these fragmentation geometries, elastic energy is less likely to contribute significantly to the breakup process. Predictions of fragment size based on the kinetic energy balance theory agree well with predictions of the earlier Mott wave analysis when fracture dissipation is included.

Energy-Horizon Theory of Fragmentation

An alternative energy-based theory sheds additional light on the dynamic fragmentation processes [32]. Like the previous several theories, it addresses only the end states of the fragmentation process, and does not speak to the evolution of transition states through the fracture process.

In the energy-horizon theory of fragmentation, two physically reasonable propositions are assumed. The first is a horizon condition (a correlation distance) that, through time and communication constraints, establishes a characteristic region within which the fracture process must be independent of the environment surrounding that region. The horizon length scale at the time of the failure event places an upper bound on the size of fragments. The second proposition is an energy condition that requires the elastic potential and kinetic energies within the region bounded by the temporal horizon to exceed the fracture energy dissipated in the failure event. These two conditions, when combined with a kinematic description of constant dilatant strain-rate loading, lead to a set of inequalities that constrain the desired spall properties. Similar relations can be posed for a body subjected to a Heaviside loading at constant and uniform tensile pressure amplitude.

Assuming a spherical region within the body of radius ct, these relations are provided by,

$$s \leq 2ct, \tag{4.177}$$

$$\frac{1}{2}\frac{P^2}{\rho\, c^2} + \frac{1}{120}\rho\dot{\varepsilon}^2\, s^2 \geq \frac{6\Gamma}{s}, \tag{4.178}$$

$$P = \rho c^2 \dot{\varepsilon} t. \tag{4.179}$$

The tensile pressure is P, while t and c are the time and elastic wave speed, respectively. Again, s is the fragment size, $\dot{\varepsilon}$ the dilatational rate of expansion, and Γ the energy required to create new fragment surface area. Equation (4.177) states that, if spall fragmentation occurs within a time t from the onset of fracture instability, then regions of diameter $2ct$ must fail independently. Fragments produced within this region must be smaller than the region itself. Equation (4.178) requires that the elastic plus kinetic energy at failure must exceed the required fragmentation energy in this same region. Equation (4.179) is the kinematic relation expressing the time history of tensile pressure leading to spall failure.

The behavior of the governing inequalities is illustrated in Fig. 4.25. From the horizon criterion in (4.177), failure at time t requires that the fragmentation energy expended in creating new fracture surface area exceed $6\Gamma/2ct$. This lower bound fragmentation energy curve is shown in Fig. 4.25. From these relations it is also observed that at time t the elastic plus kinetic energy available to fuel fragmentation is not greater than,

$$\left(\frac{1}{2} + \frac{1}{30}\right) \rho c^2 \dot{\varepsilon}^2 t^2. \tag{4.180}$$

The larger fraction of the energy available to fuel fragmentation is provided by elastic strain energy, and the smaller by the kinetic energy. The energy in (4.180) is dependent on the strain rate. Two curves plotting this energy as functions of time are illustrated in Fig. 4.25; one at higher strain rate, the other at a lower strain rate. Lastly, a minimum fracture energy value is identified. This corresponds, for example, to the constant fracture tensile pressure P_s identified in the model of Glenn and Chudnovsky [26].

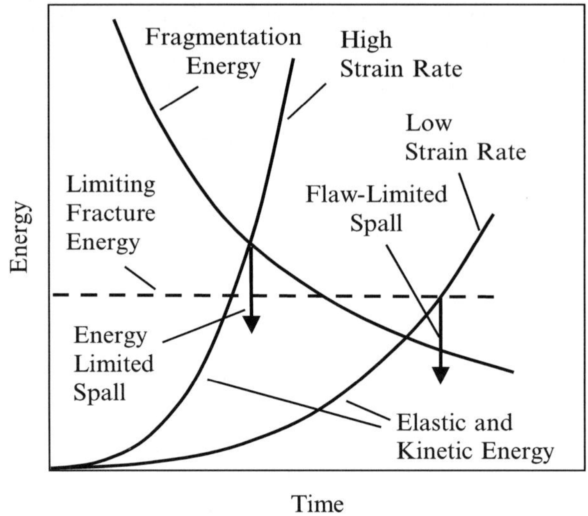

Fig. 4.25. Time dependence of the required fracture energy and the available elastic plus kinetic energy within the failure event horizon. Both energy-limited spall, and flaw-limited spall, is illustrated. The constant spall stress P_s determines the limiting fracture energy that must be exceeded before fracture failure can occur

The two energy curves driving fragmentation were selected to illustrate two distinctly different cases of fragmentation. At the higher strain rate, the minimum fracture energy (or stress) is exceeded, but further time is required to adequately fuel the fragmentation energy requirement. Fragmentation failure does not occur until the elastic and kinetic energy curve achieves the fragmentation energy curve as illustrated in the figure. This case is identified as energy-limited spall or fragmentation. The lower strain rate curve, in contrast, achieves the fragmentation energy criteria below the constant limiting fracture energy level (or constant spall strength P_s). Further tensile loading is needed to achieve this minimum fracture energy (spall stress) requirement. This case is identified as flaw limited fragmentation.

For energy-limited fragmentation, equality in the governing relations applies and specific relations for the spall strength and fragment size are obtained. For flaw-limited fragmentation, the full inequalities apply and additional features must be modeled before spall and fragmentation properties can be determined.

Energy Limited Fragmentation

If the material carried into tension is predisposed to spall through a sufficient microscopic flaw structure when the energy criteria is satisfied, then the fragmentation (4.177) through (4.179) as equalities can be solved for the spall failure and fragmentation properties [32]. These properties are for the fragment size,

$$s = \left(\frac{48\Gamma}{\rho\dot{\varepsilon}^2}\right)^{1/3}, \tag{4.181}$$

tensile spall pressure (not the Glenn–Chudnovsky constant spall stress),

$$P_s = \left(6\Gamma\rho^2 c^3 \dot{\varepsilon}\right)^{1/3}, \tag{4.182}$$

and the spall duration,

$$t_s = \left(\frac{6\Gamma}{\rho c^3 \dot{\varepsilon}^2}\right)^{1/3}. \tag{4.183}$$

In solving for the fragmentation properties in (4.181) through (4.183) the kinetic energy term (the factor of 1/30 in (4.180)) is ignored in favor of the elastic energy term to simplify the expressions. The former is less than 10% of the total driving energy.

Several further relations can be developed. By combining (4.182) and (4.183) the expression,

$$P_s^2\, t_s = 6\rho c\Gamma, \tag{4.184}$$

is obtained. This product of the spall stress and time is a constant provided Γ is itself constant independent of strain rate. The relation is useful in estimating fragmentation energies from spall data. Also, from (4.181) and (4.182), a relation for the fragment size,

$$s = \frac{2P_s}{\rho c \dot{\varepsilon}}, \tag{4.185}$$

based on the spall strength and the strain rate is obtained.

A Comparison of Energy Fragmentation Theories

The present energy-horizon theory of fragment size offers an interesting contrast with the earlier kinetic energy based theory and the later modification by Glen and Chudnovsky. In the kinetic energy model, the elastic strain energy accumulated during the tensile dilation process is ignored. In contrast, the model of Glen and Chudnovsky, and the energy-horizon model, include elastic strain energy but in uniquely different ways. Glenn and Chudnovsky introduce a material dependent tensile spall pressure P_s, which limits the magnitude of the dilatant elastic tensile strain energy that can contribute to the fracture process. The tensile pressure P in the energy-horizon theory is not limited, and continues to increase under dilatant loading until the energy-governed fragmentation failure criterion is achieved. The spall strength in the energy-limited model is not a material constant, but is a derived property

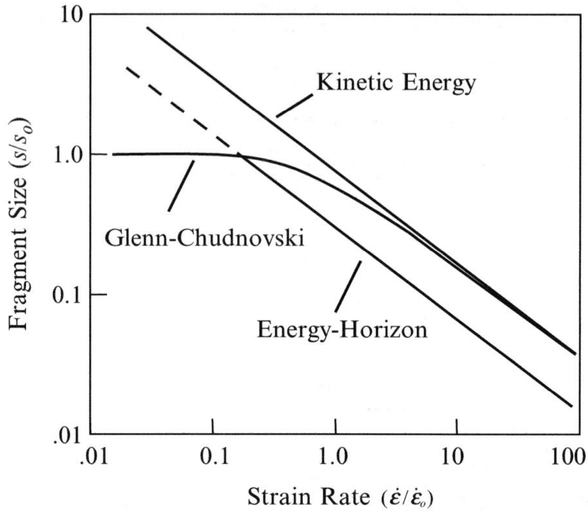

Fig. 4.26. Comparison of fragment size prediction from several energy based theories of dynamic fragmentation. Ordinate and abscissa are normalized by characteristic size and strain rate scales, respectively

from the theory (see (4.182)) that is dependent on the tensile loading rate. The fragment size predictions based on the several energy models as a function of tensile loading rate are illustrated in Fig. 4.26. Both the kinetic energy model and the energy-horizon model exhibit the same $s \sim \dot{\varepsilon}^{-2/3}$ dependence of the fragment size. Fragment size at a specified strain rate for the energy-horizon model is about a factor of 2.5 smaller due to the added contribution of elastic strain energy.

A material dependent constant spall pressure P_s in the Glenn–Chudnovski model introduces both a length scale $s_o = 12\rho c^2\Gamma/P_s^2$ and strain rate scale $\dot{\varepsilon}_o = P_s^3/\rho^2 c^3\Gamma$. These scales provide the respective normalization parameters in Fig. 4.26. The Glenn–Chudnovski model and the kinetic model converge, as expected, when the elastic energy becomes insignificant at the higher strain rates.

Differences between the Glen–Chudnovski and the energy-horizon models at all strain rates is worthy of note as both include elastic strain energy but in different ways. Neither is necessarily wrong. Each implies a different structure to the fracture-producing flaws leading to failure in the fragmentation process. The Glen–Chudnovski model suggests an avalanche of flaws, which activate when tensile pressure approaches P_s and limits the magnitude of tensile strain energy that the body can support regardless of the loading rate. The energy-horizon model, in contrast, requires a flaw structure that allows for increasingly higher levels of tensile strain energy (and spall strength) as the loading rate is increased.

A General Fragmentation Strain Rate

The rate of material straining at a point is found to be a convenient parameter in the previous several theories for capturing the intensity of fragmentation in an event. It characterizes the local kinetic energy through an expression of the form $\rho s^2 \dot{\varepsilon}^2$ where s is a length scale in the kinetic energy theories of Grady [31], and Glenn and Chudnovski [26]. Even in the energy-horizon theory the dominant elastic strain energy scales with the strain rate. In the theories described, a uniform rate of volumetric dilation was considered. This restriction does not need to be. The same energy-based theory has found successful application where this local kinetic energy is strictly distortional such as the breakup of a rapidly expanding shell or stretching jet. Consequently, a more consistent expression of the rate of deformation driving fragmentation appropriate to post-process computational simulations of fragmentation has proven necessary. The following development has provided a sensible measure of the strain rate under more general fragmentation conditions.

Consider a body in motion, characterized by a velocity field $u(x)$. Consider further, a small spherical region of volume v at an arbitrary point within the body. The kinetic energy of this region is given by,

$$T = \frac{1}{2}\rho \int\limits_{v} u^2(x) dv, \tag{4.186}$$

where ρ is considered constant over the region. The velocity can be related to the center of mass of the spherical region through a series expansion,

$$u(x_{cm} + r) = u(x_{cm}) + G \cdot r + \cdots, \tag{4.187}$$

where G is the velocity gradient tensor. Ignoring higher order terms and introducing the symmetric stretching tensor,

$$D = \frac{1}{2}\left(G + G^T\right), \tag{4.188}$$

and the antisymmetric spin tensor,

$$W + \frac{1}{2}\left(G - G^T\right), \tag{4.189}$$

gives,

$$u(x_{cm} + r) = u(x_{cm}) + W \cdot r + D \cdot r + \cdots. \tag{4.190}$$

The kinetic energy of the spherical region, neglecting higher order terms is then,

$$T = \frac{1}{2}\rho \int\limits_{v} \left(u(x_{cm}) + W \cdot r + D \cdot r\right)^2 dv. \tag{4.191}$$

Finally, the kinetic energy can be decomposed into the sum,

$$T = \frac{1}{2}\rho \int\limits_v u^2(x_{cm}) dv + \frac{1}{2}\rho \int\limits_v (W \cdot r)^2 \, dv + \frac{1}{2}\rho \int\limits_v (D \cdot r)^2 \, dv, \tag{4.192}$$

for the center-of-mass, rotational and distortional components, respectively. The first decomposition follows from the theorem in mechanics that allows separation in the center-of-mass kinetic energy plus the kinetic energy of motion about the center of mass. The decomposition into rotational and distortional kinetic energy can likewise be demonstrated.

Following the earlier kinetic energy fragmentation development, it is assumed that the center-of-mass and rotational kinetic energy at a point remains constant through the fragmentation process, and only the distortional kinetic energy fuels the failure and fragmentation process.

The distortional kinetic energy can be written as,

$$T^d = \frac{1}{2}\rho D_{ij} D_{ik} \int\limits_v r_j r_k dv, \tag{4.193}$$

where D is evaluated at the center of mass and removed from the integral. Evaluating the integral,

$$\int\limits_v r_j r_k dv = \frac{1}{3}\delta_{jk} \int\limits_v r^2 \, dv = \frac{4\pi}{15} a^5 \delta_{jk}, \tag{4.194}$$

over a sphere of radius a, the distortional kinetic energy becomes,

$$T^d = \frac{2\pi}{15}\rho a^5 \mathrm{Tr} D^2. \tag{4.195}$$

Identify the local effective strain rate with,

$$\dot{\varepsilon} = \sqrt{\mathrm{Tr} D^2} = \sqrt{\dot{\varepsilon}_x^2 + \dot{\varepsilon}_y^2 + \dot{\varepsilon}_z^2}. \tag{4.196}$$

For uniform shear-free dilation with $\dot{\varepsilon}_x = \dot{\varepsilon}_y = \dot{\varepsilon}_z$ the effective strain rate is $\dot{\varepsilon} = \sqrt{1/3}\dot{v}/v$, whereas for uniform volume-conserving extension with $\dot{\varepsilon}_y = \dot{\varepsilon}_z = -\dot{\varepsilon}_x/2$ the effective strain rate is $\dot{\varepsilon} = \sqrt{3/2}\dot{\varepsilon}_{eq}$ where $\dot{\varepsilon}_{eq}$ is the equivalent plastic strain rate.

Assuming as earlier that the distortional kinetic energy is consumed in fragment surface energy,

$$W = 4\pi a^2 \Gamma, \tag{4.197}$$

a nominal fragments size is estimated,

$$s = 2\left(\frac{30\Gamma}{\rho \dot{\varepsilon}^2}\right)^{1/3}. \tag{4.198}$$

4.3 Dynamic Fragmentation in Brittle Materials

The failure and fragmentation of brittle solids, particularly highly brittle materials such as glass, ceramic, and high-strength rock, differs starkly from that of the ductile metals. Features of brittle fracture and fragmentation are unique and warrant independent discussion. Much of the historical work in brittle solids centers around the early mining, blasting and crushing industries, which have been surveyed in the opening pages of the previous sections. In the present section several studies unique to brittle failure and fragmentation are critiqued. Particularly insightful is the seminal experimental study of Gilvarry and Bergstrom [4, 5, 28] on the catastrophic fragmentation of compressed spheres of glass and other brittle solids. Hopkinson bar techniques offer further experimental methods for assessing the character of dynamic fragmentation of brittle solids. Examination of the fractal nature of brittle fragmentation [93] offers a unique perspective on the processes of dynamic failure. Insightful comparisons are made on similarities between hydrodynamic turbulence and brittle fragmentation. These and several further topics are pursued in the present section.

4.3.1 Gilvarry–Bergstrom Brittle Spheres

An instructive study into the nature of dynamic fragmentation is provided by the extensive investigation of Gilvarry [27, 28] and Bergstrom [4, 5] into the elastic strain induced shatter of solid spheres of brittle material. These authors generated experimental fragmentation results that offer unique insight into both the energy issues governing the catastrophic breakup of a brittle solid, and the character of the resulting fragment size distributions. The supporting theoretical study of Gilvarry [27] is discussed in an earlier section of this chapter.

Distribution Characteristics of Brittle Fracture

The fragmentation tests performed consisted of spheres of brittle solids (glass, sapphire, quartz, flint) of sizes ranging from 1/8 to 1 in. in diameter compressed between platens of tungsten carbide until failure occured through explosive release of the strain energy accompanied by intense fracture and fragmentation of the material. Elastic energy stored in the samples prior to fracture was determined through measurement and assessment of compressive load and displacement of the test system. High-speed photography along with collection and detailed analysis of the fragment debris provided the principal test diagnostics.

After slow compression of the spherical samples, failure is an abrupt and explosive event. Following intense fragmentation and release of the strain energy, fragments are accelerated outward at high velocity. Large fragments originate from the outer regions of the sphere and tend toward lunar, or orange segment, in shape. These larger fragments constitute most of the mass (~90%). From high-speed photography of the ejecta velocity, the authors estimated that fragment kinetic energy was about one-half the stored elastic strain energy prior to fracture.

Fragments from the test were carefully collected and statistical distributions in size were determined through sieving techniques. The authors noted that when tests were performed in a steel collection chamber substantial secondary fragmentation occurred from impact of the high velocity fragments. Subsequent tests were performed in which a gelatin medium was introduced which precluded secondary fragmentation. Fragment distribution data from tests on 1 in. soda-lime glass spheres are plotted in Fig. 4.27 as the cumulative mass fraction $M(x)/M_o$ versus the fragment size x. The data are reasonably well described by a functional distribution of the Schuhmann type [87],

$$M(x)/M_o = (x/x_o)^n \,, \tag{4.199}$$

with a value very close to $n \simeq 1$. It is important to note that both strain energy induced (initial burst) and kinetic energy induced (secondary impact) fragmentation lead to sensibly the same functional dependence.

Sieving methods were judged to be reasonably accurate down to particle sizes of about 100 μm. Below this size, continued quantitative assessment of the fragment distribution becomes difficult. Gilvarry and Bergstrom [28] used an alternative Coulter counter method to assess the small particle characteristics of the distribution. They found that the power law distribution of the

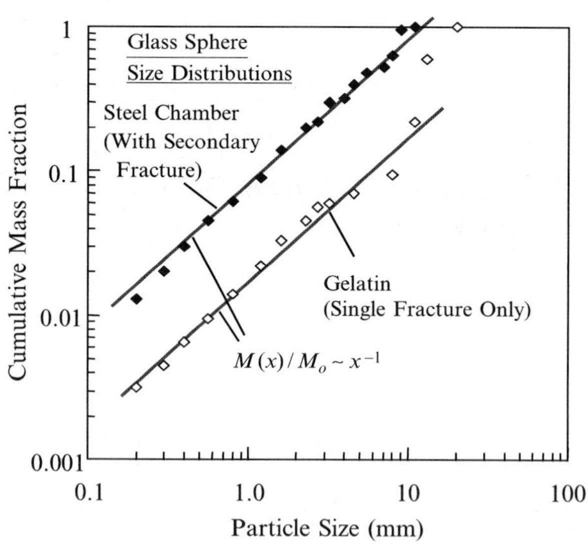

Fig. 4.27. Fragment size distribution curves for compression failure and fragmentation of 1-in. diameter soda-lime glass spheres [5]

form of (4.199), with a Schuhmann distribution index of $n \simeq 1$, continued to provide a good description of the particle size distribution of the glass sphere fragmentation tests down to sizes smaller than about 1 μm. The important observation is that a power-law distribution of the Schuhmann type with $n \simeq 1$ apparently sensibly characterizes fragment distributions in the brittle fracture of glass over at least four decades in fragment size (12 decades in fragment mass). The authors indicated that similar behavior was observed in the other brittle materials investigated.

Strain Energy Issues in the Brittle Failure Process

A key aspect of the study was the assessment of the elastic strain energy stored in the compressed sphere that fueled fragmentation during the failure process. Relations for the strain energy and the maximum stress in the test sphere are taken directly from the paper of Bergstrom [4] and are attributed, in turn, to the book on elasticity by Timoshenko and Goodier [91]. These equations are, for the specific strain energy E (total strain energy divided by the mass of the sphere),

$$\mathrm{E} = \frac{4}{5}\left(\frac{3}{4}\right)^{5/9} (\rho\pi)^{1/9} \left(\frac{1-\nu_1^2}{E_1} + \frac{1-\nu_2^2}{E_2}\right)^{2/3} \left(\frac{P}{M^{\frac{2}{3}}}\right)^{5/3}, \tag{4.200}$$

and the maximum compressive stress at the center of the sphere-platen contact zone,

$$\sigma_m = \left(\frac{40}{\pi^4}\right)^{1/5} \left(\frac{1-\nu_1^2}{E_1} + \frac{1-\nu_2^2}{E_2}\right)^{-4/5} (\rho \mathrm{E})^{1/5}. \tag{4.201}$$

In these relations ρ is the sphere density, P is the load, M is the mass of the sphere, and $\nu_1, \upsilon_2, E_1, E_2$ are Poisson's ratios and elastic moduli for the respective sphere and platen materials. The authors performed compression-to-fracture tests over a range of sphere sizes, and crucial to their study was the observed size dependence of the strain energy to fracture. It was found for the glass sphere tests that a maximum stress from (4.201) of approximately 5 GPa and 10 GPa was achieved in the 1 in. and the one-eighth inch spheres, respectively.

The specific energy at failure is plotted as a function of sphere diameter for the brittle materials tested in Fig. 4.28. Most of the data points represent an average over a number of tests. The open symbols for glass are tests performed in gelatin that precluded secondary fragmentation. All others were performed in a steel collection chamber. All materials exhibited an increase in strain energy at failure with sapphire showing somewhat less sensitivity than glass or quartz.

The corresponding Shuhmann distribution size parameter (the coefficient x_o in (4.199)) from a best fit of (4.199) to the measured cumulative mass versus

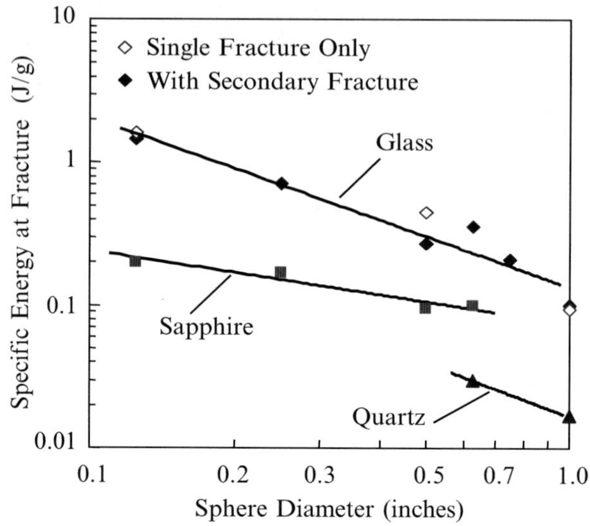

Fig. 4.28. Specific elastic strain energy at fracture for compression failure and fragmentation of soda-lime glass, single crystal sapphire and polycrystalline quartz [5]

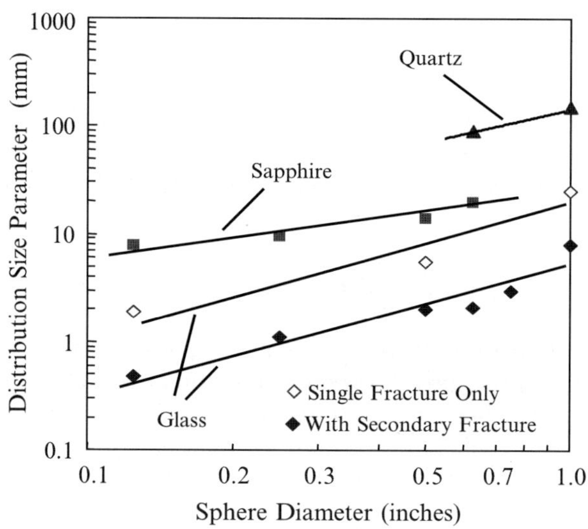

Fig. 4.29. Schuhmann distribution size parameter for compression failure and fragmentation of soda-lime glass, single crystal sapphire and polycrystalline quartz [5]

size data is shown in Fig. 4.29. The failure strain energy versus distribution size parameter is cross-plotted in Fig. 4.30. The concluding point made by the authors is that all of the data, and other data that they surveyed from the literature, did not differ with statistical significance from an inverse first power dependence,

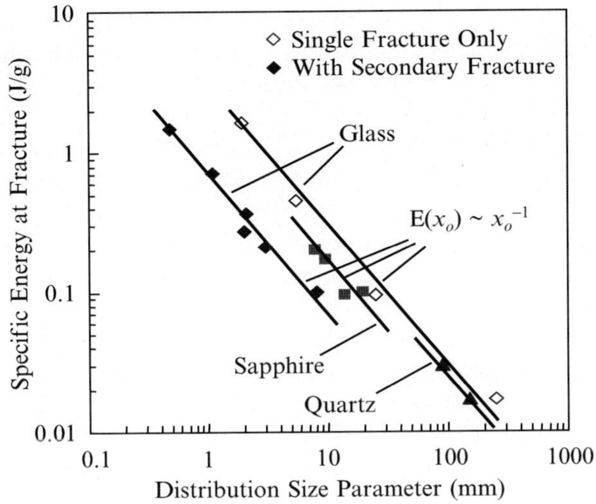

Fig. 4.30. Specific elastic strain energy at fracture versus the Schuhmann distribution size parameter for compression failure and fragmentation of soda-lime glass, single crystal sapphire and polycrystalline quartz [5]

$$\mathrm{E} \sim x_o^{-1}. \tag{4.202}$$

This result has profound significance. It can be shown that the distribution size scale x_o is inversely proportional to the new fracture surface area created in the fragmentation process. The results of these authors provide strong support for the historical Rittinger energy-based fragmentation relation [85] discussed in an earlier section.

4.3.2 Hopkinson Bar Fragmentation

Quasi-static compression of the Gilvarry–Bergstrom spheres of brittle solids leads to a catastrophic fragmentation event. Fragmentation of brittle solids can also be, and has been, explored with dynamic test methods. Such experimental methods include explosive loading, projectile impact and Hopkinson bar loading among others. The latter test method, originally developed to characterize dynamic strength properties of metals, has more recently been applied to investigate the strength, failure and fragmentation properties of brittle solids.

Early use of Hopkinson bar techniques on brittle materials focused on natural rocks and minerals (e.g., [50, 71]). Grady and Lipkin [44] have summarized the features of dynamic failure from these and other studies. More recently, Hopkinson bar methods have been extended to examine failure and fragmentation of ultra-high strength ceramics. The studies of Lankford [67], Bourne et al. [9], Chen and Ravichandran [12], Shih et al. [89], and Wang and Ramash [94] are representative.

Generally observed in these investigations is a high elastic failure stress (and concomitant elastic strain energy) followed by catastrophic fracture and fragmentation of the test sample not unlike failure of the Gilvarry–Bergstrom spheres. Unique to the Hopkinson bar experiment, however, is a transition to loading rates at which the kinetics of fracture activation and growth is not immediately able to keep pace with the rate of loading. Increasing levels of elastic stress, and strain energy, are achieved before catastrophic fracture and failure overwhelms the loading process. This increase of the effective strength of the brittle solid with the loading rate of strain is observed to initiate at strain rates in the neighborhood of $10^4\,s^{-1}$, and frequently exhibits a power law dependence on strain rate ranging up to a maximum of about one-third. Various theories, including crack activation and growth kinetics, and energy arguments, have been proposed to explain the observed behavior.

Here, we focus on one such Hopkinson bar study that investigates the failure and fragmentation of brittle solids [15, 16]. This representative study is selected for closer examination primarily because the investigations of fragment size statistics are in line with the statistical discussions of the present chapter.

Fragmentation experiments were performed on ceramic and ceramic-like materials using a Kolsky torsion bar. There was, at the time, no specific reason for using a torsion bar apparatus other than availability. Such tests could as readily have been performed on a compression Hopkinson bar. The emphasis in these tests was focused on the size and distribution of fragments produced in the failure event, and methods were devised to assess particle sizes from a few millimeters down to tens of nanometers.

The materials tested included a ferroelectric ceramic lead zirconate titanate (PZT), depleted uranium dioxide (UO_2), quartz glass (SiO_2), isotropic graphite, and four grades of oil shale with a nominal kerogen content of 10, 20, 30 and 40 gallons per ton (GPT). Test samples were cartwheel shaped with the active section 10 mm in diameter and 7.5 mm in length as shown in the Fig. 4.31 photograph. Flange sections were bonded to the input and transmission bar. An input torque pulse of 120 N-m was maintained constant over the test series providing a nominal loading strain rate of $500\,s^{-1}$. The usual strain gage measurements were performed on the input and transmission bar providing the stress and strain history throughout the test sample loading and failure.

To capture all of the fragment debris the test section of the bar was enclosed in a sealed metal chamber. Following the test, fragments were transferred to a glass jar for later analysis. For one material, the uranium dioxide, interest was also in possible airborne particulate generated in the failure event. In these tests, filtered air was circulated through the chamber with a vacuum pump and into a series of instruments designed to measure the quantity and size of the airborne particulate. These instruments included a cascade impactor, a condensation nuclei counter, and an electrostatic precipitator, allowing characterization of the mass and size distribution down to a few tens of nanometers.

Fig. 4.31. Photograph of a test sample of lead zirconate titanate ceramic (PZT) illustrating sample geometry and consequences of the torsion Kolsky-bar failure event

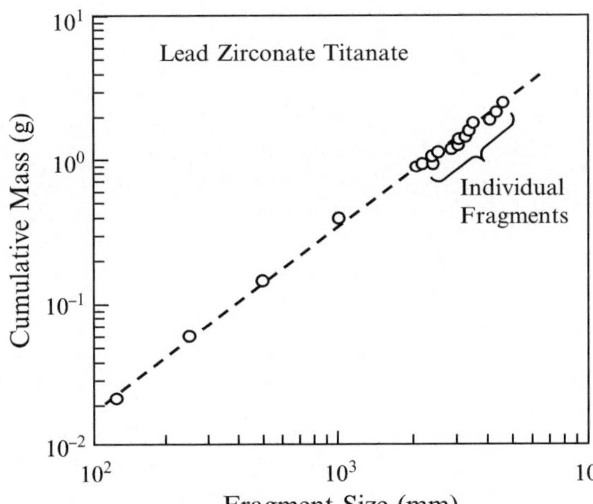

Fig. 4.32. Torsion Kolsky bar fragmentation test particle size distribution [15]

Larger particle were sorted with a set of mechanical sieves ranging from 2,000 μm down to as small as 38 μm. Particles larger than 2,000 μm were individually weight and an effective size calculated. Fragment size data were plotted in the Schuhmann cumulative mass-versus-size representation commonly used by earlier workers. One set of data for PZT ceramic is shown in Fig. 4.32, and is representative of all of the data collected excepting the graphite. Only

Table 4.1. Fragmentation properties for selected brittle solids

Material	Density ρ (kg m^{-3})	Youngs modulus E (GPa)	Fracture K_{IC} (MPa m$^{-1/2}$)	Schuhmann Index n
UO_2	10,860	192	1.6	1.5
Glass	2,200	37	0.54	1.2–1.3
Oil shale				
10 GPT	2,590	20	1.2	2.1
20 GPT	2,360	15	1.0	2.2
30 GPT	2,140	10	0.8	2.9
40 GPT	1,850	6	0.65	3.2
Graphite	1,850	12	1.4	2.6
PZT-1	7,250	110	1.22	1.3–1.4
PZT-2	7,370	114	1.38	1.4–1.5
PZT-3	7,300	112	1.47	1.25–1.35
PZT-4	7,430	110	1.59	1.4–1.5

the graphite did not plot linear in this display, exhibiting instead a bi-linear curve and hinting at more complex dynamic fracture behavior.

Pertinent properties for the materials tested are provided in Table 4.1 along with the principal experimental measurement, the distribution shape parameter, or Schuhmann index n. Several interesting features are observed. The lowest Schuhmann index of $n = 1.3$ is for glass and, although slightly higher, is not out of line with the measurements of Gilvarry and Bergstrom [28] on compression failure of glass spheres discussed earlier in the section. A trend is noted for oil shale. As the kerogen content of the oil shale increases (presumably suggesting increasingly less brittle behavior) the Schuhmann index increases. Increasing values of n indicate proportionally less mass in fine fragments relative to larger fragments and would seem in line with the increasing ductility.

Lead zirconate titanate (PZT) also shows an interesting trend. The four different formulations exhibited measurably different fracture toughness. Although scatter in the Schuhmann index n precludes any observable trend, the mass of fragments in the finer sieve sizes increased noticeably and systematically with decreasing fracture toughness [15]. Working with the smaller 125, 250 and 500 μm sieve size data, and using a constant index of $n = 1.37$ representative of all four of the materials in this size range, a Schuhmann size scale x_o, from the Schuhmann relation in the form $\ln M/M_o = n \ln x/x_o$, was determined. This size scale is plotted against fracture toughness in Fig. 4.33. An increase in fragmentation with decreasing toughness is reasonable. (Recall that new fragment surface area $A \sim 1/x_o$.) The power-law behavior (power $\sim 5/3$) has not been explained.

Sieving results for uranium dioxide (UO_2) are shown in Fig. 4.34. Of particular interest was the self-similar character of fragments in all of the sieve

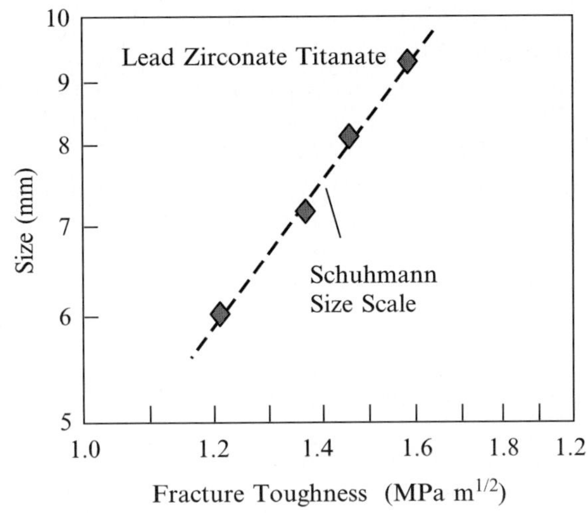

Fig. 4.33. Schuhmann size scale versus fracture toughness for four preparations of lead zirconate titanate [15]

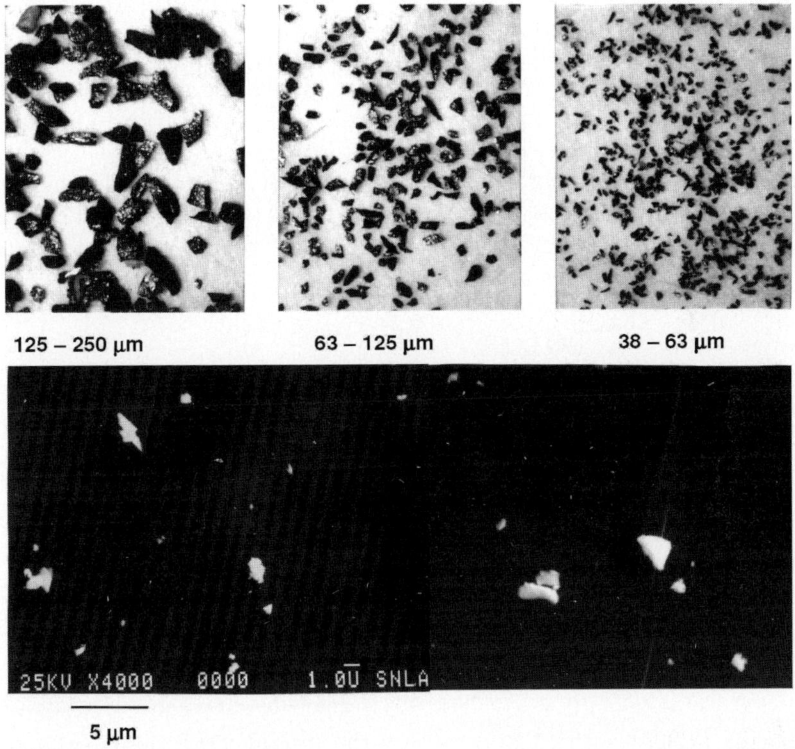

Fig. 4.34. Torsion Kolsky bar uranium dioxide fragmentation tests. Selected sieve sizes are shown in the upper photographs. Electrostatic precipitator aerosol particulate in the lower photograph

sizes. Fragment shape, angularity and surface texture were visually the same over the tested size range. Also important was the collection of substantial aerosol particulate down to the tens of nanometer size with the associated collection devices. Uranium dioxide particulate collected with the electrostatic precipitation device is illustrated in Fig. 4.34.

As noted, the graphite exhibited bi-linear behavior in a Schuhmann plot of the fragment size data. Fragments of size greater than 1,000 μm, which constituted about 95% of the mass of the sample exhibited the larger $n = 2.6$ reported in the table. The finer fragment particulate provided an n close to unity.

4.3.3 Fragmentation as a Fractal Process

Turcotte [93] has pointed out that the process of fragmentation of a solid body can result, in some cases, in a fragment distribution that is fractal. A distribution of objects is fractal if their number N greater than some linear dimension x is represented by a power-law relation,

$$N(x) \sim x^{-D}, \tag{4.203}$$

with $D > 0$. D is identified as the fractal dimension. An example is provided by the number of oceanic islands with linear dimension greater than x, equal to the square root of the area, which satisfies the relation in (4.203) with a fractal dimension of about $D \cong 1.3$ [72].

In applications dealing with the breakup of brittle solids the Rosin–Rammler distribution [86] is commonly found appropriate. The Rosin–Rammler distribution can be written in terms of the mass of fragments M with linear size smaller than x,

$$M(x) = M_o \left(1 - e^{-(x/x_o)^n}\right), \tag{4.204}$$

and is more often applied in the form attributed to Schuhmann [87],

$$M(x) = M_o (x/x_o)^n. \tag{4.205}$$

Relating the fragment number to the mass increment, $dN \simeq dM/x^3$, it is readily found that,

$$N(x) \simeq N_o (x/x_o)^{n-3}, \tag{4.206}$$

with,

$$N_o = \frac{n}{3-n} \frac{M_o}{\rho x_o^3}. \tag{4.207}$$

Comparing (4.206) with (4.203), relates the fractal dimension to the Schuhmann index through,

$$D = 3 - n, \tag{4.208}$$

and the distribution is fractal provided n is less than three. Typically $0.5 < n < 1.5$ for most rock and mineral breakage processes (e.g., [3]), and n was found to be very close to unity for the extensive compression shatter experiments of Bergstrom [4] on spheres of glass, quartz and sapphire.

Thus many breakage processes of brittle solids apparently lead to fractal-like fragment distributions with a fractal dimension D ranging between about 1.5 and 2.5. A number of examples are provided in the paper of Turcotte [93] including various rock breakage processes along with distributions of natural earth and asteroid materials. The study of Gilvarry and Bergstrom [28] on the catastrophic fracture of compressed glass spheres determined a power law dependence of the fragment size using both sieving and Coulter counter methods. Costin and Grady [15] observed a power law fragment distribution with somewhat larger n is some instances on a range of brittle materials including glass, uranium dioxide, and lead-zirconate-titanate ceramic (PZT), among others, using a torsion Kolsky bar dynamic loading methods.

The implication of the power law distribution and fractal behavior resulting from fragmentation in brittle solids is that the fracture mechanisms leading to breakup remain invariant over the span of sizes achieved in the event. This would in turn imply that fragment sizes within the range of the distribution remain large relative to the intrinsic length scales of the fracture process. The relation,

$$x_{pz} = \frac{\pi}{72}\left(\frac{K_c}{Y}\right)^2, \tag{4.209}$$

provides an estimate of the crack tip process-zone length scale [66]. Based on measured K_c values for glass and estimates of Y from Hugoniot elastic limit measurements, a measure of $x_{pz} \sim 5\,\text{nm}$ is obtained, well below the length scales considered in most brittle fracture distributions. This length scale will of course range upward for less brittle materials.

In any case, if fragmentation of brittle solids is a fractal process leading to $N \sim x^{-D}$, or, correspondingly to $M \sim x^{3-D}$, then the underlying physics of interest is in explaining the mechanisms responsible for the specific value of the fractal dimension D. Turcotte [93] has explored renormalization group methods used by others to characterize various scale invariant critical state phenomena. He identifies a parameter p_m determining the probability of fracturing of a cell of the solid body. The fractal dimension D is calculated from p_m which he in turn relates to the fragility of the brittle material. One further interesting approach is explored briefly in the last contribution to this section.

4.3.4 Unlikelihood that Brittle Fragmentation is a Poisson Process

A fragment size distribution of the form,

$$M(x) = M_o(x/x_o)^n, \tag{4.210}$$

where $M(x) \leq M_o$ is the cumulative mass of fragments less than size x, often attributed to Schuhmann [87], is known to describe the consequences of the fragmentation of brittle solids reasonable well in most instances. The distribution index n is often close to unity and commonly $0.5 \leq n \leq 1.5$. The distribution applies to both single event fragmentation such as high-velocity impact or explosive breakage, as well as to comminution processes such as crushing or grinding.

A number of the earlier authors, including Bennett [3], Lienau [70], Gilvarry [27], and Gaudin and Meloy [25], among others, attempted to rationalize (4.210) through a model of brittle fracture as a random placement of fractures through a Poisson process.

As described earlier, Lienau [70] showed that the random fragmentation of a line (a one-dimensional body) through a Poisson process leads to the cumulative number distribution,

$$N(x) = N_o(1 - e^{-x/\lambda}). \tag{4.211}$$

Bennett [3] used a relation of the form of (4.211) to distribute fracture planes in three mutually orthogonal directions and attempted to rationalize the Schuhmann distribution in (4.210). Gaudin and Meloy [25] pursued the same approach over a finite dimension body in which case (4.211) is replaced with a binomial distribution.

Difficulties in these earlier analytic studies are illustrated in the following: The mass density distribution over size is provided by the derivative of (4.210),

$$\frac{dM}{dx} \sim x^{n-1}, \tag{4.212}$$

and is observed to have a fairly weak dependence on fragment size. It would be constant if $n = 1$ (equal masses in equal size intervals).

In one dimension the Lienau relation from (4.211) yields,

$$dN \sim e^{-x/\lambda}dx, \tag{4.213}$$

and the fragment mass,

$$\frac{dM}{dx} \sim \frac{xdN}{dx} \sim xe^{-x/\lambda} \sim x. \tag{4.214}$$

The last step examines the small fragment limit of the distribution. Thus, in one dimension the mass density increases as the first power of fragment size.

Mott and Linfoot [81] were apparently the first to provide a rigorous analytic solution of the two-dimensional Poisson fragmentation process (the random partitioning of an area by the random distribution of orthogonal horizontal and vertical lines according to (4.211)). They obtained the number density distribution,

$$\frac{dN}{dx} = \frac{4x}{\lambda^2}K_o(2x/\lambda), \tag{4.215}$$

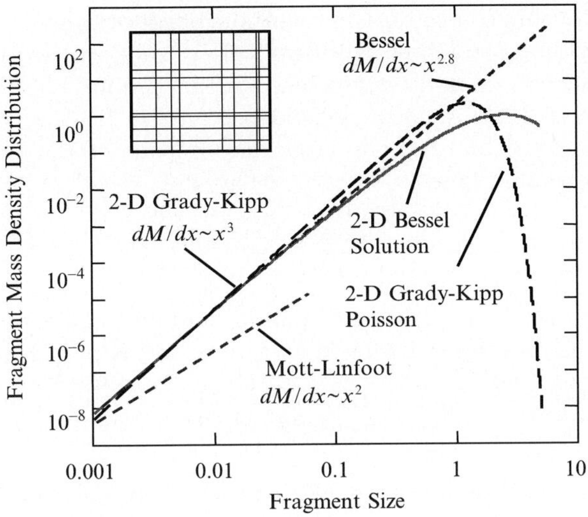

Fig. 4.35. Solutions to Poisson two dimensional (2-D) fragmentation through partitioning of an area by random orthogonal horizontal and vertical lines as illustrated in the inset

where K_o is a modified Bessel function. The Mass density in two dimensions is then,

$$\frac{dM}{dx} \sim \frac{x^2 dN}{dx} \sim x^3 K_o(x), \tag{4.216}$$

and is plotted in Fig. 4.35. A fit to the lower five decades in the plot yields approximately,

$$\frac{dM}{dx} \simeq x^{2.8}. \tag{4.217}$$

If fragment area is partitioned by a Poisson process as suggested by Grady and Kipp [42] then,

$$\frac{dM}{dx} \sim \frac{x^2 dN}{dx} \sim x^3 e^{-(x/\lambda)^2} \sim x^3, \tag{4.218}$$

with the power-law dependence applying to the small fragment portion of the distribution. Equation (4.218) is in reasonable agreement with the analytic Bessel solution in this range as illustrated in Fig. 4.35 and compared with (4.217). Note that the Mott–Linfoot assumption of the form,

$$N(m) \sim 1 - e^{-\sqrt{m/\mu}}, \tag{4.219}$$

where the mass $m \sim x^2$ leads to,

$$\frac{dM}{dx} \sim \frac{x^2 dN}{dx} \sim x^2, \tag{4.220}$$

exhibits poorer agreement with the analytic solution.

The analytic solution to random planes in three orthogonal directions has not apparently been attempted but is unlikely to support the use of the model in justifying the mass density dependence on size required by the Schuhmann distribution in (4.212). A linear dependence on size for the one-dimensional fragmentation algorithm and the cubic dependence of the two-dimensional fragmentation would suggest,

$$\frac{dM}{dx} \sim x^5, \tag{4.221}$$

for the three-dimensional orthogonal planes random fragmentation of a body. In fact, the Poisson process assumed by Grady and Kipp [42] leads to,

$$\frac{dM}{dx} \sim \frac{x^3 dN}{dx} \sim x^5 e^{-(x/\lambda)^3}. \tag{4.222}$$

In any case the preponderance of data on fragment size distributions for brittle solids suggest that brittle fragmentation is not well described by a Poisson process, at least in the simple ways that the models have been approached.

4.3.5 Scale Invariance in the Catastrophic Fracture of Brittle Solids

The statistical features of brittle fracture are not understood through any of the various probabilistic approaches discussed earlier in the present chapter. Previous discussions suggest that the efforts of earlier workers [3, 25, 27], to characterize brittle fracture through a Poisson statistical process, is probably in error. Some understanding is achieved through intriguing parallels of the dynamic fragmentation of brittle solids with hydrodynamic turbulence in fluids. Since the latter is still regarded as one of the remaining unsolved problems of classical physics, it is perhaps not surprising that a satisfactory statistical theory of brittle fracture and fragmentation continues to elude a sound theoretical basis.

Hydrodynamic Turbulence

Turbulence can occur in fluids described by the Navier–Stokes equations where governing material properties are relatively few in number. A macroscopic length scale L, a characteristic velocity V, and a viscosity η are adequate to reveal the essential features. The length L is for example the size of the structure imparting motion to the fluid while V is the nominal velocity impressed on the fluid. When the dimensionless Reynolds number $R_e = LV/\eta$ is sufficiently large turbulence arises in the fluid motion.

Turbulence in the flow emerges because large scale laminar flow is not sufficient to dissipate the energy through viscous friction. Turbulence is then the activation, the growth and the motion of irregular fluid disturbances on successively smaller length scales necessitated by the fluid dissipation requirements. The essence of turbulence is the transient and steady state motion of a hierarchy of submotions over a wide range of length scales. This cascade to successively smaller length scales in the turbulence process proceeds until velocity gradients of order V/λ, with λ the limiting substructure length scale, are achieved that are adequate to support the necessary viscous dissipation. The range of length scales between L and λ is determined by the Reynolds number and increases with increasing Reynolds number.

The range of submotions of length scales bounded by L and λ, and reasonably removed from either, is commonly called the inertial range (e.g., [23]). Invariance to scale is expected in the inertial range. Kolmogorov [65] argued for such scale invariance on dimensional grounds, and arrived at a power law dependence of features of the turbulent motions on length scale over the inertial range. Recent theoretical work on turbulence suggests that the scale invariance of Kolmogorov is not fully realized [13]. For present purposes, however, such scale invariance, or near-scale invariance, is a crucial observation of hydrodynamic turbulence, and offers a perspective for understanding the nature of dynamic brittle fragmentation

Catastrophic Fracture

Striking parallels to hydrodynamic turbulence are seen in the catastrophic fracture and fragmentation of brittle solids [39]. Consider a solid object of characteristic size L composed of brittle solid such as glass or a high-strength ceramic that is subjected to a compressive load inducing a nominal elastic strain energy per unit mass ε. Fracture of the object initiates when critical stress conditions are achieved at some site in the body. Once initiated, fracture proceeds rapidly and explosively, converting the elastic strain energy into surface fracture energy γ and kinetic energy of the ejected fragments.

Fracture in brittle solids is weakly dissipative, however, and failure through one or several through-going cracks is far from adequate to absorb the initial stored elastic strain energy. Consequently, during failure, fracture proceeds on successively finer length scales though a cascade of crack branching until length scales adequate to the dissipation of the initial elastic strain energy are achieved. This length scale is expected to scale as $\lambda \sim \gamma/\rho\varepsilon$, with ρ the material density. This limiting length scale λ is a number of decades smaller than the characteristic size L of the body.

Within the inertial range $\lambda < x < L$ there is not a length scale governing the physics of the catastrophic fracture cascade. Consequently, the fragment count within this range would be expected to exhibit a power-law dependence on fragment size. As fragment size approaches the limiting dissipation length scale λ, the functional dependence will diverge from a power-law dependence exhibiting an awareness of the dissipation limit length scale λ.

Although the specific functional form is not known for the fragment number distribution, an appropriate relation for the number distribution is readily guessed. The functional form,

$$N(x) = \frac{N_o}{1 + (x/\lambda)^{\delta}} \tag{4.223}$$

exhibits the necessary power-law dependence for $x \gg \lambda$ and the appropriate limiting behavior within the range $x \leq \lambda$. $N(x)$ is the complementary cumulative number distribution (number greater than) of fragments while N_o is the total number of fragments. The exponent δ is the fractal dimension in the power-law limit of the cumulative fragment relation. The relation readily transforms to the Schuhmann equation in a mass distribution representation of the fragments.

Similarities between hydrodynamic turbulence and catastrophic brittle fracture also are evident in the propagation of a singe crack. The laminar flow of a rapidly shearing fluid soon becomes unsettled by the emergence of insipient eddies and vortices as the intensity of shearing is increased. Again, dissipation through laminar viscosity is increasingly inadequate to balance the power input. Energy dissipated at the tip of a fast running crack in a brittle solid is known to be a strong function of the velocity of the crack. Increased dissipation is a consequence of the onset of microcrack branching instability at sufficiently high velocity [88]. Again, microcrack instability is the physical equivalent of hydrodynamic turbulence exhibited by the crack in exploring crack-tip deformations capable of the needed dissipation.

Such analogies between dynamic brittle fracture and hydrodynamic turbulence have not previously been explored, and are deserving of further attention.

4.4 Fragmentation in the Spall Process

The fracture of condensed matter through spall is a process in which the interaction of compression waves carries interior regions (away from free surfaces) into tension and fails the material if material strength is exceeded. Excepting the Hugoniot elastic limit, spall is probably the most familiar of failure processes to those in the shock compression community (e.g., [19]). The fragmentation characteristics of spall are not commonly considered, however; most likely, because the standard planar spall experiment and diagnostics do not prominantly display the fragmentation features of the event.

The processes of spall can result in intense fragmentation as the hypervelocity experiments of Piekutowski [82, 83] vividly illustrate. One radiograph from the hypervelocity impact spall experiments of Piekutowski is shown in Fig. 4.36. High-velocity normal impact of an aluminum sphere onto a thin aluminum plate results in a large-amplitude, short-duration attenuating pressure pulse which emerges at the back free surface of the sphere inducing intense

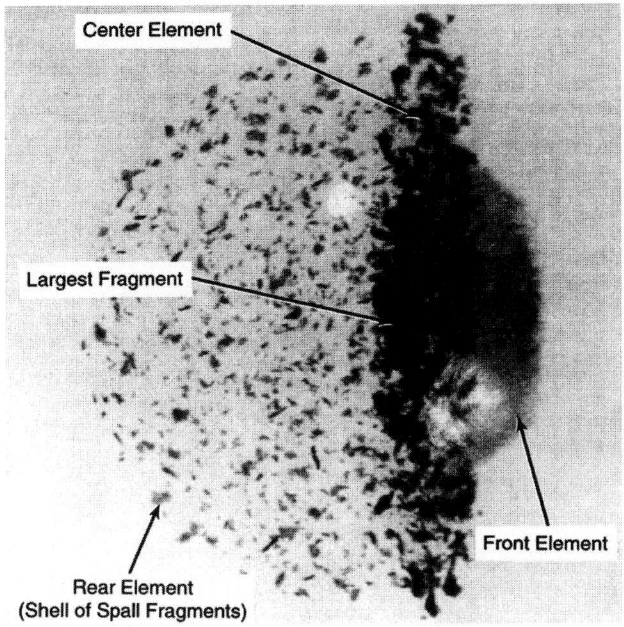

Fig. 4.36. Normal impact of 12.7 mm diameter aluminum sphere on 0.59 mm thickness aluminum plate at 6.26 kms^{-1}. Radiographic image at 46 μs after impact [82,83]

multiple spall fragmentation. The complex sequence of wave interactions leads to the total disintegration and fragmentation of the projectile.

In the present section, several analytic solutions of fragmentation in the spall process based on various spall failure criteria are explored to illustrate further issues of dynamic fragmentation and its relation to shock-wave propagation in solids. These analytic examples provide physical insight into the fragmentation processes occurring in the Piekutowski experiments and related intense spallation events.

4.4.1 Spall Fragmentation from a Pulse Emerging at a Free Surface

To explore fragmentation aspects of the spall fracture process we will investigate the consequences of a saw-toothed compressive wave propagating toward the left and incident on a free surface as illustrated in Fig. 4.37. This wave adequately approximates the impact-induced shock waves emerging at the back surface of the Piekutowski sphere. The material will be linear elastic (or linear hydrodynamic) characterized by a density ρ and a wave speed c. If time $t = 0$ corresponds to emergence and reflection of the wave at the free surface, then the compressive stress profile at any time $t < 0$ prior to arrival at the free surface is,

$$\sigma(x,t) = P - \rho c^2 \dot{\varepsilon}(t + x/c)/2, \tag{4.224}$$

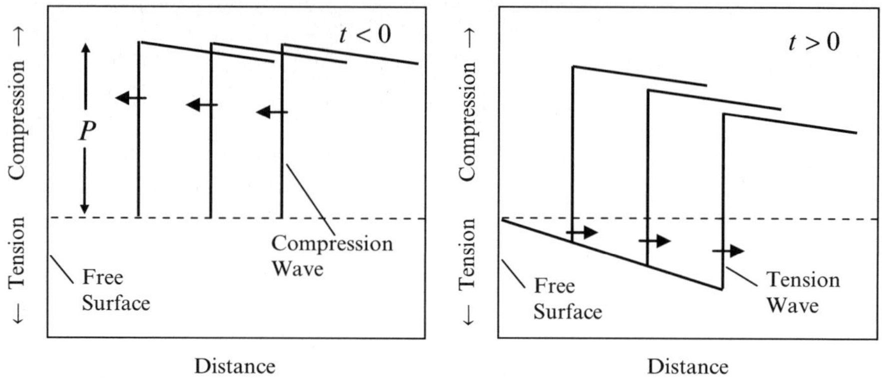

Fig. 4.37. Illustrates the compression wave (*left plot*) incident on the free surface and reflected tension wave (*right plot*)

where P is the peak compressive stress and $\dot{\varepsilon}$ is the decompression strain rate (or the velocity gradient) behind the wave peak. The peak stress P is considered here to be many times larger than the magnitude of the material tensile strength. (There will be occasional switches in sign convention for compression and tension in this section to forgo the awkward use of minus signs in the relations. The convention in use should be evident from the text.)

The incident stress wave before reflection, and the stress profile at some time after reflection, is illustrated in Fig. 4.37. Ahead of the incident shock, material is initially at rest and stress free. Reflection of the wave results in a right facing triangular wave as shown in the right hand illustration in Fig. 4.37. The stress field to the left of the leading reflected characteristic is a non-simple wave solution of the d'Alembert form (e.g., [17]),

$$\sigma(x,t) = f(t - x/c) + g(t + x/c). \qquad (4.225)$$

The stress free boundary condition leads to the solution for the reflected tensile stress for time $t > 0$ as a function of distance x from the free surface and behind the backward facing decompression shock wave,

$$\sigma(x,t) = \rho c \dot{\varepsilon}\, x, \quad x - ct < 0. \qquad (4.226)$$

The corresponding particle velocity is provided by,

$$u(x,t) = (\rho c^2 \dot{\varepsilon} t - 2P)/\rho c. \qquad (4.227)$$

The magnitude of the parameter $\dot{\varepsilon}$ provides a measure of the strain rate at which the body is carried into tension and spall.

Spall and Fragment Size Criteria

If a criterion for spall is proposed of instantaneous fracture when a constant tensile stress of magnitude σ_s (the spall stress) is achieved, then the distance

$x = x_s$ to the spall plane, is an estimate of the fragment size, and is calculated to be,

$$x_s = \frac{\sigma_s}{\rho c \dot{\varepsilon}}. \tag{4.228}$$

A subsequent identical triangular tension wave originates at the spall plane new free surface and propagates until the spall criterion is again achieved. A second spall fragment with size provided by (4.228) is again produced. The process repeats until the incident wave is exhausted, creating a fragment number of order P/σ_s.

A spall impulse criterion is derived by calculating the momentum imparted to the spalled plate, leading to the prediction of a fragment size and a corresponding spall stress that is dependent on the critical impulse I_s. The momentum expression,

$$I(x_s) = \int_0^{x_s} \rho u d\xi = \int_0^{x_s} \rho \dot{\varepsilon} \xi \, d\xi = I_s, \tag{4.229}$$

when solved provides for the fragment size,

$$x_s = (2I_s/\rho\dot{\varepsilon})^{1/2}, \tag{4.230}$$

and the spall stress at the spall plain,

$$\sigma_s = (2\rho c^2 \, I_s \dot{\varepsilon})^{1/2}. \tag{4.231}$$

Strain rate dependence of the spall stress based on a momentum criterion agrees with that suggested by Skidmore [90].

Alternatively, a spall energy criterion based on a critical work W_s follows from calculation of the elastic and kinetic energy in the plate,

$$W(x_s) = \tfrac{1}{2} \int_0^{x_s} (\rho u^2 + \sigma^2/\rho c) d\xi = \tfrac{1}{2} \int_0^{x_s} 2\rho \dot{\varepsilon}^2 \xi^2 \, d\xi = W_s. \tag{4.232}$$

A fragment size is calculated,

$$x_s = (3W_s/\rho\dot{\varepsilon}^2)^{1/3}, \tag{4.233}$$

and spall stress,

$$\sigma_s = (3\rho^2 c^3 W_s \dot{\varepsilon})^{1/3}. \tag{4.234}$$

Tuler–Butcher Spall Criteria

A spall stress that depends on details of the transient pressure load leading to spall has been observed in a number of materials (e.g., [43, 55]). Various history dependent spall criteria have been proposed. The relation proposed by Tuler and Butcher [92] has received wide application. Ignoring a threshold stress term the Tuler–Butcher fracture criterion can be written,

$$I_n = \frac{1}{\rho c} \int \sigma^n(t) dt \leq K_n. \tag{4.235}$$

For the tensile stress field provided in (4.226) the Tuler–Butcher integral can be explicitly solved for yielding,

$$I_n(x,t) = \frac{1}{\rho c} \int_{x/c}^{t} (\rho c \dot{\varepsilon}\, x)^n dt = \frac{1}{\rho c} (\rho c \dot{\varepsilon}\, x)^n (t - x/c). \tag{4.236}$$

The Tuler–Butcher integral is illustrated in Fig. 4.38. Both ordinate and abscissa are normalized to provide amplitude of unity and a propagation distance of unity at spall criticality. On the left, the shape of the Tuler–Butcher integral is illustrated for the special cases of $n = 1$, 2, and 5. This plot identifies the distance behind the tensile shock that the Tuler–Butcher integral grows the fastest. The position of the spall plane is identified for each of the cases and is seen to approach the position of the front of the tensile stress wave as n becomes large. On the right illustration in Fig. 4.38, the propagation growth of the Tuler–Butcher integral is shown for the case of $n = 2$.

At any time t the Tuler–Butcher integral is a maximum at some interior point in the interval $0 < x < ct$. The Tuler–Butcher integral at this maximum point can be equated to the critical constant K_n to establish the distance to the fracture plane, and hence fragment size,

$$x_s = \left(\frac{n \rho c^2\, K_n}{(\rho c \dot{\varepsilon})^n} \right)^{1/(n+1)}, \tag{4.237}$$

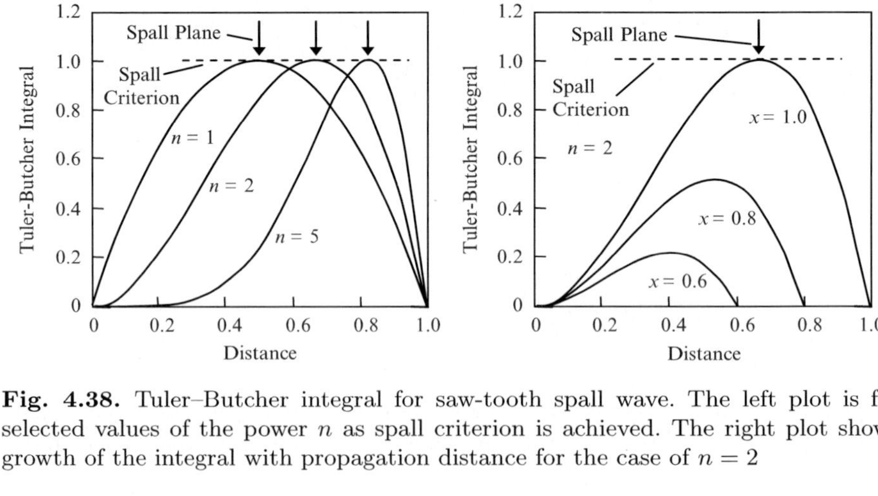

Fig. 4.38. Tuler–Butcher integral for saw-tooth spall wave. The left plot is for selected values of the power n as spall criterion is achieved. The right plot shows growth of the integral with propagation distance for the case of $n = 2$

The tensile fracture stress achieved at the fracture plane is,

$$\sigma_s = \left(n\rho^2\ c^3\ K_n\dot{\varepsilon}\right)^{1/(n+1)} . \tag{4.238}$$

The Tuler–Butcher criterion for $n = 1$ retrieves the impulse criterion of (4.230) and (4.231) with $I_s = \rho c K_1/2$ the critical impulse. A value of $n = 2$ recovers the energy criterion of (4.233) and (4.234) with $W_s = 2K_2/3$. As n becomes large the Tuler–Butcher spall criterion approaches a constant fracture stress and the fragment size dependence of (4.228).

Over a reasonable range of the Tuler–Butcher parameter n, the strain rate power for nominal fragment size ranges from 1/2 for an impulse criterion, to 2/3 for an energy criterion, to a value of 1 for a constant spall stress. The strain rate power determined for the aluminum sphere spall fragment size data of Piekutowski [83] shown in Fig. 4.39 is in the neighborhood of 0.60–0.62. For the impact velocity range of the data, the assumption that strain rate scales with the ratio of the impact velocity to the sphere diameter as is plotted in the figure is adequate [37]. Comparison with the Tuler–Butcher criteria suggests the data falls between a strain rate power of 1/2 impulse criterion and a strain rate power of 2/3 energy criterion, perhaps closer to the latter.

The Tuler–Butcher criteria also predict a strain rate dependence of the spall stress and it is of interest to provide a comparison with available experimental data. The ultra-high strain rate spall-strength data for aluminum obtained by Kanel et al. [56] is plotted in Fig. 4.40 along with their companion data for molybdenum metal. The strain rate power for aluminum and molybdenum are in the range of 0.23–0.27, somewhat below one-third for the energy-based criterion and well below one-half for the impulse criterion. The

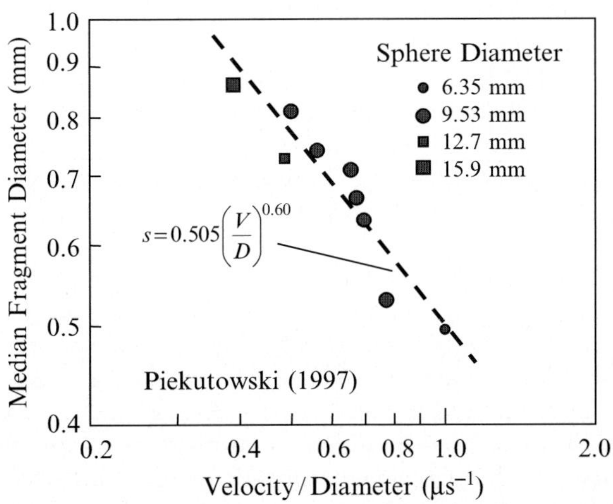

Fig. 4.39. Spall fragment size data for aluminum sphere impact experiments of Piekutowski [83] including both velocity and system scale variations

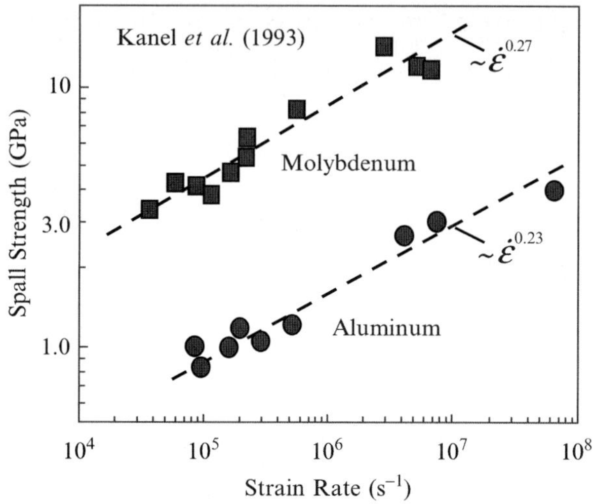

Fig. 4.40. Ultra-high strain rate spall strength data for aluminum and molybdenum from Kanel et al. [56]

data of Kanel et al. [56] are unique, however, and it is difficult to ascribe uncertainties to either the spall amplitude or the strain rate dependence.

4.4.2 Cohesive Zone Spall Solution

A further analytic solution can be pursued to explore the dependence of strength and fragmentation in the spall event. The preceding discussions have revealed two critical features governing the time dependence of fracture in the dynamic tensile fracture of solids. First, work must be performed to overcome the resistance to fracture of an opening crack and this energy must come from the local elastic or kinetic energy in the neighborhood of the crack. Second, momentum must be fluxed into the region of the crack in order to accommodate motion of the opening crack. The material and fracture properties influencing this process can be instructively explored in the one-dimensional spall of a solid in which the material response is linear elastic for all regions of the body not specifically on the plane of fracture.

Again, consider a linear elastic body, initially subjected by some impact process to a one-dimensional compressive stress of magnitude P, which is then carried into tension on the plane $x = 0$ by the symmetric interaction of two opposing release waves. Select the coordinate system such that the center of the spall plane is at rest and remains at rest throughout the spall process. A tensile shock wave illustrated in Fig. 4.41 propagates on the forward facing characteristic, $x = ct$ $(x \geq 0)$, with amplitude of the spall stress $\sigma_s = \sigma_s(P)$. Possible dependence of the spall stress on the initial shock amplitude P is assumed. Within the domain enclosed by the spall plane $x = 0$, and the tensile shock wave $x = ct$, the d'Alembert solution again applies. The flow field on left

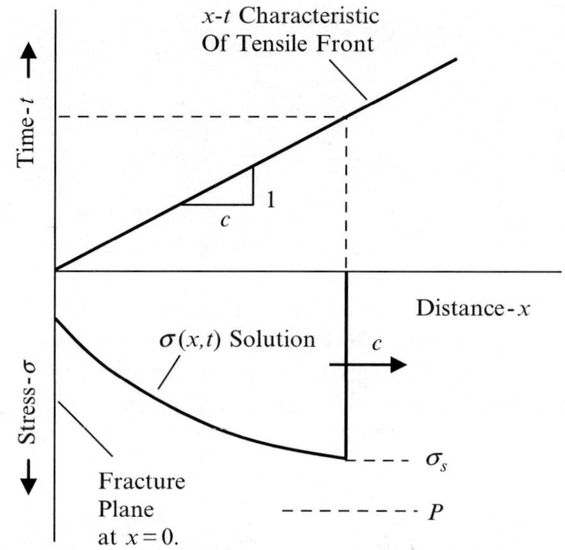

Fig. 4.41. Distance versus time (*upper*) and tensile spall stress profile (*lower*) emerging from the $x = 0$ plane as time-dependent spall fracture proceeds.

facing characteristics is constant. Consequently, stress and particle velocity, after properly matching jump conditions on the shock, have the simple wave solution,

$$\sigma(x,t) = f(t - x/c), \tag{4.239}$$

$$u(x,t) = (P - f(t - x/c))\,/\rho c. \tag{4.240}$$

The solution is completed by applying the stress boundary condition at the spall plane $x = 0$. A cohesive zone spall crack-opening resistance will be assumed of the form,

$$f(t) = \sigma(0,t) = \sigma_s(1 - y(t)/y_c). \tag{4.241}$$

The tensile stress in the spall zone decreases linearly from the spall stress $\sigma_s(P)$ to zero over the crack opening displacement $0 \le y(t) \le y_c$, and where y_c is the critical crack-opening displacement. Displacement is obtained from the particle velocity at $x = 0$,

$$y(t) = \int_0^t u(0,s)ds, \tag{4.242}$$

and leads to the differential solution for $f(t)$,

$$\frac{d}{dt}f(t) = -\frac{\sigma_s}{\rho c y_c}\left(P - f(t)\right). \tag{4.243}$$

Solutions for the stress and particle velocity field are,

$$\sigma(x,t) = P - (P - \sigma_s)\, e^{\frac{\sigma_s}{\rho c^2 y_c}(ct-x)}, \tag{4.244}$$

$$u(x,t) = \frac{1}{\rho c}(P - \sigma_s)\, e^{\frac{\sigma_s}{\rho c^2 y_c}(ct-x)}. \tag{4.245}$$

In the present cohesive zone model for the spall resistance, spall fracture energy is $\gamma_c = \sigma_s y_c/2$ and spall toughness can be defined through the expression $K_c^2 = 2\rho c^2 \gamma_c = \rho c^2 \sigma_s y_c$.

Spall strength is commonly found to depend on the shock strength P. (The amplitude of the elastic tension achieved has magnitude P if spall does not occur.) Dependence of the spall stress on the shock strength is included through assuming the form,

$$\sigma_s(P) = \sigma_{so}\left(\frac{P}{\sigma_{so}}\right)^{m-1}, 1 \le m < 2, \tag{4.246}$$

where σ_{so} is a threshold spall stress (spall does not occur if $P < \sigma_{so}$). The power m leads to a constant spall stress σ_{so} for $m = 1$ and to a spall strength approaching P as $m \to 2$.

Let $R = P/\sigma_{so}$ be the ratio of the shock strength to the spall threshold stress and let $u_o = \sigma_{so}/\rho c$. Equation (4.244) provides the time-dependent stress at the spall plane,

$$\sigma(t)/\sigma_{so} = R[1 - (1 - R^{m-2})e^{R^{2(m-1)}(ct/a_o)}], \tag{4.247}$$

where the length scale $a_o = (K_c/\sigma_{so})^2$. The stress and velocity history at any position x is provided by (4.244) and (4.245).

The time duration over which the resisting stress in the spall plane reduces to zero is calculated from (4.247),

$$-a_o R^{2(1-m)} \ln(1 - R^{m-2}) = ct_s = x_s/2. \tag{4.248}$$

The distance x_s is the minimum uninterrupted wave propagation distance required in order to flux the energy and momentum necessary to fracture completion at the spall plane. It is, for example, the minimum spacing allowing equally spaced fractures to achieve completion in a body uniformly loaded to a tension P. Equation (4.248) therefore provides a reasonable measure of the nominal fragment size as a function of the tensile amplitude P as shown in Fig. 4.42. Fragment size is plotted as a function of $R = P/\sigma_{so}$ for both $m = 1$ and $m \to 2$.

Particle velocity histories at the spall plane are illustrated in Fig. 4.43, and approximate the pullback velocity signals measured in spall experiments. Again, for a value of $m = 1$ the spall strength measured is constant. Only the pulse duration reduces with increasing amplitude of the applied tensile stress. In contrast for larger values of m (profiles for $m = 1.5$ are shown in the

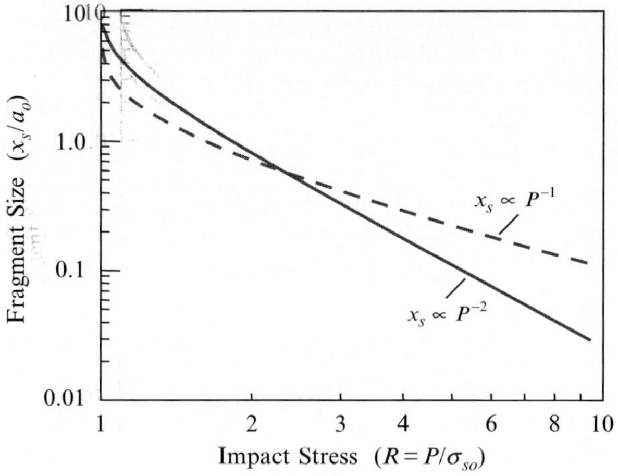

Fig. 4.42. Minimum spall or fracture plane spacing as a function of the tensile stress amplitude

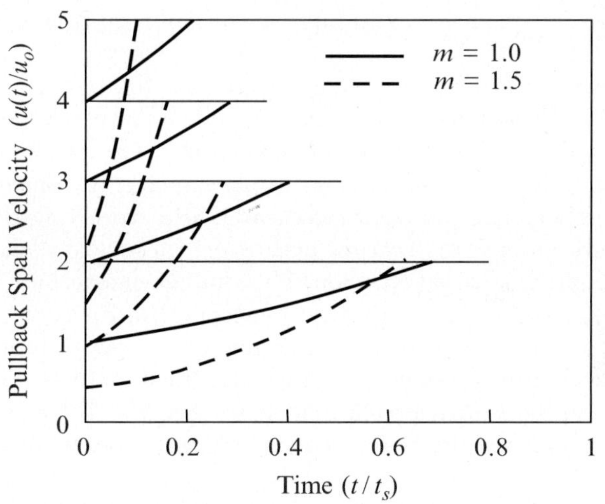

Fig. 4.43. Calculated spall velocity profiles for tensile amplitudes of $P/\sigma_{so} = 2$, 3, 4 and 5

figure) both spall stress amplitude and duration depend on the magnitude of $R = P/\sigma_{so}$.

For sensibly large R (4.248) reduces to the simpler form,

$$2ct_s = x_s = 2a_o/R^m. \tag{4.249}$$

For the special case of $m = 1$ (4.249) provides,

$$x_s = \frac{2\rho c^2}{P} y_c, \tag{4.250}$$

yielding a fragment size that depends only on the spall plane crack-opening displacement y_c, independent of the fracture energy, and with an inverse first power dependence on the shock strength P. For $m \to 2$,

$$x_s = \frac{4\rho c^2}{P^2}\gamma_c. \tag{4.251}$$

The fragment size depends only on the fracture energy γ_c, and decreases as the second power of the shock strength P. Setting $m = 2$ in (4.246) lead to solution difficulties. This issue is troublesome and perhaps other functional forms for (4.246) would be useful.

Lastly, the Tuler–Butcher criterion for the same two cases are,

$$I_1 = \frac{1}{\rho c}Pt_s = 2y_c, \tag{4.252}$$

$$I_2 = \frac{1}{\rho c}P^2 t_s = 4\gamma_c. \tag{4.253}$$

The present cohesive spall zone model offers some insight into the several common forms of the Tuler–Butcher fracture criterion. For $m = 1$ the Tuler–Butcher criterion is an impulse criterion requiring a critical crack-opening displacement, or, in one dimension, a fracture activation volume. When sufficiently above the threshold stress σ_{so}, the fracture criterion is independent of the thresholds stress and the fracture energy, requiring only a sufficient flux of momentum into the spall zone to accommodate the requisite activation volume (crack-opening displacement). In contrast, for $m \to 2$ the Tuler–Butcher criterion approaches a critical work criterion requiring a critical fracture energy, or fracture activation energy, to complete the spall process. For a critical work criterion, both sufficient energy and momentum must be transported to the spall zone to accomplish fracture.

Such interpretations of the Tuler–Butcher integral are of interest because of its relatively wide use in modeling a range of critical-state activation processes under shock and high-rate applications. A form of the Tuler–Butcher integral is used, for example, in the HVRB reaction kinetics model of Kerley [58].

4.4.3 Shock Attenuation and Spall Fragmentation in a Dissipative Medium

One further analytic solution is instructive. This solution would also apply to the impact spall of the Piekutowski aluminum sphere. It is more profitably applied to such applications as the attenuating shock and back surface spall of a concrete slab subject to a transient pressure load from a detonating explosive or impact on the opposing surface, or to the spall of a thin metal plate subjected to the impulsive loading of an intense laser or electron beam. Here, shock wave load and spall in concrete will provide the illustration. The situation is such that lateral dimensions are large relative to through-the-thickness dimensions and the analysis can be specialized to one-dimensional shock-wave propagation and spall.

Momentum and Energy Coupling into a Concrete Barrier

The coupling of momentum and energy into a slab of material caused by the planar impact of a high-velocity plate or the detonation of an in-contact plane of explosive is readily estimated by analytic means. The first is efficiently calculated through the appropriate application of pressure versus velocity diagrams [8], while the latter is readily estimated with Gurney methods [57]. The pressure loading at the input side of the slab ($x = 0$) will be approximated with the exponential form,

$$p(t) = p_o e^{-t/\tau_o}, \tag{4.254}$$

where the pulse amplitude and duration is captured by the two parameters p_o and τ_o. The corresponding particle velocity at the input interface is provided through the relation $p(t) = Zu(t)$ where Z is an appropriate constant shock impedance for the concrete at the prescribed input pressure. This correspondence becomes increasingly approximate at high shock pressures. The total impulse provided by the load is,

$$I_o = \int_0^\infty p_o e^{-t/\tau_o} dt = p_o \tau_o, \tag{4.255}$$

while the work applied and initial energy in the shock pulse is,

$$E_o = \int_0^\infty pudt = \frac{1}{2Z} p_o^2 \tau_o. \tag{4.256}$$

Let $E_x(x)$, $p_x(x)$ and $\tau_x(x)$ be the energy, pressure amplitude and pulse duration of the exponential shock pulse after propagating a distance x into the concrete barrier from the pressure loading surface, and such that $E_x(0) = E_o$, $p_x(0) = p_o$ and $\tau_x(0) = \tau_o$. The energy integral at position x corresponding (4.256) to yields,

$$E_x = \frac{1}{2Z} p_x^2 \tau_x. \tag{4.257}$$

Since momentum in the shock pulse is conserved ($I_o = p_o \tau_o = p_x \tau_x$), it follows that,

$$E_x = \frac{I_o}{2Z} p_x. \tag{4.258}$$

Thus, attenuation of the shock energy and the peak pulse pressure with propagation distance are proportional and,

$$\frac{dE_x}{dx} = \frac{I_o}{2Z} \frac{dp_x}{dx}. \tag{4.259}$$

Dissipation in Propagation of the Shock Pulse

Two forms of energy dissipation are considered in the shock pulse propagation through concrete. First, if the material exhibits a constant von Mises strength of Y, then the shock compression and subsequent pressure release will exhibit the pressure versus strain response shown on the left side of Fig. 4.44. The energy dissipation is the enclosed area as shown and is written,

$$Q_1 = \frac{4}{3} Y \varepsilon_h = \frac{4}{3} \frac{\rho_o Y}{Z^2} p_x . \tag{4.260}$$

The second equality relates the strength dissipation to the peak shock pressure of the pulse through $p_x = (Z^2/\rho_o)\varepsilon_h$.

The second source of dissipation is illustrated on the right of Fig. 4.44. The underlying physical causes are several including rate-dependent plasticity and stress wave scattering among others [36]. The loss energy is ultimately thermalized. This shock dissipation component can be written approximately as,

$$Q_2 = \frac{1}{3} \rho_o C_o^2 S \varepsilon_h^3 = \frac{1}{3} \frac{\rho_o^2 S}{Z^4} p_x^3 , \tag{4.261}$$

where C_o and S are the common linear shock velocity versus particle velocity parameters [8]. The shock dissipation model becomes,

$$\frac{dE_x}{dx} = -\frac{4}{3} \frac{\rho_o Y}{Z^2} p_x - \frac{1}{3} \frac{\rho_o^2 S}{Z^4} p_x^3 . \tag{4.262}$$

Combining the previous equations provides a relation for attenuation of the peak shock pressure,

$$I_o \frac{dp_x}{dx} = -\frac{8}{3} \frac{\rho_o Y}{Z} p_x - \frac{2}{3} \frac{\rho_o^2 S}{Z^3} p_x^3 , \tag{4.263}$$

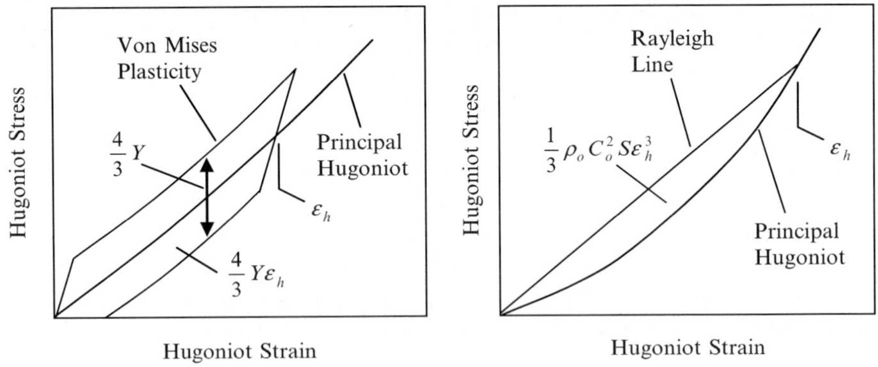

Fig. 4.44. Illustrates mechanisms of dissipation through both von Mises strength and shock dissipation in the shock propagation process

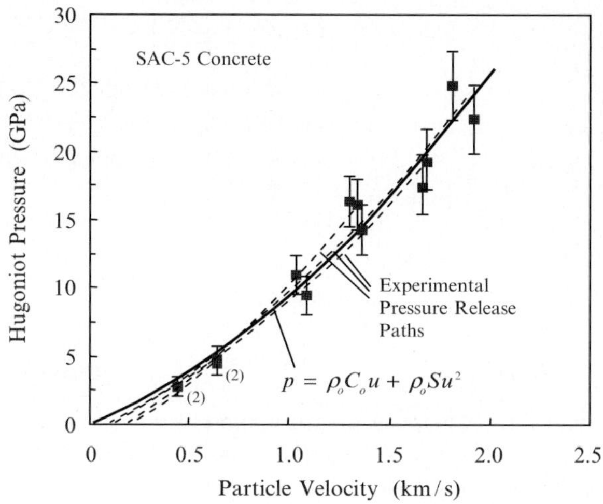

Fig. 4.45. Shock Hugoniot data for SAC-5 concrete [35] with nominal initial density of $2,300\,\text{kg m}^{-3}$. Two data points are indicated by the (2) where shown. Solid curves provide the best quadratic fit to the Hugoniot data. Dashed curves are experimental stress unloading paths from the shock states

or,

$$\frac{d\pi_x}{dx} = -\alpha_1\pi_x - \alpha_2\pi_x^3, \tag{4.264}$$

where $\pi_x = p_x/p_o$ and the coefficients α_1 and α_2 are identified in the previous equation. Equation (4.264) is readily solved for the attenuation of the peak pressure as a function of propagation distance into the concrete barrier. The solution can be performed separately for shock dissipation, which results in power-law attenuation, or for strength dissipation, leading to exponential attenuation. Generally, both dissipation mechanisms participate in the attenuation process.

The calculated shock dissipation and attenuation are carried out for a reasonably well characterized concrete [35]. Hugoniot properties are shown in Fig. 4.45. The necessary strength and equation-of-state properties are density, $\rho_o = 2,300\,\text{kg m}^{-3}$, von Mises strength, $Y = 250\,\text{MPa}$, shock velocity intercept, $C_o = 2,600\,\text{m s}^{-1}$ and shock velocity slope, $S = 1.48$. This concrete also exhibits some irreversible (pore) compaction over a portion of the Hugoniot curve [35] that is not accounted for here. An additional dissipative term for compaction, Q_3, could be readily modeled, however, and included in the dissipation and attenuation relations developed. A 1-cm thickness TNT explosive detonated in contact with the concrete slab is assume. Gurney equations provided an initial peak pressure and pulse duration of $p_o = 28.1\,\text{GPa}$ and $\tau_o = 1.46\,\mu\text{s}$.

Plots of the peak pressure attenuation, and growth of the pressure pulse duration, as a function of propagation distance into the concrete barrier are

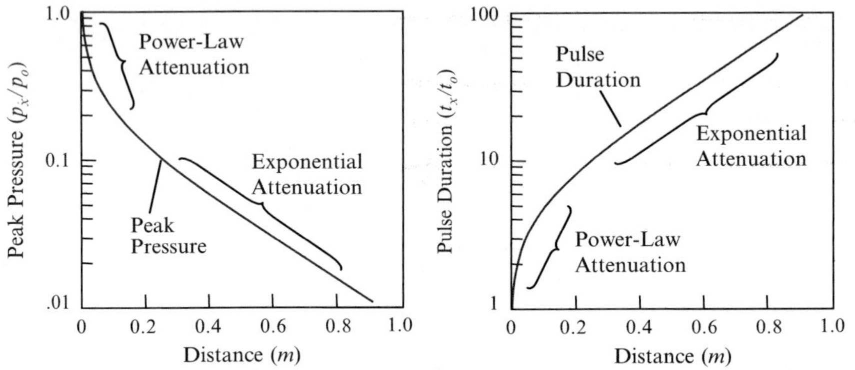

Fig. 4.46. Attenuation of the peak pressure and growth of the pressure pulse duration in concrete due to both shock dissipation (power-law attenuation) and strength dissipation (exponential attenuation)

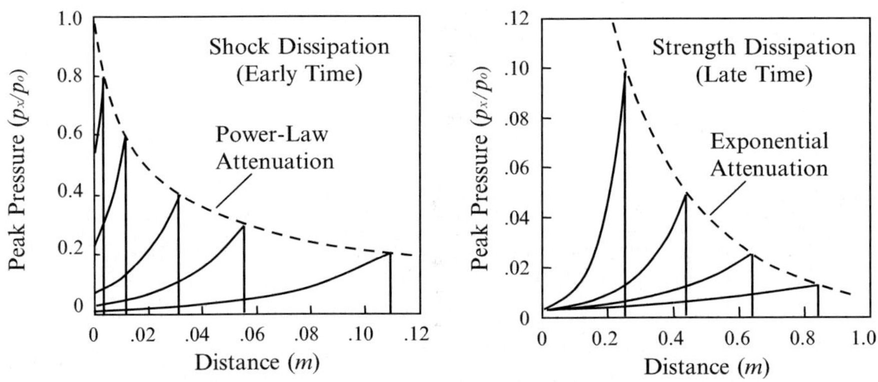

Fig. 4.47. Attenuation and propagation of the shock pressure pulse in concrete. Initial values are $p_s = 28.1\,\text{GPa}$ and $\tau_o = 1.46\,\mu\text{s}$

shown in Fig. 4.46. Calculations are based on the concrete and explosive properties provided. The calculated attenuation coefficients are $\alpha_1 = 3.34\,\text{m}^{-1}$ and $\alpha_2 = 71.8\,\text{m}^{-1}$, respectively. Shock dissipation and power-law attenuation dominates early time decay of the pressure pulse within the first 15–20 cm of concrete. Strength dissipation and exponential attenuation govern later time decay. Peak pressure attenuates from 28 GPa to less than 0.2 GPa over approximately 1 m of propagation in the concrete.

Pressure profiles within the two attenuation ranges are illustrated in Fig. 4.47. Noteworthy is the marked spreading of the pressure pulse necessary to conserving momentum as propagation and attenuation proceeds.

Solution for the Reflected Tension Wave

On emergence of the compressive pressure pulse at the back free surface, subsequent wave reflection will carry the material into tension. When the tensile spall strength is exceeded, spall fracture will occur and spall debris will be ejected at velocities commensurate with the intensity of the incident pressure wave. Frequently more than one spall plane occurs, broadening the spread in the size of fragments produced and the velocity of the fragment ejecta. The pressure wave approaching the back free surface has the functional dependence,

$$p(x,t) = p_x(x)e^{-(t-x/C_o)/\tau_x(x)}, \tag{4.265}$$

within the interval $x \le C_o t \le L$. On emergence at the back free surface following propagation through a thickness of concrete L, the pressure pulse has attenuated to,

$$p(x, L/C_o) = p_x(L)e^{-(L-x)/C_o\tau_x(L)}. \tag{4.266}$$

Let $p_x(L) = p_L$ and $\tau_x(L) = \tau_L$, for times $t > L/C_o$ write,

$$p(x,t) = p_L e^{-(C_o t - x)/C_o\tau_L}. \tag{4.267}$$

Subsequent wave propagation is reasonably approximated as linear elastic with initial condition at $t = L/C_o$ provided by (4.266).

In solving for the reflected tension wave it is convenient to introduce new distance and time variables $\bar{x} = L - x$ and $\bar{t} = t - L/C_o$ as illustrated in Fig. 4.48 providing for (4.267),

$$p(\bar{x},\bar{t}) = p_L e^{-(C_o\bar{t}+\bar{x})/C_o\tau_L}. \tag{4.268}$$

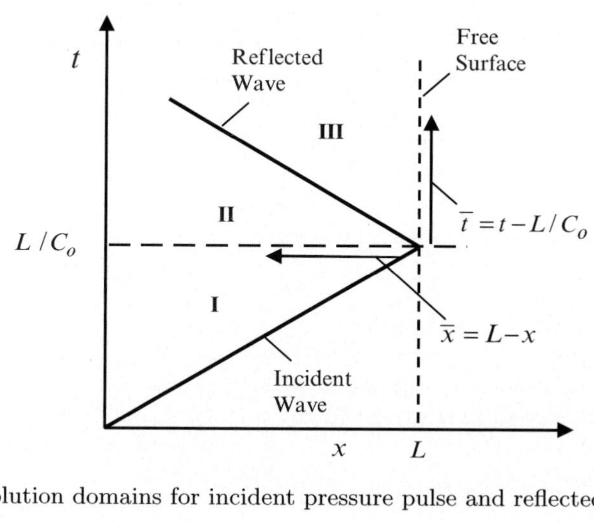

Fig. 4.48. Solution domains for incident pressure pulse and reflected tension wave

The d'Alembert solution to the elastic wave equation, in domain **III** behind the reflected tensile shock, for the pressure and particle velocity is,

$$p(\bar{x},\bar{t}) = f(C_o\bar{t}+\bar{x}) + g(C_o\bar{t}-\bar{x}), \tag{4.269}$$

$$\rho C_o u(\bar{x},\bar{t}) = f(C_o\bar{t}+\bar{x}) - g(C_o\bar{t}-\bar{x}). \tag{4.270}$$

The pressure and particle velocity are continuous on characteristics, $C_o\bar{t}+\bar{x} =$ constant, across the reflected tensile shock. Combining with the pressure free boundary condition on $\bar{x}=0$, (4.269) and (4.270) yield the solutions for tensile pressure and particle velocity in the reflected wave,

$$p(\bar{x},\bar{t}) = -2p_L e^{-\bar{t}/\tau_L}\sinh(\bar{x}/C_o\tau_L), \tag{4.271}$$

$$u(\bar{x},\bar{t}) = \frac{2p_L}{\rho C_o}e^{-\bar{t}/\tau_L}\cosh(\bar{x}/C_o\tau_L). \tag{4.272}$$

Spall Criterion and Fragment Size

The reflected wave solution can be joined with various spall criteria to determine where spall planes are located as tension and failure occurs in the pulse-loading event. The common criterion is that of abrupt spall fracture on planes where the tensile stress achieves a critical spall stress p_s. From (4.271) peak tension occurs on the characteristic $\bar{x}=C_o\bar{t}$ and the tension amplitude (a positive quantity) on this characteristic is,

$$p(\bar{x}) = p_L\left[1-e^{-2\bar{x}/C_o\tau_L}\right]. \tag{4.273}$$

Tension increases continuously with propagation distance $\bar{x}$ and spall occurs at $p(\bar{x})=p_s$. Tensile stress drops to zero at this plane and the tension again grows with continued wave propagation into the material until p_s is again achieved at a second spall plane. The process continues until the tensile wave is exhausted or until the front surface of the wall is reached. Depths of the spall planes are readily determined from the relation,

$$p(\bar{x}_n) = p_L\left[1-e^{-2\bar{x}_n/C_o\tau_L}\right] = np_s, \tag{4.274}$$

or,

$$\bar{x}_n = -\tfrac{1}{2}C_o\tau_L\ln\left(1-n\tfrac{p_s}{p_L}\right). \tag{4.275}$$

Solving for n,

$$n(\bar{x}_n) = \frac{p_L}{p_s}\left[1-e^{-2\bar{x}_n/C_o\tau_L}\right], \tag{4.276}$$

spall occurs on planes at depths $\bar{x}_n$ that yield integer values of $n(\bar{x}_n)$. The total number of spall plane is given by,

$$n_{tot} = \frac{p_L - \mathrm{mod}\,(p_L,p_s)}{p_s}. \tag{4.277}$$

Strain-Rate Dependent Fragment Size Criterion

Alternatively, when a body subjected to stress wave loading is carried into tension and experiences intense spall fracture a relation of the strain rate dependent form,

$$s = \alpha \dot{\varepsilon}^{-m}, \tag{4.278}$$

has been found appropriate for estimating the characteristic size of fragments or, for the present one-dimensional spall application, the characteristic spacing s of fractures. The rate of strain as the body is carried into tension is $\dot{\varepsilon}$. The power parameter m is typically $1/2 \leq m \leq 1$. In the present application, a calculation of the appropriate strain rate is determined from the elastic wave solution. The elastic pressure wave incident on the free surface is provided by (4.268), and the corresponding particle velocity field is,

$$u(\bar{x}, \bar{t}) = \frac{p_L}{\rho_o C_o} e^{-(C_o \bar{t} + \bar{x})/C_o \tau_L}. \tag{4.279}$$

The rate of strain is provided by the gradient of this expression on the characteristic $\bar{x} = C_o \bar{t}$,

$$\dot{\varepsilon} = \frac{\partial u}{\partial \bar{x}} = \frac{2 p_L}{\rho_o C_o^2 \tau_L} e^{-2\bar{x}/C_o \tau_L}. \tag{4.280}$$

For the case of an inverse linear dependence of the fracture spacing on strain rate,

$$s = \frac{p_s}{\rho_o C_o \dot{\varepsilon}}, \tag{4.281}$$

where again p_s is the tensile spall strength, while ρ_o and C_o are the same density and elastic wave speed, the equations yield for the fracture spacing,

$$s(\bar{x}) = \frac{C_o \tau_L p_s}{2 p_L} e^{2\bar{x}/C_o \tau_L}. \tag{4.282}$$

Location of the fractures can be determined from integer solutions of,

$$n(\bar{x}) = \int_0^{\bar{x}} \frac{d\bar{x}}{s(\bar{x})}, \tag{4.283}$$

yielding the relation,

$$n(\bar{x}) = \frac{p_L}{p_s} \left(1 - e^{-2\bar{x}/C_o \tau_L}\right), \tag{4.284}$$

or the same relation as developed through the earlier analysis.

Alternatively, consider the fracture spacing dependence on strain rate,

$$s(\bar{x}) = \alpha \dot{\varepsilon}^{-2/3}, \tag{4.285}$$

with $\alpha = (K_c/\rho_o C_o)^{2/3}$ where K_c is an appropriate spall fracture toughness for the material. The fracture spacing relation becomes,

$$s(\bar{x}) = \left(\frac{K_c C_o \tau_L}{2p_L}\right)^{2/3} e^{4\bar{x}/3C_o\tau_L}, \tag{4.286}$$

and,

$$n(\bar{x}) = \left(\frac{27}{16}\right)^{1/3} \left(\frac{p_L}{K_c/\sqrt{C_o\tau_l}}\right)^{2/3} e^{-4\bar{x}/3C_o\tau_L}. \tag{4.287}$$

Both of the spall relations provided in (4.284) and (4.287) emphasize the dependence on two key parameters in the problem. The characteristic wavelength of the incident pressure pulse $C_o\tau_L$ provides the fracture spacing length scale. The dimensionless prefactor coefficient in each case determines the fracture intensity. Both coefficients are a ratio of parameters with dimensions of stress. Fracture intensity increases with increased incident pressure amplitude p_L. Fracture intensity decreases when the fracture resistance, either p_s or $K_c/(C_o\tau_L)^{1/2}$, increases. A spall-fracture stress criterion resulting in (4.284) will exhibit replica scaling. That is, if in the present example the thickness of the concrete wall and explosive charge are scaled by some factor, the predicted spacing of spall fractures would correspondingly scale. An energy-based spall fracture criterion leading to (4.287) would not exhibit replica scaling.

Velocity of the Spall Ejecta

Emergence of the compressive wave at the free surface, and the subsequent reflected tension wave, leads to the development of successive planes of spall as described in previous paragraphs. Material ejected by the spall event (plates of various thicknesses in the present one-dimensional spall analysis) are expelled at velocities which are calculable though the present analysis. Referring to Fig. 4.48, the relevant relation from which subsequent motion of the spall ejecta can be calculated is that for the material velocity imparted by the initial compression wave in domain **II** and corresponding to the pressure relation in (4.268),

$$u(x,t) = \frac{p_L}{\rho_o C_o} e^{-(C_o t - x)/C_o\tau_L}, \tag{4.288}$$

or, in a form simpler for the present analysis,

$$u(X,T) = u_L e^{-(T+X)}. \tag{4.289}$$

The dimensionless distance and time have been introduced,

$$X = \frac{L-x}{C_o\tau_L}, \quad T = \frac{C_o t - L}{C_o\tau_L}. \tag{4.290}$$

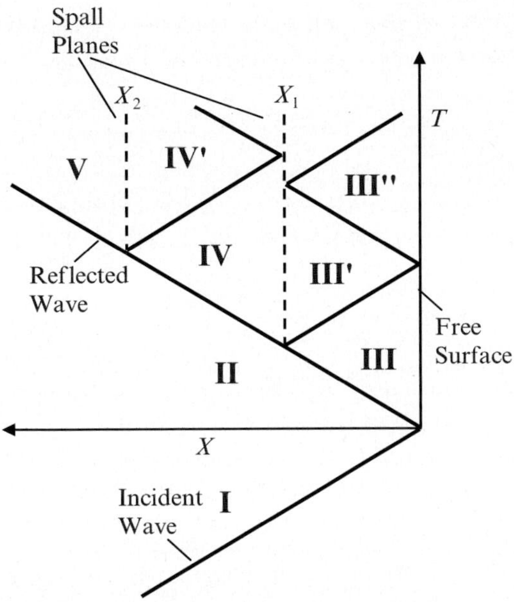

Fig. 4.49. Arrival of the incident exponential pressure pulse at the free surface and the subsequent reflected tension and spall. Scaled distance and time (see text) are on the abscissa and ordinate, respectively. Different domains are identified for which velocity solutions are obtained

The elastic wave solution provided the particle velocity in domain **III** between the free surface and the first spall plane illustrated in Fig. 4.49, given in (4.272), and written here in the simpler form,

$$u(X,T) = 2u_L e^{-T}\cosh(X). \tag{4.291}$$

The same elastic solution method provides the velocity,

$$u_n(X,T) = 2u_L e^{-2X_n} e^{-(T-X_n)}\cosh(X-X_n), \tag{4.292}$$

in the region immediately behind the nth spall plane at position X_n. The free surface and first spall plate corresponds to $n = 0$ and $X_0 = 0$. The average velocity of successive spall plates is provided by calculation of the spall plate momentum at any convenient time, say, $T = X_{n+1}$,

$$I_n = \rho_o\delta_n\bar{u}_n = 2\rho_o C_o \tau_L u_L e^{-2X_n} e^{-(X_{n+1}-X_n)} \int\limits_{X_n}^{X_{n+1}} \cosh(X-X_n)dX, \tag{4.293}$$

where $\delta_n = x_{n+1} - x_n$ is the spall plate thickness. The solution of yields for the spall plate velocity,

$$\bar{u}_n = 2u_L e^{-2\bar{x}_n/C_o\tau_L}\left(\frac{1 - e^{-2\delta_n/C_o\tau_L}}{2\delta_n/C_o\tau_L}\right), \tag{4.294}$$

where again $\bar{x}_n$ is the depth to the spall plane. When spall is intense ($p_s \ll p_L$) it follows that $\delta_n \ll C_o\tau_L$ over a reasonable range of n and,

$$\bar{u}_n \simeq 2u_L e^{-2X_n}. \tag{4.295}$$

References

1. Abramowitz, M., Stegun, I.A. (eds.): Handbook of Mathematical Function, No. 55. National Bureau of Standards Applied Mathematics Series, U. S. Government Printing Office (1954)
2. Ang, A.S., Tang, W.H.: Probability Concepts in Engineering Planning and Design, vol. 2: Decision, Risk, and Reliability. Wiley, New York (1984)
3. Bennett, J.G.: Broken coal. J. Inst. Fuel **10**, 22–39 (1936)
4. Bergstrom, B.H.: Energy and size distribution aspects of single particle crushing. In: Fairhurst, C. (ed.) Proceedings of the 5th Symposium on Rock Mechanics, pp. 155–172. Pergamon Press, Oxford (1962)
5. Bergstrom, B.H., Sollenberger, C.L., Mitchel, W.: Energy aspects of single particle crushing. Trans. AIME **220**, 367–372 (1961)
6. Bond, F.C.: Third theory of comminution. AIME Trans. **193**, 484–494 (1952)
7. Boots, B.N., Murdoch, D.J.: The spatial arrangement of random Voronoi polygons. Comput. Geosci. **9**, 351–365 (1983)
8. Boslough, M.B., Asay, J.R.: Basic principles of shock compression. In: Asay J.R., Shahinpoor, M. (eds.) High-Pressure Shock Compression of Solids, pp. 7–42. Springer, Berlin Heidelberg New York (1992)
9. Bourne, N., Millet, J., Pickup, I.: Delayed failure in shocked silicon carbide. J. Appl. Phys. **81**, 6019 (1997)
10. Brown, W.K., Wohletz, K.H.: Derivation of the Weibull distribution based on physical principles and its connection to the Rosin-Rammler and Lognormal distributions. J. Appl. Phys. **78**, 2758–2763 (1995)
11. Charles, R.J.: Energy-size reduction relationships in comminution. AIME Trans. **208**, 80–88 (1957)
12. Chen, W., Ravichandran, G.: Dynamic compressive failure of a glass ceramic under lateral confinement. J. Mech. Phys. Solids **45**, 1303–1328 (1997)
13. Chen, S.Y., Dhruva, B., Kurien, S., Sreenivasan, K.R., Taylor, M.A.: Anomalous scaling of low-order structure functions of turbulent velocity. J. Fluid Mech. **533**, 183–192 (2005)
14. Cohen, R.D.: Second law limitations on particle size distribution functions. Proc. R. Soc. Lond. **440**, 611–620 (1993)
15. Costin, L.S., Grady, D.E.: Dynamic fragmentation of ferroelectric ceramics using the torsional Kolsky bar. Sandia National Laboratories Technical Report, SAND83-2004, October (1983)

16. Costin, L.S., Grady, D.E.: Dynamic fragmentation of brittle materials using the torsional Kolsky bar. Institute of Physics Conference Ser. No. 70, Oxford, pp. 321–328 (1984)
17. Courant, R., Hilbert, D.: Methods of Mathematical Physics, vol. 2. Interscience, New York (1962)
18. Crain, I.K.: The Monte-Carlo generation of random polygons. Comput. Geosci. **4**, 131–141 (1978)
19. Curran, D.R., Seaman, L., Shockey, D.A.: Dynamic failure of solids. Phys. Rep. **147**, 253–288 (1987)
20. Drugan, W.J.: Dynamic fragmentation of brittle materials: analytical mechanics-based models. J. Mech. Phys. Solids **49**, 1181–1208 (2001)
21. Englman, R., Rivier, N., Jaeger, Z.: Fragment size distribution in disintegration by maximum entropy formalism. Phil. Mag. B **56**, 6, 751–759 (1987)
22. Epstein, B.: The mathematical description of certain breakage mechanisms leading to the logarithmico-normal distribution. J. Franklin Inst. **244**, 471–477 (1947)
23. Falkovich, G., Sreenivasan, K.R.: Lessons from hydrodynamic turbulence. Phys. Today **59**, 4, 43–49 (2006)
24. Gaudin, A.M.: An investigation of crushing phenomena. AIME Trans. **73**, 253–316 (1926)
25. Gaudin, A.M., Meloy, T.P.: Model and a comminution distribution equation for single fracture. AIME Trans. **223**, 40–43 (1962)
26. Glenn, L.A., Chudnovsky, A.: Strain-energy effects on dynamic fragmentation. J. Appl. Phys. **59**, 1879–1380 (1986)
27. Gilvarry, J.J.: Fracture of brittle solids. I. Distribution function for fragment size in single fracture (theoretical). J. Appl. Phys. **32**, 391–399 (1961)
28. Gilvarry, J.J., Bergstrom, B.H.: Fracture of brittle solids. II. Distribution function for fragment size in single fracture (experimental). J. Appl. Phys. **32**, 400–410 (1961)
29. Grady, D.E.: Fragmentation of solids under impulsive stress loading. J. Geophys. Res. **86**, 1047–1054 (1981)
30. Grady, D.E.: Application of survival statistics to the impulsive fragmentation of ductile rings. In: Meyers M.A., Murr, L.E. (eds.) Shock Waves and High-Strain-Rate Phenomena, pp. 181–192. Plenum, New York (1981)
31. Grady, D.E.: Local inertial effects in dynamic fragmentation. J. Appl. Phys. **53**, 322–325 (1982)
32. Grady, D.E.: The spall strength of condensed matter. J. Mech. Phys. Solids **36**, 3, 353–384 (1988)
33. Grady, D.E.: Particle size statistics in dynamic fragmentation, J. Appl. Phys. **68**, 12, 6099–6105 (1990)
34. Grady, D.E.: Spall and fragmentation in high-temperature metals. In: Darison, L., Grady, D.E., Shahinpoor, M. (eds.) High-Pressure Shock Compression of Solids II – Dynamic Fracture and Fragmentation, pp. 219–236. Springer (1996)
35. Grady, D.E.: Shock equation of state properties of concrete. In: Jones, N., Brebbia, C.A., Watson, A.J. (eds.) Structures under Shock and Impact IV, Computational Mechanics Publications, pp. 405–414. Southampton, UK (1996)
36. Grady, D.E.: Scattering as a mechanism for structured shock waves in metals. J. Mech. Phys. Solids **46**, 2017–2032 (1998)

37. Grady, D.E.: Comparison of hypervelocity fragmentation and Spall experiments with Tuler–Butcher Spall and fragment size criteria. Int. J. Impact Eng. **33**, 305–315 (2005)
38. Grady, D.E.: Fragmentation of Rings and Shells. Springer, Berlin Heidelberg New York (2006)
39. Grady, D.E.: Fragment size distributions from the dynamic fragmentation of brittle solids. J. Impact Eng. (in press, 2007)
40. Grady, D.E., Hightower, M.M.: Natural fragmentation of exploding cylinders. In: Meyers, M.A., Murr, L.E., Staudhammer, K.P. (eds.) Shock-Wave and High-Strain-Rate Phenomena in Materials, pp. 713–721. Marcel Dekker, New York (1992)
41. Grady, D.E., Kipp, M.E.: Continuum Modeling of Explosive Fracture in Oil Shale, Int. J. Rock Mech. Mining Sci., **17**, 149–150 (1980)
42. Grady, D.E., Kipp, M.E.: Geometric statistics and dynamic fragmentation. J. Appl. Phys. **58**, 1210–1222 (1985)
43. Grady, D.E., Kipp, M.E.: Dynamic rock fragmentation. In: Atkinson, B.K. (ed.) Fracture Mechanics of Rock. Academic Press, San Diego, CA (1987)
44. Grady, D.E., Lipkin, J.: Criteria for impulsive rock fracture. Geophys. Res. Lett. **7**, 255–258 (1980)
45. Grady, D.E., Olsen, M.L.: A statistical and energy based theory of dynamic fragmentation. Int. J. Impact Eng. **29**, 293–306 (2003)
46. Grady, D.E., Winfree, N.A.: Impact fragmentation of high-velocity compact projectiles on thin plates: a physical and statistical characterization of fragment debris. J. Impact Eng. **26**, 249–262 (2001)
47. Grady, D.E., Kipp M.E., Benson, D.A.: Energy and statistical effects in the dynamic fragmentation of metal rings. Proceedings of the Conference of the Mechanical Properties of High Rates of Strain. Oxford, Inst Phys. Conf. Ser. No. **70**, 315–320 (1984)
48. Grady, D.E., Parsons, D.A., Winfree, N.A.: Physics-based models and analysis tools for debris formation in meteoroid and orbital debris impact. Applied Research Associates Report, for National Aeronautics and Space Administration, March (2000)
49. Grady, D.E., Wilson, L.T., Reedal, D.R., Kuhns, L.D., Kipp, M.E., Black, J.W.: Comparing alternative approaches in the scaling of naturally fragmenting munitions. In: Crewther, (ed.) Proceedings 19th International Symposium on Ballistics, pp. 591–598. Interlaken, Switzerland, May 7–11, I. R, (2001)
50. Green, S.J., Perkins, R.D.: Uniaxial compression tests at varying strain rates on three geological materials. In: Gray, K.E. (ed.) Proceeding of 10th Symposium on Rock Mechanics, pp. 35–54. Austin, Texas (1968)
51. Griffith, L.: A theory of the size distribution of particles in a comminuted system. Can. J. Res. **21A**, 6, 57–64 (1943)
52. Hahn, G.J., Shapiro, S.S.: Statistical Models in Engineering. Wiley, New York (1967)
53. Jeans, J.H.: The Dynamic Theory of Gases, 3rd edn. Cambridge University Press, Cambridge (1921)
54. Johnson, W.A., Mehl, R.F.: Reaction kinetics in processes of nucleation and growth. Trans. AIMME **135**, 414–458 (1939)
55. Kanel, G.I.: The resistance of, metals to spall fracture. Combust. Explo. Shock **18**, 329–335 (1982)

56. Kanel, G.I. et al.: Spallation near the ultimate strength of solids. In: Schmidt, S.C. et al. (eds.) Shock Compression of Condensed Matter-1993, pp. 1043–1046. Elsevier, Amsterdam, The Netherlands (1994)
57. Kennedy, J.E.: The Gurney model for explosive output for driving metal. In: Zukas, J.A., Walters, W.P. (eds.) Explosive Effects and Applications, pp. 221–258. Springer, Berlin Heidelberg New York (1997)
58. Kerley, G.I.: CTH EOS Package: Porosity and Reactive Burn Models. Sandia National Laboratories Report, SAND92-0553, April (1992)
59. Kiang, T.: Mass distributions of asteroids, stars and galaxies. Zeitschrift für Astrophysik **64**, 426–432 (1966)
60. Kiang, T.: Random fragmentation in two and three dimensions. Zeitschrift für Astrophysik **64**, 433–439 (1966)
61. Kick, F.: Das Gesetz der proportionalem Wederstand and Seine Anwendung. Leipsig (1985)
62. Kipp, M.E., Grady, D.E.: Dynamic fracture growth and interaction in one dimension. J. Mech. Phys. Solids **33**(4), 399–415 (1985)
63. Kipp, M.E., Grady, D.E.: Random flaw nucleation and interaction in one dimension. In: Murr, L.E., Staudhammer, K.P., Meyers, M.A. (eds.) Metallurgical applications of shock-wave and high-strain-rate phenomena, pp. 781–791. Marcel Dekker, New York (1986)
64. Kittel, C.: Introduction to Solid State Physics. Wiley, New York (1971)
65. Kolmogorov, A.N.: Doklady Akad. Nauk., SSSR; **30**, 9; **31**, 99; **32**, 16 (1941)
66. Lawn, B.R., Wilshaw, T.R.: Fracture of Brittle Solids. Cambridge University Press, London (1975)
67. Lankford, J.: Uniaxial compressive damage in α-SiC at low homologous temperatures. J. Am. Cer. Soc. **62**, 310 (1979)
68. Lee, E.H.: The continuum mechanics aspect of material properties determination. In: Mueller W., Shaw, M. (eds.) Energetics III, pp. 85–122. Gordon and Breach, New York (1967)
69. Levesey, J.C.: Maxwellian distribution derived from N-body disintegration theory. Am. J. Phys. **48**, 222–225 (1980)
70. Lienau, C.C.: Random fracture of a brittle solid. J. Franklin Inst. **221**, 485–494, 674–686, 769–787 (1936)
71. Lundberg, B.: A split Hopkinson Bar study of energy absorbtion in dynamic rock fragmentation. Int. J. Rock Mech. Min. Sci. **13**, 187–197 (1976)
72. Mandelbrot, B.B.: The Fractal Geometry of Nature. W.C. Freeman, San Francisco, CA (1982)
73. Matthews, J., Walker, R.L.: Mathematical Methods of Physics. W.A. Benjamin, New York, pp. 228–234 (1964)
74. Maxwell, J.C.: Proc. Cambridge Philos. Soc. **12**, 713, XCIV (1879)
75. Mock, W., Holt, W.H.: Fragmentation behavior of Armco iron and HF-1 steel explosive filled cylinders. J. Appl. Phys. 54, 2344–2351 (1983)
76. Mott, N.F.: Fragmentation of H. E. Shells: A Theoretical Formula for the Distribution of Weights of Fragments, United Kingdom Ministry of Supply. AC3642, March (1943)
77. Mott, N.F.: A Theory of the Fragmentation of Shells and Bombs., United Kingdom Ministry of Supply AC4035, May (1943)
78. Mott, N.F.: A Theory of Fragmentation. Application to Wire Wound Bombs such as the American 20 lb. F. United Kingdom Ministry of Supply, AC6338, May (1944)

79. Mott, N.F.: Fragmentation of shell cases. Proc. R. Soc. A **189**, 300–308 (1947)
80. Mott, N.F.: Brittle fracture in mild steel plates. Engineering **165**, 16 (1948)
81. Mott, N.F., Linfoot, E.H.: A Theory of Fragmentation, United Kingdom Ministry of Supply AC3348, February (1943)
82. Piekutowski, A.J.: Formation and Description of Debris Clouds Produced by Hypervelocity Impact. NASA Contractor Report, 4704, February (1996)
83. Piekutowski, A.J.: Effects of scale on debris cloud formation. Int. J. Impact Eng. **20**, 639–650 (1997)
84. Reedal, D., Wilson, L., Grady, D., Chhabildas, L., Reinhart, W.: Impact and explosion induced failure and fragmentation studies on tungsten. Proceedings of the 15th U.S. Army Symposium on Solid Mechanics, pp. 569–585. Myrtle Beach, South Carolina, April 12–14, (1999)
85. von Rittinger, P.R.: Lehrbuch der Aufbereitungskunde. Verlag, Ernst. und Korn, Berlin (1867)
86. Rosin, P., Rammler, E.: The laws governing the fineness of powdered coal. J. Inst. Fuel **7**, 29–36 (1933)
87. Schuhmann, R.: Principles of comminution, I., size distribution and surface calculations. AIME Tech. Publ. 1189, Mining Technology, 1–11 (1940)
88. Sharon, E., Gross, S.P., Fineberg, J.: Energy dissipation in dynamic fracture. Phys. Rev. Lett. **76**, 2117–2120 (1996)
89. Shih, C.J., Nesterenko, V.F., Meyers, M.A.: High-strain-rate deformation and comminution of silicon carbide. J. Appl. Phys. **83**, 4660–4671 (2000)
90. Skidmore, I.C.: Appl. Mat. Res. **4**, 131–147 (1965)
91. Timoshenko, S., Goodier, J.N.: Theory of Elasticity, 2nd edn. McGraw Hill, New York, pp. 372–384 (1951)
92. Tuler, F.R., Butcher, B.M.: A criterion for the time dependence of dynamic fracture. Int. J. Fracture Mech. **4**, 431–437 (1968)
93. Turcotte, D.L.: Fractals and fragmentation. J. Appl. Phys. **91**, 1921–1926 (1986)
94. Wang, H., Ramash, K.T.: Dynamic strength and fragmentation of hot-pressed silicon carbide under uniaxial compression. Acta Met. **52**, 355–367 (1997)
95. Watson, D.F.: Computation the n-dimensional Delaunay tessellation with application to Voronoi polytypes. Comput. J. **24**, 167–172 (1981)
96. Waza, T., Matsui, T., Kani, K.: Laboratory simulations of planetesimal collision II, – ejecta velocity distribution. J. Geophys. Res. **90**, 1995–2011 (1985)
97. Weaire, D., Rivier, N.: Soap, cells and statistics – random patterns in two dimensions. Contemp. Phys. **25**(1), 59–99 (1984)
98. Wesenberg, D.L., Sagartz, M.J.: Dynamic fracture of 6061-T6 aluminum cylinders. J. Appl. Mech. **44**, 643–646 (1977)
99. Zhou, F., Molinari, J.F., Ramesh, K.T.: Effects of material properties and strain rate on the fragmentation of brittle materials. Appl. Phys. Lett. **88**, 261918 (2006)

Index

Shock Wave Science and Technology Reference Library, Volume 3
Solids II

About the Authors

Chapter 1

Stephen A. Sheffield

Los Alamos National Laboratory, Los Alamos NM 87545, USA

Stephen A. Sheffield received a B.S. in Mechanical Engineering from the University of Utah, an M.S. in Mechanical Engineering from the University of New Mexico, and a Ph.D. in Engineering Science (Shock Wave Physics) from Washington State University. He has been a technical staff member doing explosives-related work, first at Sandia National Laboratories for 18 years and since at Los Alamos National Laboratory for 22 years. He is intimately familiar with the design of gas-gun-driven shock experiments and has innovated and used the magnetic *in-situ* gauging technique for over 20 years. He has also used laser-based velocity interferometry for over 30 years (co-developer of the interferometer technique called ORVIS). His research has centered on experimental studies associated with high explosive initiation, detonation propagation, detonation reaction zones, as well as the study of shock-induced chemistry in organic liquids. He has been instrumental in developing important new understanding relating to 1) shock initiation of several homogeneous liquid explosives (including a modification to the classical homogeneous initiation theory), 2) shock initiation of solid heterogeneous explosives, 3) detonation reaction-zone measurements, and 4) the effect of chemistry changes on initiation and detonation properties of liquid nitromethane. His work on shock chemistry in organic liquids forms the basis for future studies in this area.

Ray Engelke

Los Alamos National Laboratory, Los Alamos NM 87545, USA

Ray Engelke received B.S. and M.A. degrees in Physics from Long Beach State College and a Ph.D. in Physics from the University of New Mexico. He has worked as a technical staff member, contractor, and guest scientist at Los Alamos National Laboratory over a period of 36 years, doing research related to explosives initiation, detonation propagation, shock-driven chemistry, and ab-initio quantum chemical calculations. He pioneered the work of studying chemically-sensitized homogeneous nitromethane (NM) and physically-sensitized heterogeneous NM-based explosive materials; this work has led to significant understanding related to the initiation and propagation characteristics of both homogeneous and heterogeneous explosives. This work includes studies of critical diameter as a function of chemical and physical sensitizations, diameter-effect curves with sensitization, and two-dimensional detonation shock wave shape as a function of charge diameter. He has done a large amount of experimental work to understand the first steps of shock chemistry in explosive and non-explosive materials. His quantum-chemical studies have produced evidence that extremely powerful pure nitrogen structures (e.g., N_8 cubane) may exist as metastable structures.

Shock Wave Science and Technology Reference Library, Volume 3 Solids II

Chapter 2

Fan Zhang

Defence R&D Canada – Suffield, PO Box 4000
Medicine Hat
Alberta, T1A 8K6
Canada
Fan.zhang@drdc-rddc.gc.ca

Fan Zhang is a Senior Scientist in the Department of National Defence at Defence Research and Development Canada – Suffield and an adjunct Professor at the University of Waterloo in the Department of Mechanical Engineering. He specializes in shock waves, detonations and explosions, more specifically in multiphase reactive flow and high energy density systems. He earned his doctoral degree in science in 1989 from the University of Technology Aachen (RWTH), Germany, and received a Borchers Medal, a Friedrich-Wilhelm Prize and several best paper awards. He has published more than a hundred refereed journal and proceedings papers, book chapters and special issues in journals.

Anguang Hu

University of Ottawa
Department of Chemistry
Ottawa, Ontario
K1N 6N5
Canada

Dr. Anguang Hu's main area of scientific research is *ab initio* electronic structure techniques on modeling of chemistry, physics, and materials science with high-performance computing. He is one of developers for several academic and commercial electronic structure programs.

Saman Alavi

University of Ottawa
Department of Chemistry
Ottawa, Ontario
K1N 6N5
Canada

Dr. Saman Alavi is a chemist with a research background in theoretical/computational chemistry. He has been working in the field of molecular simulations of materials with applications in the field of materials and environmental engineering. His research interests include simulations of structure and dynamics of inclusion compounds (clathrates and calixarenes), energetic materials, and green solvents, and the study of proton transfer dynamics in acid-base complexes.

Chapter 2, cont.

Tom Woo

University of Ottawa
Centre for Catalysis
Research and Innovation
Department of Chemistry
Ottawa, Ontario
K1N 6N5
Canada
twoo@uottawa.ca
http://titan.chem.
uottawa.ca

Dr. Tom Woo is an Associate Professor of Chemistry and Canada Research Chair in Catalyst Modelling and Computational Chemistry, at the University of Ottawa and the Centre for Catalysis Research and Innovation. He earned his Ph.D. degree in 1998 at the University of Calgary under Professor Tom Ziegler and worked as a post-doctoral fellow at the ETH, in Zurich. He joined the University of Western Ontario in 2000 as an Assistant Professor and in 2005 he joined the University of Ottawa. His research interests include the development and application of molecular simulation methods to study chemical reaction mechanisms, energetic materials, and catalysis.

Chapter 3

John B. Aidun

Multiscale Dynamic
Materials Modeling
Sandia National
Laboratories
Albuquerque
New Mexico, USA

John B. Aidun (Sandia National Laboratories, Albuquerque, NM) John Aidun is Manager of the Multiscale Dynamic Materials Modeling Department in the Computation, Computers, Information, and Mathematics Center at Sandia National Laboratories (SNL). He received a Ph.D. degree in physics from Washington State University in 1989, where he conducted experimental research on shock wave induced solid state phase transitions. He was a post doctoral researcher (1990–1992) and then staff member (1993–1994) in the Equations of State and Mechanics of Materials Group (T-1) at Los Alamos National Laboratory. He joined the Material Mechanics Department at SNL in 1995 and has been a manager since 2001. Research interests include multiscale materials simulation methods development with emphasis on materials chemistry and reactivity; equations of state; constitutive modeling; shock wave physics; philosophy of science.

Chapter 3, cont.

Zhiping Tang

University of Science and Technology of China
Department of Modern Mechanics
Hefei, Anhui 230026
China
zptang@ustc.edu.cn

Professor Zhiping Tang's main research areas are impact dynamics, DEM and multi-scale numerical method, and laser propulsion. He is the vice chairman of the Chinese Committee of Explosion Mechanics, associate editor of Explosion and Shock Waves.

Chapter 4

Dennis Grady

Applied Research Associates
4300 San Mateo Blvd. NE, A-220
Albuquerque, NM 87110
USA
dgrady@ara.com

Dennis Grady is an Associate and Principal Scientist with Applied Research Associates headquartered in Albuquerque, New Mexico. He received his Ph.D. in physics from Washington State University in 1971. Following three years at SRI International he joined Sandia National Laboratories, retiring in 1996. His research interests are focused on mechanical and thermodynamic effects of the intense shock environment. He is a member of the American Physical Society, International Hypervelocity Society, and the American Geophysical Union.